# INTERACTIONS AMONG CELL SIGNALLING SYSTEMS

The Ciba Foundation is an international scientific and educational charity. It was established in 1947 by the Swiss chemical and pharmaceutical company of CIBA Limited — now CIBA-GEIGY Limited. The Foundation operates independently in London under English trust law.

The Ciba Foundation exists to promote international cooperation in biological, medical and chemical research. It organizes about eight international multidisciplinary symposia each year on topics that seem ready for discussion by a small group of research workers. The papers and discussions are published in the Ciba Foundation symposium series. The Foundation also holds many shorter meetings (not published), organized by the Foundation itself or by outside scientific organizations. The staff always welcome suggestions for future meetings.

The Foundation's house at 41 Portland Place, London W1N 4BN, provides facilities for meetings of all kinds. Its Media Resource Service supplies information to journalists on all scientific and technological topics. The library, open five days a week to any graduate in science or medicine, also provides information on scientific meetings throughout the world and answers general enquiries on biomedical and chemical subjects. Scientists from any part of the world may stay in the house during working visits to London.

Ciba Foundation Symposium 164

# INTERACTIONS AMONG CELL SIGNALLING SYSTEMS

*A Wiley-Interscience Publication*

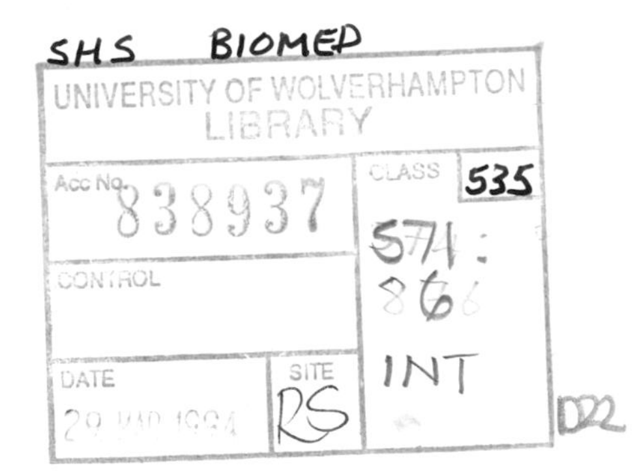


1992

JOHN WILEY & SONS

Chichester · New York · Brisbane · Toronto · Singapore

Published in 1992 by John Wiley & Sons Ltd.
Baffins Lane, Chichester
West Sussex PO19 1UD, England



*Other Wiley Editorial Offices*

John Wiley & Sons, Inc., 605 Third Avenue,
New York, NY 10158-0012, USA

Jacaranda Wiley Ltd, G.P.O. Box 859, Brisbane,
Queensland 4001, Australia

John Wiley & Sons (Canada) Ltd, 22 Worcester Road,
Rexdale, Ontario M9W 1L1, Canada

John Wiley & Sons (SEA) Pte Ltd, 37 Jalan Pemimpin #05-04,
Block B, Union Industrial Building, Singapore 2057

Suggested series entry for library catalogues:
Ciba Foundation Symposia

Ciba Foundation Symposium 164
xi + 268 pages, 49 figures, 5 tables

*Library of Congress Cataloging-in-Publication Data*

Interactions among cell signalling systems.
p. cm.—(Ciba Foundation symposium ; 164)
Editors, Ryo Sato, Gregory R. Bock (organizers), and Kate Widdows.
A Wiley-Interscience publication.
Includes bibliographical references and index.
ISBN 0 471 93073 3
1. Cellular signal transduction—Congresses. 2. Cell interaction—Congresses I. Satō, Ryō, 1923- . II. Bock, Gregory. III. Widdows, Kate. IV. Series.
[DNLM: 1. Cell Communication—physiology—congresses. 2. Signal Transduction—physiology—congresses. W3 C161F v. 164]
QP517.C45I55 1992
574.87—dc20
DNLM/DLC
for Library of Congress 92-95
CIP

*British Library Cataloguing in Publication Data*

A catalogue record for this book is available
from the British Library

ISBN 0 471 93073 3

Phototypeset by Dobbie Typesetting Limited, Tavistock, Devon.
Printed and bound in Great Britain by Biddles Ltd., Guildford.

# Contents

*Symposium on Interactions among cell signalling systems, held in collaboration with the CIBA-GEIGY Foundation (Japan) for the Promotion of Science, at the Portopia Hotel on Port Island, Kobe, Japan, 23–25 April 1991*

*Editors: Ryo Sato, Gregory R. Bock (Organizers) and Kate Widdows*

# Participants

**N. Akaike** Department of Neurophysiology, Tohoku University School of Medicine, 1-1 Seiryo-machi Aoba-ku, Sendai 980, Japan

**D. A. Cantrell** Lymphocyte Activation Laboratory, Imperial Cancer Research Fund, PO Box 123, 44 Lincoln's Inn Fields, London WC2A 3PX, UK

**G. Carpenter** Department of Biochemistry, Vanderbilt University School of Medicine, Nashville, TN 37232-0146, USA

**J.-P. Changeux** Institut Pasteur, Neurobiologie Moléculaire, Bâtiment des Biotechnologies, 25 rue du Dr Roux, 75015, Paris, France

**G. L. Collingridge** Department of Pharmacology, University of Birmingham, The Medical School, Edgbaston, Birmingham B15 2TT, UK

**J. W. Daly** Laboratory Bioorganic Chemistry, NIDDK, Building 8 Room 1A15, National Institutes of Health, Bethesda, MD 20892, USA

**J. Exton** Department of Molecular Physiology & Biophysics, Room 802, Light Hall, Vanderbilt University School of Medicine, Nashville, TN 37232, USA

**E. H. Fischer** Department of Biochemistry, SJ-70, University of Washington, Seattle, WA 98195, USA

**B. Fredholm** Department of Pharmacology, Karolinska Institute, PO Box 60400, S-104 01 Stockholm, Sweden

**T. Hunter** Molecular Biology & Virology Laboratory, The Salk Institute, PO Box 85800, San Diego, CA 92186-5800, USA

**H. Kanoh** Department of Biochemistry, Sapporo Medical College, West-17 South-1, Sapporo 060, Japan

**M. Kasuga** Department of Internal Medicine, Kobe University School of Medicine, Kobe, Japan 650

**M. Kato-Homma** Department of Hygiene & Oncology, Faculty of Medicine, Tokyo Medical & Dental University, 1-5-45 Yushima, Bunkyo-ku, Tokyo 113, Japan

**E. G. Krebs** Department of Pharmacology, University of Washington School of Medicine, Mail Stop SL-15, Seattle, WA 98195, USA

**R. H. Michell** School of Biochemistry, University of Birmingham, PO Box 363, Birmingham B15 2TT

**K. Mikoshiba** Institute for Protein Research, Osaka University, 3-2 Yamadaoka, Suita, Osaka 565, Japan

**A. C. Nairn** Laboratory of Molecular & Cellular Neuroscience, The Rockefeller University, 1230 York Avenue, New York, NY 10021, USA

**Y. Nishizuka** Department of Biochemistry, Kobe University School of Medicine, Kobe 650, Japan

**P. Parker** Protein Phosphorylation Laboratory, Imperial Cancer Research Fund, PO Box 123, 44 Lincoln's Inn Fields, London WC2A 3PX, UK

**H. Rasmussen** Division of Endocrinology, Department of Internal Medicine, Yale University School of Medicine, Fitkin 1, PO Box 3333, New Haven, CT 06510-8056, USA

**H-l. Su** Department of Biochemistry, Institute of Endocrinology, Tianjin Medical College, 22 Qi Xiang Tai Road, Heping District, Tianjin 300070, People's Republic of China

**C. Tanaka** Department of Pharmacology, Kobe University School of Medicine, Kobe 650, Japan

**G. Thomas** Friedrich Miescher Institute, PO Box 2543, CH-4002 Basle, Switzerland

**K. Toyoshima** Department of Oncogene Research, Research Institute for Microbial Diseases, Osaka University, 3-1 Yamadaoka, Suita City, Osaka 565, Japan

**R. W. Tsien** Department of Molecular & Cellular Physiology, Arnold & Mabel Beckman Center, Stanford University Medical Center, Stanford, CA 94305-5425, USA

**M. Ui** Department of Physiological Chemistry, Faculty of Pharmaceutical Sciences, University of Tokyo, Tokyo 113, Japan

**J. Warner** Department of Physiology & Pharmacology, University of Southampton, Bassett Crescent East, Southampton SO9 3TU, UK

**C-c. Yang** Institute of Life Sciences, National Tsing Hua University, Hsinchu, Taiwan 30043

# Preface

In September 1987, CIBA-GEIGY Ltd set up a new foundation in Japan—the CIBA-GEIGY Foundation (Japan) for the Promotion of Science. The new Foundation was established with the approval of the Japanese Ministry of Education, Science and Culture, and this was the first such approval ever given to a foreign company in Japan. The goal of the Foundation is to promote creative research in Japan by offering competitive research grants and awards for exchange visits between Japan and Europe.

Most of the Ciba Foundation's symposia are held at its London headquarters, but the Foundation has a policy of holding at least one meeting each year outside the UK. In 1987, before the establishment of the new CIBA-GEIGY Foundation, the Ciba Foundation held a meeting in Kyoto, Japan, on Applications of Plant Cell and Tissue Culture. When the new Foundation was set up, and its Managing Trustee appointed, the possibility was discussed of the Ciba Foundation holding another symposium in Japan, and the two Foundations agreed to organize a joint meeting in the traditional Ciba Foundation symposium format. We chose the area of cell signalling as being one in which there have been recent important advances, and where Japanese scientists have made outstanding contributions. The meeting took place in Kobe on 23–25 April 1991.

We would like to thank Professor Yasutomi Nishizuka for chairing the symposium, and for his tireless and energetic help throughout all stages of the planning. Without his efforts, the meeting would not have been held successfully. We would also like to thank CIBA-GEIGY (Japan) Ltd for their generosity in sponsoring an open meeting after the symposium, which provided an opportunity for a large number of local scientists to hear some of the participants in the symposium present their latest data.

The organization of this symposium in Japan was entirely consistent with the goals of the Ciba Foundation, and the CIBA-GEIGY Foundation (Japan) for the Promotion of Science. We sincerely hope that our joint activities will help to foster better communication and better understanding between scientists inside and outside Japan.

Gregory R. Bock
*Assistant Director, The Ciba Foundation*

Ryo Sato
*Managing Trustee, The CIBA-GEIGY Foundation (Japan) for the Promotion of Science*

# Introduction

Yasutomi Nishizuka

*Department of Biochemistry, Kobe University School of Medicine, Kobe 650, Japan*

The molecular basis of the transduction of extracellular signals into intracellular events and cell proliferation has long been a subject of great interest, and our knowledge of the mechanisms of cell signalling pathways has been expanding rapidly. It is now clear that signalling pathways for many hormones, neurotransmitters, growth factors and other biologically active substances consist of a series of proteins, including specific receptors, GTP-binding proteins, second messenger-generating enzymes, protein kinases, and the target functional proteins. Some growth factor receptors themselves have protein kinase activity. External signals can be transmitted directly to the cell interior through ion channels, and cellular responses are mediated by effector proteins, such as calcium-binding proteins. Recent molecular cloning analysis has revealed that some of these signalling proteins show considerable heterogeneity and tissue-specific expression, and it is becoming evident that an elaborate network of various signalling pathways operates within a single cell.

In April 1991, the Ciba Foundation and the CIBA-GEIGY Foundation (Japan) for the Promotion of Science jointly organized a symposium on cell signalling, with particular emphasis on interactions among various signalling pathways. There are diverse interactions, including negative and positive feedback control, counteraction, potentiation, cooperation, co-transmission, synergism and antagonism, which are all important to our understanding of the dynamic aspects of cellular regulation.

This volume contains the papers presented at the symposium by leading scientists in this field of research. As with all Ciba Foundation symposia, considerable time was devoted to the discussion of experimental methods and results and their implications. The areas covered, which are all currently under intensive study and are developing rapidly, include endocrinology, neuroscience (memory in particular), immunology, gene expression, cell proliferation and differentiation, and certain aspects of the cardiovascular system.

# Inositol lipids and phosphates in the proliferation and differentiation of lymphocytes and myeloid cells

Robert H. Michell*, Louise A. Conroy*, Michael Finney*, Philip J. French*, Christopher M. Bunce‡, Kim Anderson*, Michael A. Baxter°, Geoffrey Brown‡, John Gordon‡, Eric J. Jenkinson†, Janet M. Lord‡, Christopher J. Kirk* and John J. T. Owen†

*Departments of Biochemistry*, Anatomy†, Immunology‡ and Medicine°, The University of Birmingham, Edgbaston, Birmingham B15 2TT, UK*

*Abstract.* It is established that receptor-stimulated hydrolysis of phosphatidylinositol 4,5-bisphosphate is an essential signalling reaction in the responses of many haemopoietic cells to stimuli: examples include platelet activation, antigen-driven initiation of cell proliferation in mature B and T lymphocytes and histamine release by mast cells, and chemotaxis and oxygen radical generation by neutrophils. However, the roles of inositol lipids and phosphates in the development of haemopoietic and immune cells are less well understood. This paper discusses three such situations: the sequential employment of phosphatidylinositol 4,5-bisphosphate hydrolysis and cyclic AMP accumulation as two signals essential to the action of the B lymphocyte-stimulatory cytokine interleukin 4; the involvement of antigen receptor-triggered inositol lipid hydrolysis in apoptotic elimination of immature anti-self T lymphocytes in the fetal mouse thymus; and the possible role of changes in the levels of abundant inositol polyphosphates in the differentiation of HL-60 promyelocytic cells and of normal human myeloid blast cells.

*1992 Interactions among cell signalling systems. Wiley, Chichester (Ciba Foundation Symposium 164) p 2–16*

Since the pioneering work of Fisher & Mueller (1968) showed that the polyclonal mitogen phytohaemagglutinin very rapidly provokes a stimulation of phosphatidylinositol (PtdIns) turnover in lymphocytes, these cells have probably provided the best evidence for an essential link between receptor-stimulated inositol lipid hydrolysis and long-term changes in cell state. Despite widespread acceptance of the resulting view that antigen-activated phosphatidylinositol 4,5-bisphosphate hydrolysis, yielding the second messengers inositol 1,4,5-trisphosphate [Ins(1,4,5)$P_3$] and 1,2-diacylglycerol, is an essential step in rousing both B and T cells from quiescence, many problems relating to the role of inositol lipids and phosphates in controlling the proliferation and differentiation of haemopoietic cells remain unresolved. In particular, the signalling mechanisms used by some growth factors and cytokines clearly involve activation of PtdIns(4,5)$P_2$ hydrolysis and/or the activation of a 3-kinase which catalyses the formation of one or more of the 3-phosphorylated lipids PtdIns3P, PtdIns(3,4)$P_2$ and PtdIns(3,4,5)$P_3$ (Carpenter & Cantley 1990, Stephens et al 1991), but many more of these key agents control cells through as yet unknown mechanisms (see, for example, Benton 1991).

Studies we have recently undertaken suggest that interleukin 4 (IL-4) uses a novel signalling mechanism, they broaden the significance of PtdIns(4,5)$P_2$ hydrolysis as a signalling reaction involved in a variety of haemopoietic processes, and they focus attention on an urgent need to understand the biological role in haemopoietic and other cells of abundant inositol phosphates that are formed from non-lipid sources.

**Signalling by human interleukin 4**

Interleukin 4 acts in synergy with small quantities of anti-immunoglobulin in rousing quiescent human B cells into proliferation, and it helps to determine the immunoglobulin isotypes secreted by the resulting plasma cells, particularly as an essential factor in inducing IgE synthesis.

We have investigated the signals generated by interleukin 4 in quiescent human tonsillar B cells and the role of these signals in regulating the subsequent expression of the B cell-specific (glyco)protein CD23 (Finney et al 1990, 1991). From these studies has come what appears to be the first evidence that an extracellular stimulus acting through a cell-surface receptor can use inositol lipid hydrolysis and cyclic AMP (cAMP) accumulation as sequential signals, with both being essential for the activation of B cell-specific genes. The key observations are summarized in Fig. 1. When a population of quiescent ($G_0$) human tonsillar B cells was challenged with recombinant human interleukin 4 (hIL-4), a small, transient accumulation of Ins(1,4,5)$P_3$ was observed, which peaked within about 30 seconds and had returned to the pre-stimulus level within one minute or so. As expected, this was accompanied by a transient increase of cytosolic [$Ca^{2+}$], as detected in cell populations loaded with fura-2. The [$Ca^{2+}$] rise was, however, very variable—large, small or sometimes not detectable. We hope to explore the spatial and temporal details of this response soon by single-cell $Ca^{2+}$ imaging. Although it has yet to be directly demonstrated, we assume that a third transient signal generated during this early phase must be a 1,2-diacylglycerol-mediated activation of protein kinase C.

A few minutes later there was a progressive rise in the cellular cAMP level, usually to 150–300% of the control level: this rise was maintained for at least 30 min (Fig. 1). As a working hypothesis, we assume that this accumulation of cAMP is the result of a delayed activation of adenylate cyclase, but we cannot at present rule out other explanations such as an unchanged adenylate cyclase

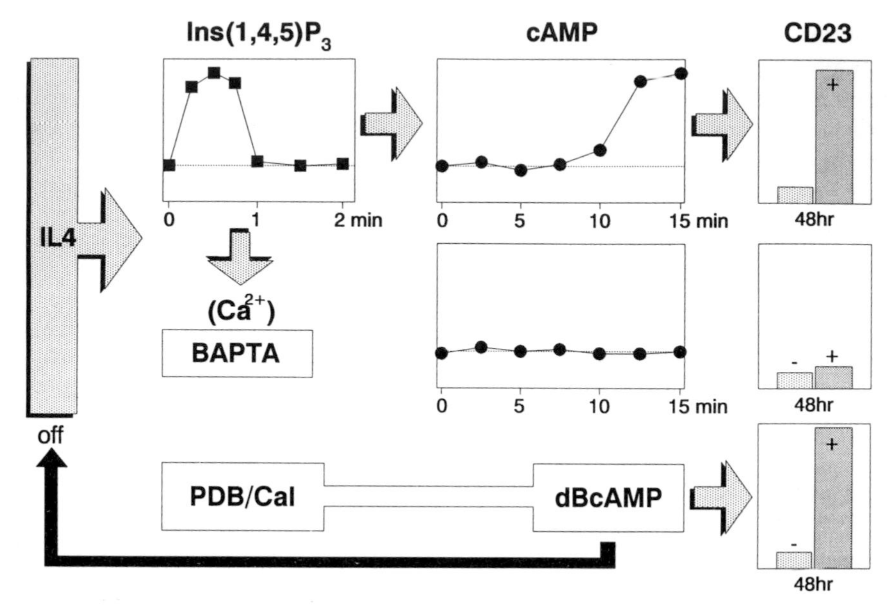


FIG. 1. A diagrammatic summary of the results obtained by Finney et al (1990) relating to the mechanism of action of recombinant human interleukin 4 (hIL-4) on quiescent human tonsillar B cells. The top panel illustrates the sequence of events seen during stimulation of quiescent human tonsillar B lymphocytes by hIL-4. Within the first minute, there is a transient rise in Ins(1,4,5)$P_3$ level, as assayed by a specific protein-binding assay (the dotted line represents the level in unstimulated cells). A few minutes later, the level of cyclic AMP (cAMP) (also assayed by a protein-binding assay) shows a sustained rise (typically 150–300%). Finally, and much later, there is an increase in the expression of CD23: the − and + symbols on the bars of the CD23 histograms indicate, respectively, cells incubated without and with hIL-4 (or, in the bottom panel, the sequence of pharmacological agents designed to mimic IL-4). The middle panel illustrates the fact that ablation of the initial Ins(1,4,5)$P_3$-associated $Ca^{2+}$ rise by intracellular buffering of $Ca^{2+}$ with BAPTA [1,2-bis(2-aminophenoxy)ethane *N,N,N′,N′*-tetraacetic acid] prevents both the subsequent cAMP rise and the CD23 expression. The bottom panel shows that imitation of the IL-4-induced signal sequence with the pharmacological agents phorbol dibutyrate (PDB), ionomycin (CaI, a $Ca^{2+}$ ionophore) and dibutyryl cAMP (dBcAMP) can induce a CD23 expression similar to that evoked by IL-4. As indicated by the filled arrow at the bottom, this pharmacological treatment induces CD23 expression only if the agents are presented in an order that mimics the order of signals generated by IL-4, and pretreatment of cells with dibutyryl cAMP prevents them from responding successfully to hIL-4.

activity occurring in concert with inhibition of cAMP phosphodiesterase activity. Possible routes to this delayed activation of adenylate cyclase could include either intracellular mechanisms or the secretion of an ‘autocrine’ agent in response to hIL-4: failure to abolish this response with aspirin suggests that this autocrine agent, if it exists, is not a classical prostaglandin.

A major functional response of quiescent B cells to treatment with IL-4 is activation of the expression of several ‘cell activation’ surface (glyco)proteins.

One of these is CD23, whose expression provides an excellent index of cell activation—the extracellular domain of this molecule, when cleaved and released, acts on its parent cell as an autocrine stimulatory factor, driving B cells further towards activation and differentiation. We therefore adopted assay of the expression of CD23, which is potently activated by hIL-4 (Fig. 1), as a method by which to determine the role in IL-4 action of the striking sequence of hIL-4-induced cellular messenger changes described above. When the hIL-4-driven rise in cytosolic [$Ca^{2+}$] was buffered out by pre-loading cells with the chelator BAPTA [1,2-bis(2-aminophenoxy)ethane-*N*,*N*,*N'*,*N'*-tetraacetic acid], both of the later changes (cAMP elevation and CD23 induction) were ablated, indicating that the transient increases in $Ins(1,4,5)P_3$ and $Ca^{2+}$ levels that are evoked by IL-4 are essential for both the early and late components of the train of events that characterizes B cell activation.

To explore further the roles of $Ca^{2+}$, protein kinase C and cAMP in the signalling pathway leading to activation of characteristic B cell genes we attempted to reconstruct the observed signalling sequence pharmacologically, by the application, singly and in combination, of a $Ca^{2+}$ ionophore (ionomycin), phorbol dibutyrate and dibutyryl cyclic AMP. Pharmacological mimicry of the same temporal sequence as that seen during the response to hIL-4 produced an essentially identical induction of CD23, confirming that this sequence of three signals can provide sufficient information to provoke the selective expression of this B cell-specific protein. Moreover, omission of any of the three pharmacological agents from the above sequence led to failure of CD23 expression, as did their presentation in reverse order. In particular, transient application of ionophore plus phorbol dibutyrate did not elicit either a subsequent accumulation of cAMP or the eventual expression of CD23. We therefore conclude that at least one additional signal, which has not yet been identified, must be present for some or all of the 5–10 min period before the initiation of the cyclic AMP rise. This signal must be an essential contributor to the cellular response to hIL-4: a novel tyrosine kinase-mediated pathway is one obvious possibility.

Having established several of the essential features of this pathway, we are now in a position to explore its inter-relationships with other signals controlling B cell behaviour. Amongst the stimuli that tend to nullify, at least in part, the responses of B cells to IL-4 are $\gamma$-interferon, transforming growth factor $\beta$ and antibodies to the B cell surface protein CD19 (the function of which is unknown). To date, we have shown that the subsets of IL-4-mediated responses that are inhibited by each of these three 'IL4-antagonistic' agents are subtly different, and also that the counter-IL-4 actions of these agents are at least partially additive: both observations indicate that the modes of action of these agents differ, and our aim is now to determine at what levels they exert their antagonistic actions on the stimulation of B cells.

## $PtdIns(4,5)P_2$ hydrolysis appears to be a signal involved in antigen-driven T cell elimination and in antigen-driven T cell activation

It has recently become clear that T lymphocytes use the same antigen receptor to initiate two quite different responses at different stages in their life-span. In most cells, this receptor (the T cell receptor, TCR) consists of an $\alpha\beta$ heterodimer of clonally variable polypeptides, which recognizes a complex of self-MHC (major histocompatibility complex) and antigenic peptide on the surface of antigen-presenting cells. The activated TCR transmits its activation to the cell interior via the associated multisubunit signal-generating complex CD3. The TCR is first used, soon after the TCR gene rearrangement that generates a polyclonal population of immature T cells, to provoke the apoptotic 'suicide' of potentially anti-self cells before the completion of their development in the thymus (Smith et al 1989): apoptosis is a form of positively regulated cell elimination, an important and readily assayable element of which is an endonuclease-mediated fragmentation of genomic DNA into oligonucleosomal pieces (Wyllie et al 1980, Williams 1991). Later, the TCR mediates the recognition step that initiates the activation of mature T cells by a foreign antigen. A key question is whether: (a) the same signals are always generated by ligation of the antigen receptor, whatever the developmental stage of the lymphocytes, but the developing cells change in ways that allow them to mount different responses; or (b) cells respond in different ways at these two developmental stages because the signal or signals emanating from the receptors change in their nature or balance as the cells mature.

One recent body of evidence has suggested that the antigen-recognizing $\alpha\beta$ heterodimers in the TCR–CD3 complexes of immature and mature lymphocytes are equally effective at recognizing antigen presented on self-MHC, but that in immature cells they might not effectively couple antigen activation to stimulation of the signal-generating CD3 component of the receptor complex (Finkel et al 1991). However, it is not clear how this 'faulty coupling' model of signalling in immature T cells could fully explain two features of the antigen-driven apoptotic elimination of potentially anti-self T cells. Firstly, the 'faulty coupling' model puts its emphasis on the apparently diminished $Ca^{2+}$ signal that is seen when immature T cells are stimulated by antigen. However, an antigen-initiated $Ca^{2+}$ signal seems to be essential for apoptosis: modest treatment of thymocytes with a $Ca^{2+}$ ionophore is a perfectly adequate stimulus to provoke extensive apoptosis. Secondly, any hypothesis that focuses primarily on reduced signalling efficacy, instead of on a change in the nature of the signals generated or of the response systems within the cell, fails to accommodate in any obvious way the fact that immature and mature T cells show equally striking, though very different, responses to antigen stimulation, rather than one of the pair being responsive and the other unresponsive.

A key signal in the antigen-induced activation of mature T cells is PtdIns(4,5)$P_2$ hydrolysis, as has recently been elegantly confirmed by the observation that transfection into the Jurkat J-HM1-2.2 T cell line of an $M_1$ muscarinic receptor that is coupled to phosphoinositidase C activation through the G protein $G_q$ confers on these cells the ability to grow in response to cholinergic stimulation (Desai et al 1990). It therefore seemed essential to determine to what degree PtdIns(4,5)$P_2$ hydrolysis is associated with, and possibly important in, the antigen-driven elimination of immature thymic T cells. To do this, we used thymus lobes from fetal mice. If taken at Day 14 of gestation and cultured for one week or less, the thymic T cell progenitors in these lobes rearrange their antigen receptor genes and the great majority of the resulting T cells remain immature for the duration of the experiment (Smith et al 1989).

To investigate inositol lipid signalling in these lobes, we labelled them with [2-$^3$H]inositol during culture. Because the radiosensitivity of the proliferating T cell population limits the acceptable isotope load, most data have been assembled as 'within-lobe' (InsP + $InsP_2$)/$InsP_5$ + $InsP_6$) ratios: these represent levels of labelled inositol monophosphates plus bisphosphates accumulated in the presence of $Li^+$ (which blocks InsP phosphatase and Ins(1,4)$P_2$/Ins(1,3,4)$P_3$ 1-phosphatase; Berridge et al 1989), divided by the amount of labelled inositol pentakisphosphates plus hexakisphosphates (a reference value representing compounds whose labelling does not rapidly alter in response to stimulation in these cells). Because of the relatively slow access of antibodies to the cells in the cultured lobes, and the likelihood that the resulting activation of individual cells will be unsynchronized, we collected Ins(1,4,5)$P_3$-derived inositol phosphates over a prolonged period: the figures generated represent an integrated signal over the entire incubation period rather than reflecting the kinetics of cell activation.

Addition of either of two different (hamster and rat) monoclonal antibodies against the TCR-associated CD3 signalling complex (Table 1), each capable of inducing extensive apoptosis of the immature T cells present in these cultured lobes, led to a clear-cut accumulation of InsP/$InsP_2$, confirming that activation of CD3 can initiate inositol lipid hydrolysis in these immature T cells (Table 1; see also Michell et al 1989, Conroy et al 1991). HPLC analyses on a limited number of extracts from such stimulated lobes confirmed that there was accumulation within minutes of Ins(1,3,4)$P_3$, a unique metabolite of Ins(1,4,5)$P_3$, indicating that stimulated hydrolysis of PtdIns(4,5)$P_2$ (rather than, for example, PtdIns hydrolysis) is the sole or major source of the accumulated inositol phosphates. Assays done in parallel showed that incubation with the same anti-CD3 antibodies provoked extensive T cell elimination by apoptosis, as detected either microscopically or by the appearance of oligonucleosomal ladders, characteristic of endonuclease activation, in gels of extracted DNA.

**TABLE 1 Stimulation of the accumulation of inositol phosphates at a $Li^+$ block in cultured fetal mouse thymus lobes incubated with agents that activate the TCR–CD3 complex**

| *Additions to lobes*[a] *(in 10 mM $Li^+$)* | *Ratio $InsP + InsP_2/InsP_5 + InsP_6$*[b] *[mean ± SD (n)]* |
|---|---|
| Control (no antibody) | 0.87 ± 0.09 (11) |
| Anti-CD3 (hamster monoclonal antibody) | 1.95 ± 0.36 (7) |
| Anti-CD3 (purified rat monoclonal antibody) | 1.86 ± 0.19 (4) |
| Anti-αβ | 1.20 ± 0.16 (6) |
| Staphylococcal enterotoxin B (SEB) | 0.97 ± 0.12 (18) |

[a]Thymus lobes from 14-day gestation mouse embryos were grown in organ culture for up to 10 days with [$^3H$]inositol. Antibodies or SEB were added to the lobes, in the presence of 20 mM $Li^+$, for two hours.
[b]Inositol phosphates were analysed using Dowex-1 anion-exchange mini-columns. Data are expressed as ratios of the responsive ($InsP + InsP_2$) divided by the unresponsive ($InsP_5 + InsP_6$).

Although exposure of lobes to a $Ca^{2+}$ ionophore provoked apoptosis, presumably as a result of the involvement of a $Ca^{2+}$-activated endonuclease, it did not provoke $InsP/InsP_2$ accumulation. Conversely, cyclosporin, whose site of inhibition is thought to be downstream of both PtdIns(4,5)$P_2$ hydrolysis and [$Ca^{2+}$] elevation, did not change the accumulation of $InsP/InsP_2$ driven by anti-CD3, but did prevent apoptosis. These observations are compatible with a model in which PtdIns(4,5)$P_2$-derived Ins(1,4,5)$P_3$ provokes a $Ca^{2+}$ signal that is essential for the initiation of apoptosis, with cyclosporin interfering with the downstream signalling pathway at a point distal to $Ca^{2+}$.

Antigen is recognized by the αβ dimer of the T cell receptor complex, and one hypothesis is that apoptotic elimination of immature thymocytes occurs because of ineffective coupling in these cells between TCR-αβ and the CD3 complex, resulting in inefficient signalling (see above). We therefore analysed the inositol phosphates generated in response to an agonistic broad-spectrum antibody to TCR-αβ. We also examined the response to staphylococcal enterotoxin B (SEB), a 'superantigen' which activates only those cells whose TCRs include certain types of $V_{\beta 8}$ subunits. The anti-αβ antibody produced effective apoptosis and induced an accumulation of inositol phosphates that was substantial, but smaller than that induced by anti-CD3 (Table 1). SEB caused apoptosis mainly of certain $V_{\beta 8}$ clones of T cells (about one fifth of the population eliminated by anti-CD3), and it led to a detectable but much smaller stimulation of inositol phosphates in the entire lobes. This response was shown, by cell sorting with Dynabeads coated with an anti-$V_{\beta 8}$ antibody, to be occurring selectively in the SEB-responsive $V_{\beta 8}$ population.

As a result of these experiments, we conclude that the TCR-mediated elimination of immature T cells that occurs in the fetal thymus involves as an early signal the hydrolysis of PtdIns(4,5)$P_2$, with this presumably giving rise to the increase in cytosolic $Ca^{2+}$ that is one of the cellular initiators of apoptosis. It therefore appears that this major signalling pathway is activated in a similar manner in immature T cells when they are induced to initiate their own suicide and in mature T cells when they are activated by an antigen–MHC complex. Whether the different responses of the cells at these different developmental stages arise (a) from the generation of different 'second signals' alongside PtdIns(4,5)$P_2$ hydrolysis or (b) from differences in the downstream response elements possessed by the two cell populations remains to be determined.

## Inositol polyphosphate levels change dramatically during differentiation of myeloid cells

Although Ins(1,4,5)$P_3$ has now been assigned a key role in cellular control, as a receptor-generated second messenger that opens a transmembrane $Ca^{2+}$ channel in the boundary of an intracellular $Ca^{2+}$ reservoir, the functions of a variety of other inositol polyphosphates which are present in cells at much higher abundance are not known (e.g. Hanley 1988, Irvine 1989, Berridge & Irvine 1989, Shears 1989). Several years ago we and others realized that $InsP_4$, $InsP_5$ and $InsP_6$ are all relatively abundant in HL-60 promyeloid cells: these are bipotent cells capable of either neutrophilic or monocytic differentiation (Michell et al 1988, French et al 1988, Pittet et al 1989). We have now analysed in some detail the complement of inositol metabolites, primarily inositol polyphosphates, that is formed in HL-60 cells when they grow, and the ways in which the levels of these compounds change when the cells are induced to differentiate towards either neutrophils or monocytes.

HL-60 cells were adapted to grow in a serum-free medium containing 1 mg/L inositol, in which they differentiated normally towards neutrophils in 0.9% (v/v) dimethylsulphoxide (DMSO) or towards monocytes in 10 nM 12-*O*-tetradecanoyl phorbol-13-acetate (TPA). Cells that had been equilibrium-labelled with [2-$^3$H]*myo*-inositol during growth contained a complex pattern of inositol metabolites, several of which were at relatively high concentrations. These included $InsP_5$ and $InsP_6$, which were present at concentrations of about 25 μM and 60 μM, respectively. Striking and different changes occurred in the levels of some of the inositol polyphosphates as the cells differentiated towards either neutrophils or monocytes. Most notable were a large but gradual accumulation of Ins(1,3,4,5,6)$P_5$ as HL-60 cells decreased in size and acquired characteristics of neutrophils, and much more rapid and sequential declines in $InsP_4$, $InsP_5$ and $InsP_6$ as the cells started to take on monocytic character (French et al 1991). There was a marked accumulation of free inositol and of PtdIns in the cells during neutrophil differentiation, probably caused at least in part by an increased rate of inositol uptake providing an increased intracellular inositol supply (French et al 1991, Baxter et al 1990). The accumulation of Ins(1,3,4,5,6)$P_5$ that occurred during neutrophil differentiation was the same whether it was induced by DMSO or by a combination of retinoic acid and a differentiation factor produced by a T lymphocyte cell line. Ins(1,4,5)$P_3$, a physiological intracellular mediator of $Ca^{2+}$ release from membrane stores, was present at much lower concentrations (about 0.4 μM) and its concentration did not change appreciably during these differentiation processes.

An obvious concern was that these observations in HL-60 cells, a leukaemia-derived cell line, might not accurately reflect the behaviour of normal myeloid precursor cells. We have now confirmed that largely similar patterns of inositol polyphosphates are found in equivalent extracts of primary cultures of normal myeloid precursor cells maintained in the presence of interleukin 3, and also that TPA provokes declines in Ins(1,3,4,5)$P_4$ and Ins(1,3,4,5,6)$P_5$ levels in these cells that are similar to those seen in HL-60 cells (Bunce et al 1990, 1992).

These observations, which have provided the clearest examples yet reported of biological regulation of the levels of these fairly abundant intracellular compounds, suggest that inositol polyphosphates such as Ins(1,3,4,5,6)$P_5$ and Ins(1,2,3,4,5,6)$P_6$ must play some important, though not yet understood, role in cells. In particular, our observations suggest some form of participation either in the processes of haemopoietic differentiation or in the expression of differentiated cell character in myeloid cells, and they provide a unique opportunity for careful studies directed at identifying the function(s) of these compounds. These inositol polyphosphates are potent bivalent cation chelators, they appear to be located mainly in the cytosol compartment (unpublished data), and they are formed from non-lipid precursors. We therefore aim to undertake experiments to test the hypothesis that long-term regulation of the cystolic levels of the inositol polyphosphates Ins(3,4,5,6)$P_4$, Ins(1,3,4,5,6)$P_5$ and Ins$P_6$ plays some role in controlling the complex spatial and temporal patterns of receptor-stimulated $Ca^{2+}$ signalling that have been discovered in recent years (e.g. Cuthbertson & Cobbold 1991).

## *Acknowledgements*

We are grateful to the Medical Research Council, The Royal Society, the Leukaemia Research Fund and the British Diabetic Association for financial support. K. Anderson, L. A. Conroy, M. Finney and P. J. French were supported by MRC and SERC research studentships.

## References

Baxter MA, Bunce CM, Lord JM, French PJ, Michell RH, Brown G 1990 Changes in inositol transport during DMSO-induced differentiation of HL60 cells towards neutrophils. Biochem Biophys Acta 1091:158–164

Benton HP 1991 Cytokines and their receptors. Curr Opin Cell Biol 3:171–175

Berridge MJ, Irvine RF 1989 Inositol phosphates and cell signalling. Nature (Lond) 341:197–205

Berridge MJ, Downes CP, Hanley MR 1989 Neural and developmental actions of lithium: a unifying hypothesis. Cell 59:411–419

Bunce CM, Patton WN, Pound JD, Lord JM, Brown G 1990 Phorbol myristate acetate treatment of normal myeloid blast cells promotes monopoiesis and inhibits granulopoiesis. Leuk Res 14:1007–1017

Bunce CM, French PJ, Patton WM, Scott SA, Michell RH, Brown G 1992 Levels of inositol metabolites within normal myeloid blast cells and changes during their differentiation. Proc R Soc Lond B Biol Sci 247:27–33

Carpenter CL, Cantley L 1990 Phosphoinositide kinases. Biochemistry 29:11147–11156

Conroy LA, Jenkinson EJ, Owen JJT, Michell RH 1991 The role of inositol lipid hydrolysis in the selection of immature thymocytes. Biochem Soc Trans 19:90S

Cuthbertson KSR, Cobbold PH (eds) 1991 Oscillations in cell calcium. Cell Calcium 12:61–268

Desai DM, Newton ME, Kadlecek T, Weiss A 1990 Stimulation of the phosphatidylinositol pathway can induce T-cell activation. Nature (Lond) 348:66–69

Finkel TH, Kubo RT, Cambier JC 1991 T-cell development and transmembrane signaling: changing biological responses through an unchanging receptor. Immunol Today 12:79–85

Finney M, Guy GR, Michell RH et al 1990 Signalling by human interleukin 4 involves sequential activation of phosphatidylinositol 4,5-bisphosphate hydrolysis and adenylate cyclase. Eur J Immunol 20:151–156

Finney M, Michell RH, Gillis S, Gordon J 1991 Regulation of the interleukin-4 signal in human B lymphocytes. Biochem Soc Trans 19:287–291

Fisher DB, Mueller GC 1968 An early alteration in the phospholipid metabolism of lymphocytes by PHA. Proc Natl Acad Sci USA 60:1396–1402

French PJ, Bunce CM, Brown G, Creba JA, Michell RH 1988 Inositol phosphates in growing and differentiating HL60 cells. Biochem Soc Trans 16:985–986

French PJ, Bunce CM, Stephens LR et al 1991 Changes in the levels of inositol lipids and phosphates during the differentiation of HL60 promyelocytic cells towards neutrophils or monocytes. Proc R Soc Lond B Biol Sci 245:193–201

Hanley MR, Jackson TR, Vallejo M et al 1988 Neural functions: metabolism and actions of inositol metabolites in mammalian brain. Philos Trans R Soc Lond B Biol Sci 320:381–398

Irvine RF 1989 Functions of inositol phosphates. In: Michell RH, Drummond AH, Downes CP (eds) Inositol lipids in cell signalling. Academic Press, New York p 135–161

Michell RH, King CE, Piper CE et al 1988 Inositol lipids and phosphates in erythrocytes and HL60 cells. In: Gunn RB, Parker JC (eds) Cell physiology of blood. (41st Annual Symp Soc General Physiologists) Rockefeller University Press, New York p 345–356

Michell RH, Conroy LA, Finney M et al 1989 Inositol lipids and phosphates in the regulation of the growth and differentiation of haemopoietic and other cells. Philos Trans R Soc Lond B Biol Sci 327:193–207

Pittet D, Schlegel W, Lew DP, Monod A, Mayr GW 1989 Mass changes in inositol tetrakis- and pentakisphosphate isomers induced by chemotactic peptide stimulation in HL60 cells. J Biol Chem 264:18489–18493

Shears SB 1989 Metabolism of the inositol phosphates produced upon receptor activation. Biochem J 260:313–324

Smith CA, Williams GT, Kingston R, Jenkinson EJ, Owen JJT 1989 Antibodies to CD3/T cell receptor complex induce death by apoptosis in immature T cells in thymic cultures. Nature (Lond) 337:181–184

Stephens L, Hughes KT, Irvine RF 1991 Pathway of phosphatidylinositol(3,4,5)-trisphosphate synthesis in activated neutrophils. Nature (Lond) 351:33–39

Williams GT 1991 Programmed cell death: apoptosis and oncogenesis. Cell 65:1–2

Wyllie AH, Kerr JFR, Curries AR 1980 Cell death—the significance of apoptosis. Int Rev Cytol 68:251–305

## DISCUSSION

*Tsien:* You have shown that there are several messenger systems that participate in the activation of B cells. Do you know whether and how protein kinases are involved?

*Michell:* We know that if we leave phorbol dibutyrate out of the pharmacological reconstruction sequence we don't get the response. Similarly, I would expect that both the calcium part of the signal and the cAMP part of the signal involve protein kinase activation. We do see some characteristic changes of inositol phosphates in the cells during activation by IL-4, but these changes have not been very informative (Finney et al 1991). This is partly because these quiescent B cells, which are selected by density gradient centrifugation from tonsils, are metabolically rather quiet cells, so it's difficult to do good labelling experiments without pre-activation, and we obviously don't want that. I would guess that many of the key events are protein kinase-mediated, but we have no direct evidence.

*Tsien:* Which kinds of protein kinase C exist in these cells? Is there only one particular subtype?

*Michell:* I don't think that has been looked at.

*Kanoh:* You used phorbol dibutyrate to mimic the effect of diacylglycerol liberation. Is it not dangerous, in view of the pleiotropic effects of phorbol esters, to mimic protein kinase C activation in this way, rather than by using cell permeant diacylglycerols?

*Michell:* We used phorbol dibutyrate because we wanted to make a transient application and we could wash most of it away. I would expect to get the same results with a small amount of a permeant diacylglycerol, but it would be difficult to judge the quantity that would be needed to give only a minute or two of stimulation before its removal by metabolism. The fact that phorbol dibutyrate works, and that it works only if it is presented in a transient manner, reassured us that our approach was probably reasonable.

*Kanoh:* Have you ever checked the time course of diacylglycerol or phosphatidic acid liberation during the experiment?

*Michell:* We attempted to do that for diglyceride by labelling studies, but were not successful. We are now trying to apply an enzymic assay.

*Fredholm:* Could you be generating an agonist during the first phase, during the 10-minute delay before the cAMP rise? Possible candidates in this type of cell are adenosine or a prostanoid, either of which could easily be produced by the proliferative stimulus.

*Michell:* We haven't attempted to do experiments to check on adenosine, but we know that blockade of prostanoid synthesis with indomethacin does not prevent the cAMP rise.

*Parker:* In the response of B lymphocytes to IL-4 you saw a lag of about 10 min before the cAMP accumulation. If you treat the cells first with phorbol dibutyrate and ionomycin and then with IL-4, can you get the cAMP response without a time lag? In other words, is that pathway priming a cAMP response that may be directly IL-4-dependent?

*Michell:* We have not done that particular experiment.

*Thomas:* How important is the timing of events in these experiments? Is the signal a long-term one or is it 'washed out' with the phorbol dibutyrate? If you add the phorbol dibutyrate first, how long can you wait after the washout before you add the dibutyryl cAMP and still observe the response?

*Michell:* We have not done that experiment. As you will appreciate, there are an enormous variety of experiments we could do. We have focused mainly on trying to mimic the sequence of events with the timing that is seen normally in the cell, and on changing the order.

*Hunter:* Do you know whether IL-4 induces tyrosine phosphorylation events in human tonsil B cells?

*Michell:* I don't know of any data which indicate that it does.

*Hunter:* That might provide the secondary parallel pathway. Is the phospholipase C that's activated by IL-4 sensitive to pertussis toxin? If not, the enzyme might be PLC-$\gamma$, which can be activated by tyrosine phosphorylation.

*Michell:* We haven't done the pertussis toxin experiment. The obvious candidate for the secondary signal is tyrosine phosphorylation.

*Hunter:* Do you plan to use tyrosine kinase inhibitors to see whether they block the IL-4 response?

*Michell:* Yes; we have just begun such experiments.

*Ui:* Adenylate cyclase is a calmodulin-dependent enzyme. Might calmodulin be involved in the cAMP rise?

*Michell:* The experiment with BAPTA indicates that the calcium transient is necessary for the cAMP rise, but the rise in cAMP is temporally separated from the calcium increase so I doubt whether it is a direct response to calcium–calmodulin activation. It's certainly possible that a calcium–calmodulin-mediated event might be an intermediate step along the way to the cyclic AMP rise, but I don't think there can be a direct coupling.

*Ui:* Cyclic GMP could first be produced, and activate a phosphodiesterase.

*Michell:* We haven't assayed cyclic GMP.

*Cantrell:* IL-4 regulates activation of lots of different cells, including some T cells. Has the signalling pathway you have observed in human B cells been found in any other type of cell?

*Michell:* The original work looking for effects of IL-4 on inositol lipids was done in mouse B cells by Kevin Rigley in Gerry Klaus's laboratory, and he failed to see any effects (O'Garra et al 1987). John Gordon plans to investigate some other IL-4-sensitive cell populations.

*Fredholm:* My question relates to the T cell apoptosis experiments. David McConkey, Pier-Luigi Nicotera, Mikoel Jondal and Sten Orrenius have done a series of experiments which indicate that there are many ways to induce apoptosis in immature $CD4^+/CD8^+$ thymocytes; it can be achieved by raising cAMP (McConkey et al 1990), or by using corticosteroids (McConkey et al 1989a), agonistic anti-CD3 antibodies (McConkey et al 1989b) or agents such as dioxins (McConkey et al 1988). They have postulated that the apoptosis is brought about by 'unbalanced signalling'—that very strong stimulation of one particular pathway is deleterious to the cells. Your's was a rather more balanced scheme.

*Michell:* You are right that there are a number of ways in which you can provoke this response. We felt that we should do it by trying to go through the antigen receptor rather than using various pharmacological manipulations. We hoped that in this way we would harness the process that is going on *in situ*. One of the disadvantages of the mouse thymic culture system, despite the fact that it's a good reflection of T cell selection *in vivo*, is that there are undoubtedly additional factors in the thymus that play a role but which are not visible in our experiment. In contrast, many of the experiments that McConkey and co-workers have done have been on isolated thymocytes, released from the thymus and studied in culture. Our experience with the mouse thymus is that when the cells are taken from the thymic environment in the cultured lobe they undergo spontaneous apoptosis. That is why very few of our experiments have been attempted on isolated thymocytes. The *Staphylococcus* enterotoxin B experiment was done on isolated cells, but this involved gathering several people together to work very hard and very fast before the cells underwent spontaneous apoptosis.

We don't know what those very early and rapid spontaneous apoptotic events represent, but they are associated with a strange alteration in inositol phosphate metabolism. Inositol pentaphosphate and hexaphosphate are normally very metabolically stable in cells; although they show some relatively slow changes during myeloid cell differentiation (French et al 1991, Bunce et al 1992), they are excellent metabolically stable compounds against which to measure changes in other inositol phosphates during experiments on thymic lobes. However, if you take the mouse thymocytes out of their thymic environment, $InsP_5$ and $InsP_6$ disappear extremely rapidly by a completely unknown mechanism. This is the only situation in which we have seen those compounds changing so rapidly in any type of cell.

*Fischer:* The differentation of $CD4^+/CD8^+$ cells into helper cells and cytotoxic T lymphocytes goes through the T cell receptor–CD3 complex. Is there any evidence for participation of lymphocyte-specific tyrosine kinases in these reactions?

*Michell:* We have only done some preliminary experiments with inhibitors of tyrosine kinases.

*Fischer:* Can you block differentiation with herbimycin or other inhibitors of tyrosine kinases?

*Michell:* We have some indications that we can, but they are very preliminary.

*Fischer:* Are there changes in the level of phosphorylation of the ζ chains?

*Michell:* We haven't even contemplated doing that experiment on these almost invisible pieces of tissue.

*Toyoshima:* After activation of T lymphocytes with anti-CD3 ε chain antibodies, expression of *fyn* is increased 10-fold two hours after stimulation and gradually decreases (Katagiri et al 1989). Those T cells which express the *fyn* gene product probably escape apoptosis. $CD4^-/CD8^-$ T cells in mice with *lpr/lpr* or *gld/gld* mutations also overexpress *fyn*.

*Cantrell:* Professor Michell, you said that you think there is hydrolysis of $PtdInsP_2$ in anti-CD23-treated cells. Are you *sure* that it is hydrolysis of $PtdInsP_2$ and not hydrolysis of PtdIns?

*Michell:* Not entirely. Louise Conroy has done a very limited number of experiments in which she has analysed the inositol phosphates from cells after relatively brief stimulation; she managed to do HPLC by pooling the extracts from many lobes. She saw an appreciable accumulation of Ins(1,3,4)$P_3$ during 30 min of stimulation, presumably as a secondary metabolite of Ins(1,4,5)$P_3$. These results reassured us that $PtdInsP_2$ breakdown is occurring, but we cannot rule out some hydrolysis of PtdIns4P and/or PtdIns in the stimulated lobes.

*Cantrell:* PtdIns metabolism can be induced by CD3 agonists in T cells with no calcium changes induced, suggesting that there is hydrolysis of PtdIns in situations in which $PtdInsP_2$ is not hydrolysed.

*Michell:* Recently, John Owen's group have been imaging thymic lobes loaded with fluo-3 in a confocal microscope and stimulating them with anti-CD3. They have seen substantial numbers of cells lighting up with a calcium signal. This suggests, but does not prove, that Ins(1,4,5)$P_3$ is being generated.

*Tsien:* In the T cells, ionomycin stimulated apoptosis without an alteration of the levels of inositol phosphates. Did you try to prevent apoptosis with BAPTA?

*Michell:* No. This experiment was done essentially to confirm conclusions obtained by others. Although we have no reason to doubt the interpretation generally made from this result, I agree that this would have been a good experiment to do.

## References


Bunce CM, French PJ, Patton WP, Scott SA, Michell RH, Brown G 1992 Levels of inositol metabolites within normal myeloid blast cells and changes during their differentiation towards monocytes. Proc R Soc Lond B Biol Sci 247:27–33

Finney M, Michell RH, Gillis S, Gordon J 1991 Regulation of the interleukin-4 signal in human B lymphocytes. Biochem Soc Trans 19:287–291

French PJ, Bunce CM, Lord JM et al 1991 Changes in the levels of inositol lipids and phosphates during the differentiation of HL60 promyeolcytic cells towards neutrophils or monocytes. Proc R Soc Lond B Biol Sci 245:193–201

Katagiri T, Urakawa K, Yamanashi Y et al 1989 Overexpression of *src* family tyrosine kinase $p59^{fyn}$ in $CD4^-/CD8^-$ T cells with lympho-proliferative disorder. Proc Natl Acad Sci USA 86:10064–10068

McConkey DJ, Hartzell P, Duddy SK, Hakansson H, Orrenius S 1988 2,3,7,8-tetrachlorodibenzo-*p*-dioxin kills immature thymocytes by a $Ca^{2+}$-mediated endonuclease activation. Science (Wash DC) 242:256–259

McConkey DJ, Nicotera P, Hartzell P, Bellomo G, Wyllie AH, Orrenius S 1989a Glucocorticoids activate a suicide process in thymocytes through an elevation of cytosolic $Ca^{2+}$ concentration. Arch Biochem Biophys 269:365–370

McConkey DJ, Hartzell P, Amados-Pérez JF, Orrenius S, Jondal M 1989b Calcium dependent killing of immature thymocytes by stimulation via the CD3/T cell receptor complex. J Immunol 143:1801–1806

McConkey DJ, Orrenius S, Jondal M 1990 Agents that elevate cAMP stimulate DNA fragmentation in thymocytes. J Immunol 145:1227–1230

O'Garra A, Rigley KP, Holman M, McLaughlin JB, Klaus GGB 1987 B-cell-stimulatory factor 1 reverses Fc receptor-mediated inhibition of B-lymphocyte activation. Proc Natl Acad Sci USA 84:6254–6258

# The inositol 1,4,5-trisphosphate receptor

Katsuhiko Mikoshiba*‡, Teiichi Furuichi†, Atsushi Miyawaki*†, Shingo Yoshikawa*†, Nobuaki Maeda*, Michio Niinobe*, Shinji Nakade*, Toshiyuki Nakagawa*, Hideyuki Okano* and Jun Aruga‡

*Institute for Protein Research, Osaka University, 3-2 Yamadaoka, Suita, Osaka 565, †National Institute for Basic Biology, 38 Nishigonaka, Myodaiji, Okazaki 444 and ‡The Institute of Medical Science, The University of Tokyo, 4-6-1, Shirokane-dai, Minato-ku, Tokyo 108, Japan

*Abstract*. Inositol 1,4,5-trisphosphate ($InsP_3$) is a second messenger that releases $Ca^{2+}$ from its intracellular stores. The $InsP_3$ receptor has been purified and its cDNA has been cloned. We have found that the $InsP_3$ receptor is identical to $P_{400}$ protein, first identified as a protein enriched in cerebellar Purkinje cells. We have generated an L-fibroblast cell transfectant that produces cDNA-derived $InsP_3$ receptors. The protein displays high affinity and specificity for $InsP_3$. $InsP_3$ induces greater $Ca^{2+}$ release from membrane vesicles from transfected cells than from those from control L-fibroblasts. After incorporation of the purified $InsP_3$ receptor into lipid bilayers $InsP_3$-induced $Ca^{2+}$ currents were demonstrated. These results suggest that the $InsP_3$ receptor is involved in physiological $Ca^{2+}$ release. Immunogold labelling using monoclonal antibodies against the receptor showed that it is highly concentrated on the smooth-surfaced endoplasmic reticulum and slightly on the outer nuclear membrane and rough endoplasmic reticulum; no labelling of Golgi apparatus, mitochondria and plasmalemma was seen. Cross-linking experiments showed that the receptor forms a homotetramer. The approximately 650 N-terminal amino acids are highly conserved between mouse and *Drosophila*, and this region contains the critical sequences for $InsP_3$ binding. We have investigated the heterogeneity of the $InsP_3$ receptor using the polymerase chain reaction and have found novel subtypes of the mouse $InsP_3$ receptor that are expressed in a tissue-specific and developmentally specific manner.

*1992 Interactions among cell signalling systems. Wiley, Chichester (Ciba Foundation Symposium 164) p 17–35*

Many of the receptors for neurotransmitters in the nervous system are coupled to the phosphoinositide cycle. In this cycle, phosphatidylinositol bisphosphate is hydrolysed in response to receptor stimulation, producing inositol 1,4,5-trisphosphate ($InsP_3$) and diacylglycerol. Diacylglycerol activates protein kinase C, which then exerts its physiological function through phosphorylation (Nishizuka 1988). $InsP_3$ functions as a second messenger, releasing calcium ions from their storage site. By mobilizing intracellular calcium ions $InsP_3$ plays diverse roles in many physiological processes. Although it had been shown that

$InsP_3$ releases calcium ions from calcium storage sites, the molecular mechanism behind the calcium release was unknown. Here, we briefly describe the biochemical and molecular characterization of the $InsP_3$ receptor and describe the cross-talk between this and other signalling systems.

## Purkinje cell-enriched $P_{400}$ protein as an $InsP_3$ receptor

The mouse cerebellum contains five types of neuron. The Purkinje cell, one of these, plays an important role in information processing, because the sole output from the cerebellar cortex is the axons of Purkinje cells. In the course of biochemical analysis a protein termed $P_{400}$ was found to be enriched in Purkinje cells, but greatly reduced in the cerebella of Purkinje cell-deficient ataxic mutants (Mikoshiba et al 1979, 1985). An $InsP_3$-binding protein was characterized biochemically using $InsP_3$-binding activity as a marker; this protein was also enriched in cerebellar Purkinje cells (Supatapone et al 1988a). It has been confirmed recently that these two proteins are identical through experiments using specific monoclonal antibodies against $P_{400}$ (Maeda et al 1990).

The apparent relative molecular mass in SDS–PAGE of the receptor is 250 000 in the mouse (Maeda et al 1988, 1989, 1990) and 260 000 in the rat, although from the amino acid sequence an $M_r$ of 313 000 is predicted (Furuichi et al 1989a). Mannosyl-glycoprotein endo-β-*N*-acetylglucosaminidase F digestion abolishes binding of concanavalin A, suggesting that the protein has asparagine-linked oligosaccharide chains (Maeda et al 1988).

## The primary structure of the $InsP_3$ receptor

Two mouse cerebellum cDNA libraries, synthesized by priming with random hexamers or oligo(dT), were constructed in a bacteriophage λgt11 expression vector and screened with the three anti-$P_{400}$ monoclonal antibodies. From the cDNA sequence (Furuichi et al 1989a) the $InsP_3$ receptor is expected to be composed of 2749 amino acids, with a relative molecular mass of 313 000; the apparent $M_r$ is 250 000 (Furuichi et al 1989b). The $InsP_3$ receptor is an integral membrane protein; it lacks a definable signal sequence with a hydrophobic stretch at the N-terminus, which indicates that an N-terminal hydrophilic region is located on the cytoplasmic face of the membrane.

Recently, the primary structure of the ryanodine receptor, which mobilizes calcium from the calcium storage site (the sarcoplasmic reticulum) in skeletal muscle, has been described (Takeshima et al 1989). The $InsP_3$ receptor has fragmentary sequence identity with the ryanodine receptor (Furuichi et al 1989a, Mignery et al 1989). In particular, the putative transmembrane domains show remarkable similarity to the transmembrane segments of the ryanodine receptor. The similarity does not extend to the putative binding sites for modulators (calcium, nucleotides and calmodulin) of the ryanodine receptor proposed by

Takeshima et al (1989). The two receptors thus resemble each other in their proposed transmembrane topologies, their large cytoplasmic N-terminal regions, their short C-terminal regions and their transmembrane regions near the C-terminus. These similarities imply that the $InsP_3$ receptor is also involved in calcium mobilization.

## The biochemical properties of the $InsP_3$ receptor

The $InsP_3$ receptor can be phosphorylated by the purified catalytic subunit of cyclic AMP-dependent protein kinase and slightly by $Ca^{2+}$/calmodulin-dependent protein kinase II (Supattapone et al 1988b, Yamamoto et al 1989). Protein kinase C also phosphorylates the receptor slightly. Protein kinase A phosphorylates only seryl residues in the $InsP_3$ receptor (Yamamoto et al 1989). These results suggest that $InsP_3$-induced $Ca^{2+}$ release is regulated by protein kinase A and also slightly by $Ca^{2+}$/calmodulin-dependent protein kinase II and protein kinase C.

Cross-linkage of the $InsP_3$ receptor was done to investigate its subunit composition. Agarose–PAGE analysis after cross-linkage of the purified receptor revealed four distinct bands of $M_r$ 320 000, 650 000, 1 000 000 and 1 250 000 (Fig. 1) (Maeda et al 1991). No higher molecular weight cross-linked products were observed at any of the concentrations of cross-linker used. The same pattern of cross-linking was found with microsome-bound receptors. To investigate whether the native receptor exists in a covalently or non-covalently coupled state we denatured the purified $InsP_3$ receptor using SDS under reducing conditions; after this the receptor showed an $M_r$ of 320 000 in agarose–PAGE. When it was denatured under non-reducing conditions, four bands were observed on gels, as with the cross-linked samples. However, longer incubation of the sample resulted in a disappearance of the higher molecular mass bands and a corresponding increase in the lower molecular mass bands. From these results, it is proposed that the native cerebellar $InsP_3$ receptor exists as a non-covalently coupled homotetramer of $M_r$ 320 000 subunits (Maeda et al 1991).

## The $InsP_3$ receptor is localized on smooth-surfaced endoplasmic reticulum

Immunohistochemical studies of the $InsP_3$ receptor have shown that it is expressed at high levels in Purkinje cells (Ross et al 1989, Furuichi et al 1989a) and that it is also expressed in many regions of the mouse brain, such as the hippocampus, striatum and cerebral cortex (Nakanishi et al 1991). *In situ* hybridization has revealed that it is ubiquitously distributed in various tissues other than brain (Furuichi et al 1990). Immunogold experiments using three monoclonal antibodies (mAb 10A6, 4C11 and 18A10) have been used to investigate the localization of the $InsP_3$ receptor in mouse cerebellar Purkinje cells. The epitopes recognized by the three antibodies were detected on the smooth

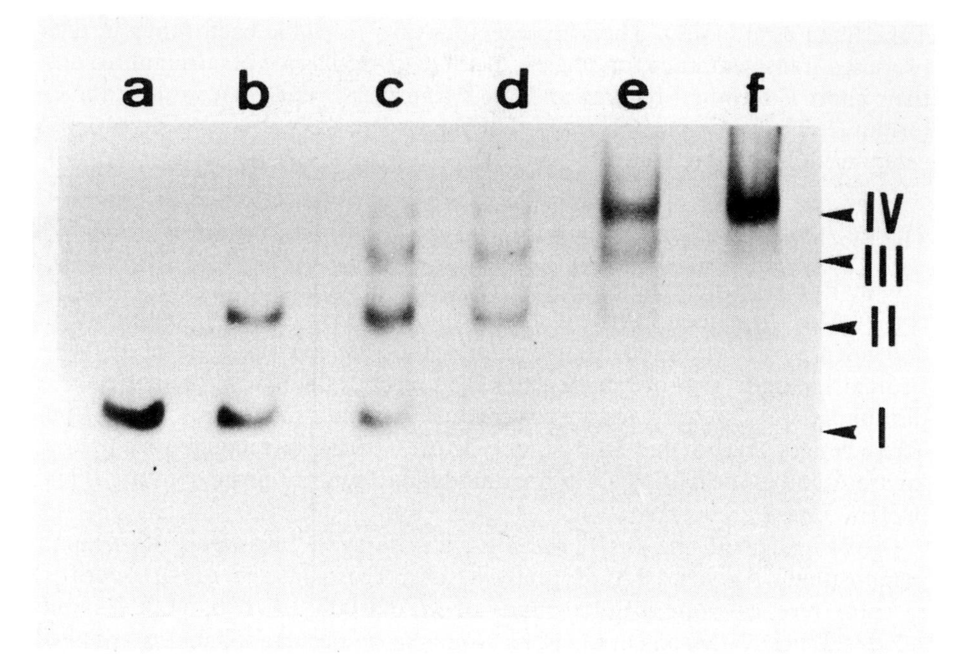


FIG. 1. Cross-linking of cerebellar microsomal fractions shows that the $InsP_3$ receptor forms a homotetrameric structure. The cross-linkage was done with disuccinimidyl tartrate at concentrations of 0 (lane a), 0.06 (b), 0.6 (c), 1.7 (d), 5 (e) and 10 (f) mM. $M_r$ values for bands labelled I–IV are, respectively, 320 000, 650 000 1 000 000 and 1 250 000.

endoplasmic reticulum, particularly on the stacks of flattened smooth endoplasmic reticulum, subsurface cisternae and spine apparatus; rough endoplasmic reticulum and the outer nuclear membrane were scarcely stained (Otsu et al 1990). There was a very low level of reaction with the plasma membrane, synaptic densities, mitochondria and Golgi apparatus. The three antibodies recognized the $InsP_3$ receptor at the cytoplasmic side of the membrane, suggesting that both N and C-termini are on the cytoplasmic side. Gold particles are usually localized on the fuzzy structure of the cytoplasmic surface of smooth endoplasmic reticulum (Fig. 2). Because the fuzzy structure is observed even in the control experiment, in which the primary monoclonal antibody is omitted, it has been suggested that the fuzzy structure corresponds to the feet structures of ryanodine receptors. Satoh et al (1990) have made similar observations with antisera raised against the purified $InsP_3$ receptor.

## The expressed $InsP_3$ receptor has $Ca^{2+}$-releasing activity

Complementary DNA with a β-actin promoter at the 5′ end has been transfected into L-fibroblast cells to examine whether the clone endodes an $InsP_3$-binding

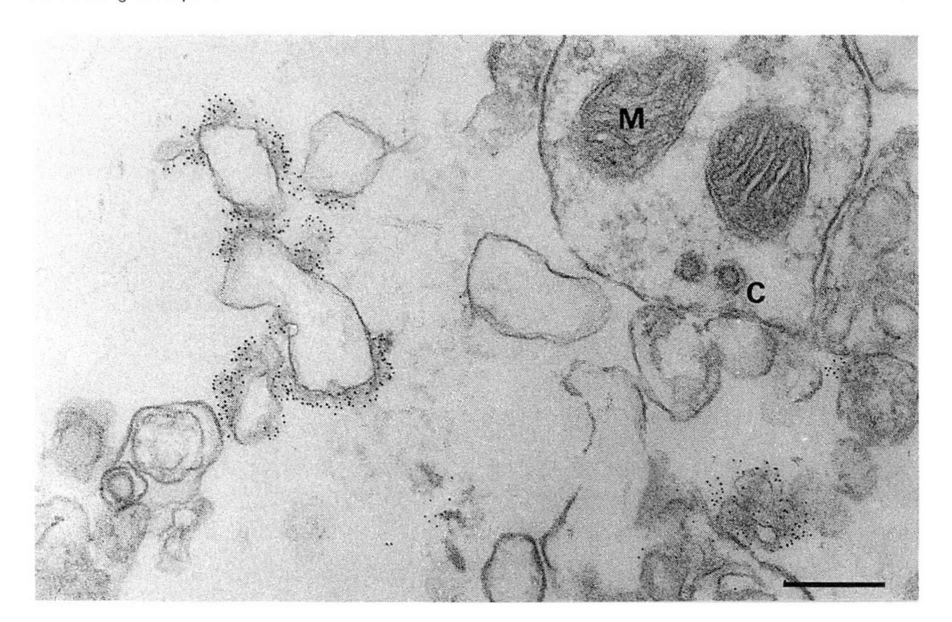


FIG. 2. Immunogold staining of the $InsP_3$ receptor in the fuzzy structures associated with the cytoplasmic surface of the smooth vesicotubular structures. These are presumably derived from the dendritic spine apparatus. M, mitochondrion; C, coated vesicle. Scale bar, 0.2 μm.

sequence and a $Ca^{2+}$ release channel. The cells contain endogenous protein immunoreactive against anti-$P_{400}$ monoclonal antibodies (Fig. 3) (Furuichi et al 1989a, Miyawaki et al 1990). The $M_r$ of the endogenous $InsP_3$ receptor is smaller than that of the receptor from mouse cerebellum. The cells transfected with cDNA produced a large amount of protein immunoreactive against all three of the monoclonal antibodies. The expressed protein displays high affinity and specificity for $InsP_3$ (in comparison with other inositol phosphates), and a high binding capacity, as does the $InsP_3$ receptor in cerebellar microsomes.

Thus, stable transformants that express the cerebellar $InsP_3$ receptor have been established in an L-fibroblast cell line. Direct evidence for calcium ion-releasing activity of the cerebellar $InsP_3$ receptor expressed in natural membranes of non-neuronal cells has been demonstrated (Miyawaki et al 1990). $InsP_3$ releases only a fraction of the $Ca^{2+}$ within cells, indicating that only some of the $Ca^{2+}$ pools are sensitive to $InsP_3$. It is believed that $InsP_3$-sensitive pools possess $InsP_3$ receptors, whereas $InsP_3$-insensitive pools do not. Calcium ion release experiments have revealed that the $InsP_3$-sensitive calcium ion pools in L-fibroblast transformant cells are larger in size than those in non-transfected cells (Fig. 4).

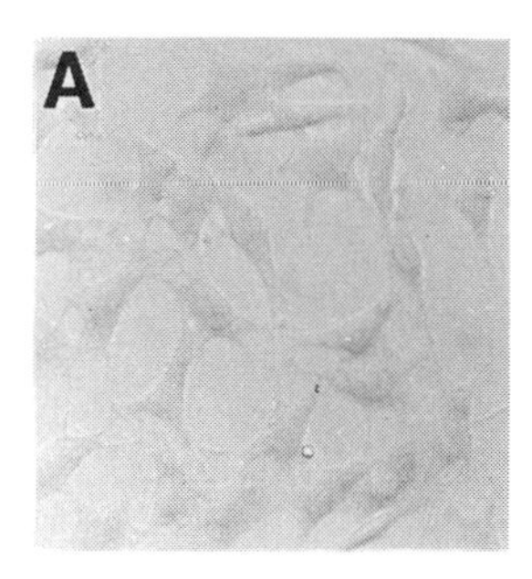


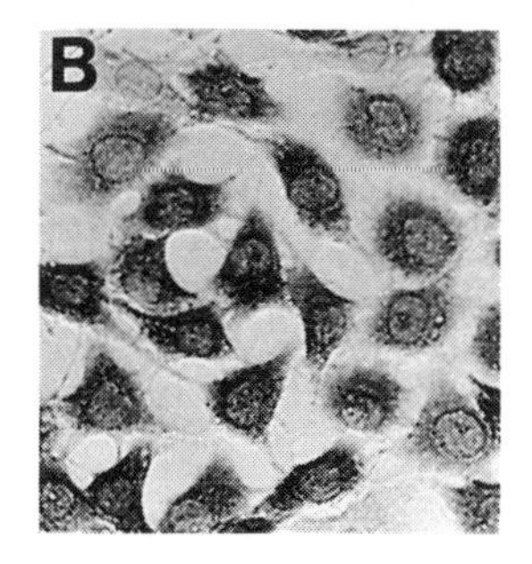


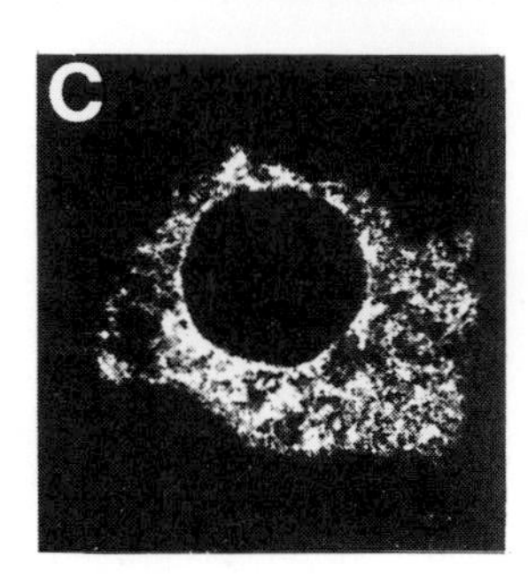


FIG. 3. $InsP_3$ receptors expressed after transfection of $InsP_3$ receptor cDNA into L-fibroblasts. The figure shows immunohistochemical staining of the receptor on L-fibroblasts (using anti-$P_{400}$ monoclonal antibody 4C11, visualized with fluorescein isothiocyanate-labelled goat anti-rabbit IgG) before (A) and after (B) transfection. (C) A confocal section of a single transfected cell strongly fluorescently labelled. Original magnifications: A, ×210; B, ×250; C, ×100.

The expression of the cDNA-derived $InsP_3$ receptor may cause some $InsP_3$-insensitive calcium ion pools in L-fibroblasts to be converted into $InsP_3$-sensitive pools in transfected cells. This may account for the larger size of the $InsP_3$-sensitive $Ca^{2+}$ pool in the transfected cells. We do not know in which organelle the cDNA-derived $InsP_3$ receptors accumulate. Although there might be sorting mechanisms that cause synthesized $InsP_3$ receptors to be preferentially localized in the $InsP_3$-sensitive $Ca^{2+}$ pools, the increase in maximal $Ca^{2+}$ release suggests that the cDNA-derived $InsP_3$ receptors are distributed not only into $InsP_3$-sensitive pools, but also into $InsP_3$- insensitive calcium ion pools in transfected cells. The $EC_{50}$ value for $InsP_3$-induced calcium ion release is about 10-fold lower than that before transfection (Fig. 4). The greater sensitivity of the transformant may be a result of several factors. The measured $EC_{50}$ value is not necessarily a reflection of the actual $K_m$ value for $InsP_3$-induced calcium ion release; it probably results from the greatly increased number of $InsP_3$ receptors. cDNA-derived $InsP_3$ receptors may actually differ from endogenous L-fibroblast receptors in the relative efficiency of coupling between receptor occupancy and channel opening. The increased sensitivity may be due to a considerably decreased relative effectiveness of proteins that normally control the activity of the receptor, such as protein kinase A and calmedin (Supattapone et al 1988a).

### Functional regulation of the $InsP_3$ receptor as analysed by reconstitution experiments—cross talk with phosphorylation systems

Reconstitution of the $InsP_3$ receptor purified from mouse cerebellum into planar lipid bilayers has shown that the receptor contains a cation-selective

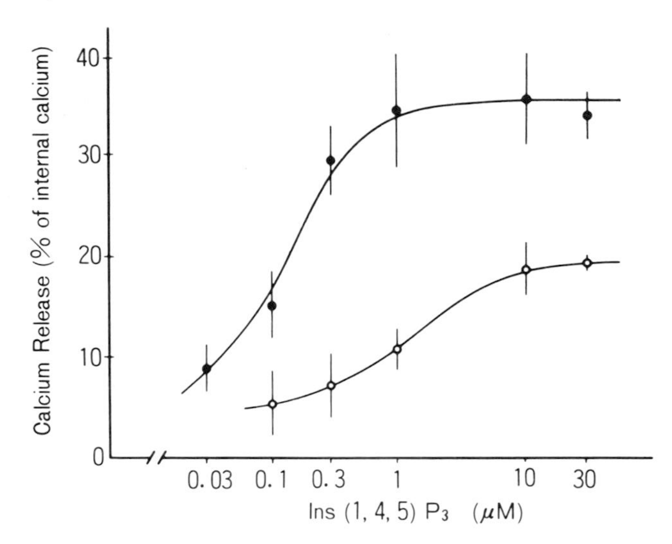


FIG. 4. Dose–response curves for $InsP_3$-induced $Ca^{2+}$ release from membrane fractions of L-fibroblast cells expressing the $InsP_3$ receptor (closed circles) and from fibroblasts in which the expression vector alone had been transfected (open circles). Results are shown as % of the $Ca^{2+}$ content before addition of $InsP_3$ and are mean values ± SD from six assays. The membrane fractions were incubated in medium containing $^{45}Ca^{2+}$ and EGTA was added to buffer free $Ca^{2+}$ to 200–400 nM. $Ca^{2+}$ uptake was initiated by addition of ATP (5 mM) and an ATP-regenerating system (creatine phosphate and creatine kinase). After 17 min incubation at 30 °C $InsP_3$ was added. Samples were filtered at 16 min and at 17 min 40 s, and the amount of $^{45}Ca^{2+}$ bound to the filters was assessed by liquid scintillation counting. A slight difference in $Ca^{2+}$-pumping activity was detected between the two cell types. In each experiment the level of $^{45}Ca^{2+}$ accumulated after 17 min in the membrane fractions from cells expressing the receptor was about 85% of that in the fractions from cells containing only the expression vector.

channel that is opened by $InsP_3$ (Fig. 5). The channels show several conductance states. Addition of ATP in the presence of $InsP_3$ generated large conductance currents, probably resulting from a change in the full open conductance level or a shift to a greater conductance state. Even in the absence of ATP the same high conductance level was occasionally observed. Thus, it is likely that ATP modifies the channel to allow it to reach a state of greater conductance. A photoaffinity labelling experiment with [$\alpha$-$^{32}$P]-8-azide-ATP has proved that ATP binds to the cerebellar $InsP_3$ receptor (Maeda et al 1991). Scatchard analysis of [$\alpha$-$^{32}$P]ATP binding to the $InsP_3$ receptor purified from mouse cerebellum indicated there was a single ATP-binding site with a $K_d$ value of 17 μM and a $B_{max}$ of 2.3 pmol/μg of protein. Scatchard analysis of binding of tritiated $InsP_3$ to the purified receptor gave a $B_{max}$ value of 2.1 pmol/μg of protein. It seems, therefore, that there are the same number of ATP- and $InsP_3$-binding sites on the receptors. A nucleotide-binding consensus sequence,

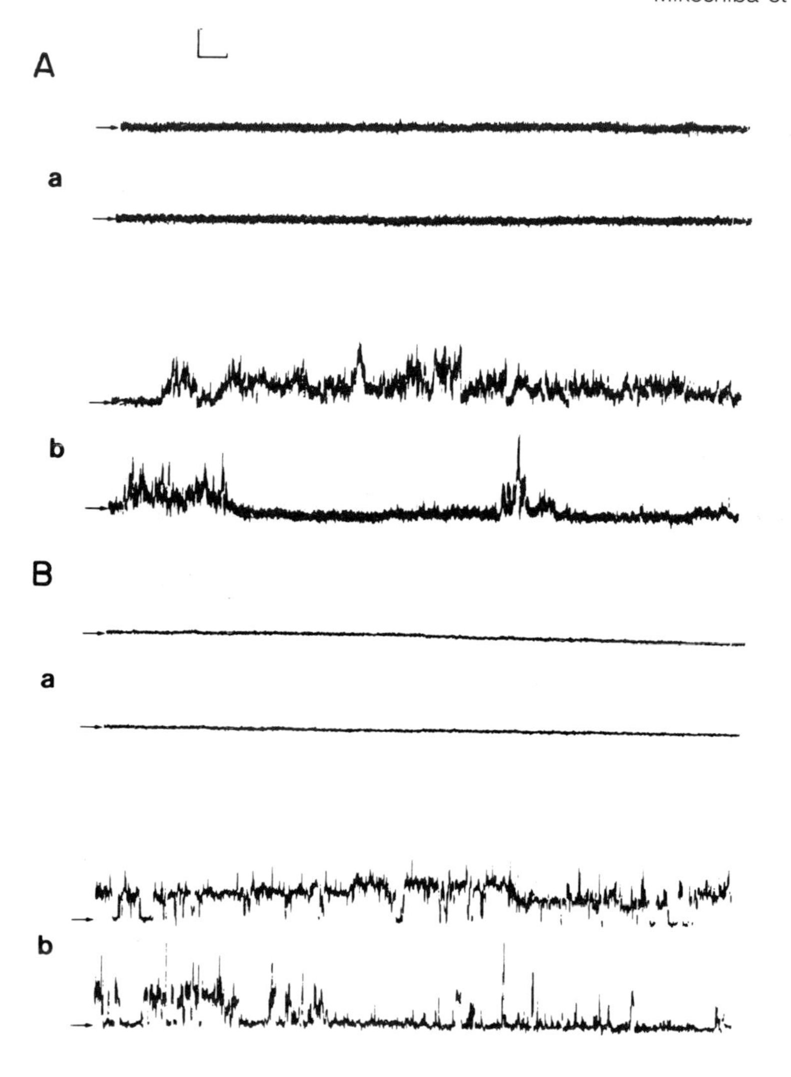


FIG. 5. Recordings of single-channel currents mediated by the purified $InsP_3$ receptor incorporated into planar lipid bilayers. (A) $Ca^{2+}$ currents were recorded after the addition of 2 µg of the $InsP_3$ receptor to one chamber, termed cis, containing 125 mM Tris, 250 mM Hepes (pH 7.4) and 0.1 µM free $Ca^{2+}$. The other chamber, designated trans, contained 53 mM $Ca(OH)_2$ and 250 mM Hepes (pH 7.4). No fluctuation in current was observed before the addition of $InsP_3$ (a). Channel opening was observed after addition of 4.8 µM $InsP_3$ to the cis chamber (b). (B) $Na^+$ currents mediated by the $InsP_3$ receptor were recorded in asymmetric NaCl solutions. The cis chamber contained 0.1 M NaCl, 0.1 µM free $Ca^{2+}$ and 5 mM Tris–Hepes (pH 7.4). Recordings are (a) before and (b) after addition of 4.8 µM $InsP_3$ to the cis chamber. Vertical calibration, 0.5 pA; horizontal calibration, 5 s. Arrows to the left indicate the closed state.

Gly-X-Gly-X-X-Gly, is found in the N-terminal cytoplasmic domain of the mouse cerebellar $InsP_3$ receptor (amino acid residues 2016–2021). The binding is selective for adenine nucleotides and the affinity is in the order ATP > ADP > AMP. It has been reported by Smith et al (1985) and Suematsu et al (1985) that ATP stimulates $InsP_3$-induced calcium release in smooth muscle cells. Ehrlich & Watras (1988) reported that 100 μM AMP-PCP, a non-hydrolysable analogue of ATP, increased the open probability of the $InsP_3$-gated channels of the aortic sarcoplasmic reticulum two-fold. The cerebellar receptors and the smooth muscle $InsP_3$ receptors are regulated similarly by adenine nucleotides. Ferris et al (1989) demonstrated $InsP_3$-induced $Ca^{2+}$ flux from lipid vesicles containing cerebellar $InsP_3$ receptors. This flux was stimulated by ATP in a concentration-dependent manner (Ferris et al 1990). ATP increased $InsP_3$-induced calcium flux at 1–10 μM, but the effect diminished between 0.1 and 1.0 mM. As the intracellular concentration of ATP is about 1 mM Ferris et al (1990) speculated that rapid filling of $Ca^{2+}$ stores by the $Ca^{2+}$–ATPase may deplete the local ATP concentration to submillimolar levels, with the $InsP_3$ receptor channel then being activated, facilitating $InsP_3$-induced $Ca^{2+}$ release. Our experiments in lipid bilayers showed that ATP stimulated channel opening most effectively at a concentration of 0.6 mM. This discrepancy may be due to differences in the lipids used for reconstitution. It is likely that the reduction of the enhancing effect of ATP at higher concentrations reported by Ferris et al (1990) is mediated by the inhibition of $InsP_3$ binding; ATP inhibits $InsP_3$ binding to the purified $InsP_3$ receptor with an $IC_{50}$ value of 2 mM (Maeda et al 1991).

Calcium inhibits binding of $InsP_3$ to microsomal fractions of the cerebellum, but sensitivity to $Ca^{2+}$ is lost in the purified receptor, suggesting that the calcium-regulating molecule calmedin is present in microsomal fractions (Supatapone et al 1988a).

The calmodulin antagonists W7, W13 and CGS-934313 inhibited $InsP_3$-induced calcium mobilization in rat liver epithelial cells. It was therefore suggested that $InsP_3$-induced $Ca^{2+}$ release is modulated by calmodulin. The $InsP_3$ receptor binds to calmodulin in the presence of calcium ions (Maeda et al 1990), but the addition of calmodulin did not affect $InsP_3$ binding to the cerebellar $InsP_3$ receptor.

## Tissue-specific and developmentally regulated diversity of the $InsP_3$ receptor

Novel subtypes of the $InsP_3$ receptor were found through the polymerase chain reaction (PCR). Different subtypes of the mouse $InsP_3$ receptor are expressed in a tissue-specific and developmentally specific manner (Nakagawa et al 1991a,b). The newly discovered subtypes of $InsP_3$ receptor differ from that previously reported in the structure of two small variably spliced segments (SI and SII; Fig. 6). Some of the subtypes are specific to the central nervous system;

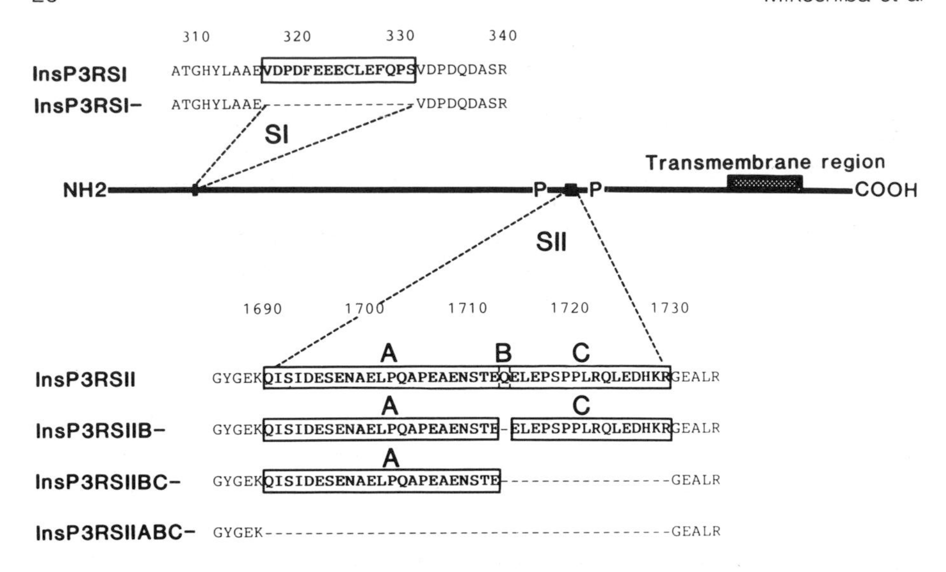


FIG. 6. Heterogeneity of $InsP_3$ receptors produced from RNA splicing of the SI and SII regions. SI is localized near the $InsP_3$-binding site and SII is located at the centre of two phosphorylation sites, indicated by P.

these include a domain consisting of 40, 23 or 39 amino acids located in the centre of the protein (SII). Peripheral tissues exclusively express a form in which the variable central domain (SII) is absent (Danoff et al 1991, Nakagawa et al 1991a). Another variant identified contains a 15-amino acid domain (SI) located near the amino terminus (Nakagawa et al 1991a, Mignery et al 1990). Because these variations in sequence are close to sites of phosphorylation or $InsP_3$ binding, it is speculated that selective expression of $InsP_3$ receptor subtypes generates tissue-specific and developmentally specific, functionally distinct channels.

## Structure–function relationship of the $InsP_3$ receptor

A series of internal deletion or C-terminal truncation mutant proteins were expressed in NG108-15 cells to analyse the structure–function relationships of the receptor (Fig. 7) (Miyawaki et al 1991). We have recently succeeded in cloning cDNA of the *Drosophila* $InsP_3$ receptor. The approximately 650 N-terminal amino acids are highly conserved between the mouse and *Drosophila* receptors, and are within the large cytoplasmic portion of the receptor. These results, combined with those of Mignery & Sudhof (1990), suggest that this region contains the critical sequences for $InsP_3$ binding, which probably form the three-dimensionally restricted binding site. Each $InsP_3$ receptor subunit binds one $InsP_3$ molecule through the N-terminal region. Cross-linking experiments

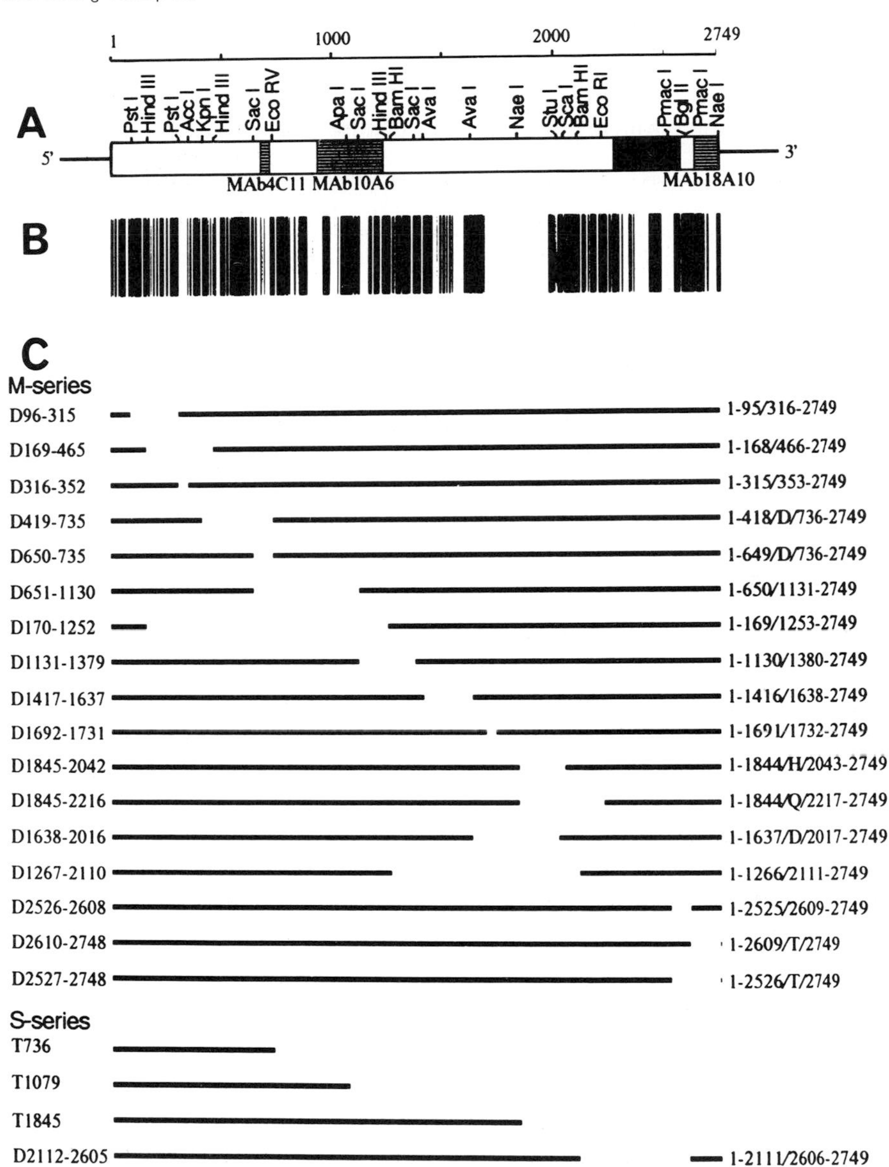


FIG. 7. Strategies for the construction of mutant $InsP_3$ receptors. (A) A restriction map of $InsP_3$ receptor cDNA. (B) Distribution of the amino acids identical in mouse and *Drosophila* $InsP_3$ receptors. (C) Internal deletion and C-terminal truncation mutants. Approximately 650 N-terminal amino acids are conserved between the mouse and *Drosophila* receptors; these play an important role in $InsP_3$ binding. The mutants that lack the transmembrane region also show $InsP_3$-binding activity as soluble proteins.

have revealed that $InsP_3$ receptors form intermolecular associations at the transmembrane domains and/or the C-termini. The interaction between the receptor subunit and $InsP_3$ may cause a conformational change in the tetrameric complex which results in the opening of the $Ca^{2+}$ channel.

## References

Danoff SK, Ferris CD, Donath C et al 1991 Inositol 1,4,5-trisphosphate receptors: distinct neuronal and nonneuronal forms derived by alternative splicing differ in phosphorylation. Proc Natl Acad Sci USA 88:2951–2955

Ehrlich BE, Watras J 1988 Inositol 1,4,5-trisphosphate activates a channel from smooth muscle sarcoplasmic reticulum. Nature (Lond) 336:583–589

Ferris CD, Huganir RL, Supattapone S, Snyder SH 1989 Purified inositol 1,4,5-trisphosphate receptor mediates calcium flux in reconstituted lipid vesicles. Nature (Lond) 342:87–89

Ferris CD, Huganir RL, Snyder SH 1990 Calcium flux mediated by purified inositol 1,4,5-trisphosphate receptor reconstituted into lipid vesicles is allosterically regulated by adenine nucleotides. Proc Natl Acad Sci USA 87:2147–2151

Furuichi T, Yoshikawa S, Miyawaki A, Wada K, Maeda N, Mikoshiba K 1989a Primary structure and functional expression of the inositol 1,4,5-trisphosphate-binding protein $P_{400}$. Nature (Lond) 342:32–38

Furuichi T, Yoshikawa S, Mikoshiba K 1989b Nucleotide sequence of cDNA encoding $P_{400}$ protein in the mouse cerebellum. Nucleic Acids Res 17:5385–5386

Furuichi T, Shiota C, Mikoshiba K 1990 Distribution of inositol 1,4,5-trisphosphate receptor mRNA in mouse tissues. FEBS (Fed Eur Biochem Soc) Lett 267:85–88

Maeda N, Niinobe M, Nakahira K, Mikoshiba K 1988 Purification and characterization of $P_{400}$ protein, a glycoprotein characteristic of Purkinje cells, from mouse cerebellum. J Neurochem 51:1724–1730

Maeda N, Niinobe M, Inoue Y, Mikoshiba K 1989 Developmental expression and intracellular location of $P_{400}$ protein characteristic of Purkinje cells in the mouse cerebellum. Dev Biol 133:67–76

Maeda N, Niinobe M, Mikoshiba K 1990 A cerebellar Purkinje cell marker $P_{400}$ protein is an inositol 1,4,5-trisphosphate ($InsP_3$) receptor protein—purification and characterization of $InsP_3$ receptor complex. EMBO (Eur Mol Biol Organ) J 9:61–67

Maeda N, Kawasaki T, Nakade S et al 1991 Structure and functional characterization of inositol 1,4,5-trisphosphate receptor channel from mouse cerebellum. J Biol Chem 266:1109–1116

Mignery GA, Sudhof TC 1990 The ligand binding site and transduction mechanism in the inositol 1,4,5-trisphosphate receptor. EMBO (Eur Mol Biol Organ) J 9:3893–3898

Mignery GA, Sudhof TC, Takei K, De Camilli P 1989 Putative receptor for inositol 1,4,5-trisphosphate similar to ryanodine receptor. Nature (Lond) 342:192–195

Mignery GA, Newton CL, Archer BT, Sudhof TC 1990 Structure and expression of the rat inositol 1,4,5-trisphosphate receptor. J Biol Chem 265:12679–12685

Mikoshiba K, Huchet M, Changeux J-P 1979 Biochemical and immunological studies on the $P_{400}$ protein, a protein characteristic of the Purkinje cell from mouse and rat cerebellum. Dev Neurosci 2:254–275

Mikoshiba K, Okano H, Tsukada Y 1985 $P_{400}$ protein characteristic to Purkinje cells and related proteins in cerebella from neuropathological mutant mice: autoradiographic study by $^{14}C$-leucine and phosphorylation. Dev Neurosci 7:179–187

Miyawaki A, Furuichi T, Maeda N, Mikoshiba K 1990 Expressed cerebellar-type inositol 1,4,5-trisphosphate receptor, $P_{400}$, has calcium release activity in a fibroblast L cell line. Neuron 5:11–18
Miyawaki A, Furuichi T, Ryou Y et al 1991 Structure–function relationships of the mouse inositol 1,4,5-trisphosphate receptor. Proc Natl Acad Sci USA 88:4911–4915
Nakagawa T, Okano H, Furuichi T, Aruga J, Mikoshiba K 1991a The subtypes of the mouse inositol 1,4,5-trisphosphate receptor are expressed in a tissue-specific and developmentally specific manner. Proc Natl Acad Sci USA 88:6244–6248
Nakagawa T, Shiota C, Okano H, Mikoshiba K 1991b Differential localization of alternative spliced transcripts encoding inositol 1,4,5-trisphosphate receptors in mouse cerebellum and hippocampus: *in situ* hybridization study. J Neurochem 57:1807–1810
Nakanishi S, Maeda N, Mikoshiba K 1991 Immunohistochemical localization of an inositol 1,4,5-trisphosphate receptor, $P_{400}$, in neural tissue: studies in developing and adult mouse brain. J Neurosci 11:2075–2086
Nishizuka Y 1988 The molecular heterogeneity of protein kinase C and its implications for cellular regulation. Nature (Lond) 334:661–665
Otsu H, Yamamoto A, Maeda N, Mikoshiba K, Tashiro Y 1990 Immunogold localization of inositol 1,4,5-trisphosphate ($InsP_3$) receptor in mouse cerebellar Purkinje cells using three monoclonal antibodies. Cell Struct Funct 15:163–173
Ross CA, Meldolesi J, Milner TA, Satoh T, Supattapone S, Snyder SH 1989 Inositol 1,4,5-trisphosphate receptor localized to endoplasmic reticulum in cerebellar Purkinje neurons. Nature (Lond) 339:468–470
Satoh T, Ross CA, Villa A 1990 The inositol 1,4,5-trisphosphate receptor in cerebellar Purkinje cells: quantitative immunogold labeling reveals concentration in an ER subcompartment. J Cell Biol 111:615–624
Smith JB, Smith L, Higgins BL 1985 Temperature and nucleotide dependence of calcium release by myo-inositol 1,4,5-trisphosphate in cultured vascular smooth muscle cells. J Biol Chem 260:14413–14416
Suematsu E, Hirata M, Sasaguri T, Hashimoto T, Kuriyama H 1985 Roles of $Ca^{2+}$ on the inositol 1,4,5-trisphosphate-induced release of $Ca^{2+}$ from saponin-permeabilized single cells of the porcine coronary artery. Comp Biochem Physiol 82:645–649
Supattapone S, Worley PF, Baraban JM, Snyder SH 1988a Solubilization, purification, and characterization of an inositol trisphosphate receptor. J Biol Chem 263:1530–1534
Supattapone S, Danoff SK, Theibert A, Joseph SK, Snyder SH 1988b Cyclic AMP-dependent phosphorylation of a brain inositol trisphosphate receptor decreases its release of calcium. Proc Natl Acad Sci USA 85:8747–8750
Takeshima H, Nishimura S, Matsumoto T et al 1989 Primary structure and expression from complementary DNA of skeletal muscle ryanodine receptor. Nature (Lond) 339:439–445
Yamamoto H, Maeda N, Niinobe M, Miyamoto E, Mikoshiba K 1989 Phosphorylation of $P_{400}$ protein by cyclic AMP-dependent protein kinase and $Ca^{2+}$/calmodulin-dependent protein kinase II. J Neurochem 53:917–923

## DISCUSSION

*Yang:* Do your monoclonal antibodies affect the binding of $InsP_3$?

*Mikoshiba:* We tested three antibodies, 4C11, 10A6 and 18A10; two of these (4C11 and 10A6) recognize the N-terminal portion of the receptor and have no effect on $InsP_3$-binding activity or calcium-releasing activity. The third antibody, 18A10, which recognizes the C-terminal portion, blocks calcium-releasing activity but not $InsP_3$-binding activity.

*Changeux:* Is there cooperative binding of $InsP_3$? What is the Hill coefficient?

*Mikoshiba:* The Hill coefficient is almost one, so there seems to be no cooperativity in $InsP_3$ binding.

*Tsien:* What do you know about cooperativity in calcium-releasing activity?

*Mikoshiba:* We have been studying a lipid bilayer system and have found that the $InsP_3$ receptor $Ca^{2+}$ channel exhibits four conductance levels, suggesting that $InsP_3$ molecules open the channel in an additive manner. Ehrlich & Watras (1988) also find this. However, others (Meyer & Stryer 1988, Meyer et al 1990) report that opening of the calcium channel requires binding of four $InsP_3$ molecules. This discrepancy might be attributed to the difference in tissues used—cerebellum versus basophilic leukaemia cells—or to methodological differences—we use a lipid bilayer system and they use permeabilized cells. To determine if there is cooperativity in $Ca^{2+}$-releasing activity, we need to do statistical analysis of single-channel recordings from our lipid bilayer system.

*Changeux:* In the bilayer system there may be association of tetramers and cooperative intermolecular interactions between tetramers. With the nicotinic acetylcholine receptor, interactions *between* oligomers take place in lipid bilayers.

*Tsien:* Professor Mikoshiba, you contrasted your results on stoichiometry with those of Meyer & Stryer (1988). Initially there does seem to be a difference, because you observe different conductance levels depending on the amount of $InsP_3$ that is present, and presumably on the amount of $InsP_3$ binding there is. The Meyer and Stryer model for oscillations, which seems to have some support, requires only that the degree of calcium flux is steeply dependent on $[InsP_3]$. If the probability of the high conductance state is greater than the probability of the low conductance state there could still be overall cooperativity even if one $InsP_3$ molecule is sufficient to generate a small amount of conductance. There could be an overall steepening of $Ca^{2+}$ flux as a function of $[InsP_3]$, which is all that the Meyer and Stryer model really requires. One needs to be cautious about how one makes the comparison and more information is needed, not only about the conductance levels, but also with respect to the opening probabilities.

*Changeux:* Does the $InsP_3$ receptor become desensitized?

*Mikoshiba:* In our studies of $InsP_3$-induced $Ca^{2+}$ release using microsome fractions we have not detected desensitization.

*Changeux:* So the channel remains open as long as $InsP_3$ is present.

*Mikoshiba:* The channel closes when $InsP_3$ is hydrolysed.

*Changeux:* Is there any effect of ATP on $InsP_3$ binding?

*Mikoshiba:* ATP inhibits $InsP_3$ binding. The ATP-binding site is very close to the phosphorylation site.

*Changeux:* I am not clear about this; are you talking about a reversible ATP-binding site or is it a phosphorylation site?

*Mikoshiba:* The phosphorylation site is different from the ATP-binding site, but they are close to each other.

*Krebs:* The receptor is phosphorylated by cyclic AMP-dependent protein kinase. How many phosphorylation sites are there?

*Mikoshiba:* There are two consensus sequences for cyclic AMP-dependent phosphorylation. Both serine residues in these sequences have been found to be phosphorylated. Recently, Sol Synder's group have shown that in the CNS the C-terminal phosphorylation site is highly phosphorylated; however, in peripheral tissues the N-terminal site is highly phosphorylated (Danoff et al 1991).

*Carpenter:* Were the sites of phosphorylation determined by *in vitro* or *in vivo* experiments?

*Mikoshiba:* We did the experiments *in vitro*, using microsome fractions. We have not used an *in vivo* system yet.

*Thomas:* Does phosphorylation at either site affect ATP binding, and do the monoclonal antibodies interfere with ATP binding?

*Mikoshiba:* We have not yet done these experiments. The sequence of the ATP-binding site, which conforms to the consensus sequence Gly-X-Gly-X-X-Gly (residues 2016–2021; Furuichi et al 1989), is different from the sequences recognized by our monoclonal antibodies, so the antibodies will probably not have a direct effect on ATP binding.

*Parker:* Is the affinity of the receptor for ATP such that you wouldn't expect it to dynamically control the function of the channel? If that is the case, does ADP have an affinity for that site? Could the ATP : ADP ratio control channel function?

*Mikoshiba:* The binding is selective for adenine nucleotides and the order of affinity amongst these is ATP>ADP>>AMP. ATP actually promotes conduction in the reconstituted cerebellar $InsP_3$ receptor channel in the presence of $InsP_3$. We have not tested whether the ATP:ADP ratio controls channel functioning.

*Parker:* Might you expect the energy status of the cell to affect channel function through the competition of ADP or AMP for the ATP site?

*Mikoshiba:* That is possible, but we have no evidence.

*Akaike:* A decrease in intracellular ATP concentration below 2.5 mM decreases the affinity of the $GABA_A$ receptor in pyramidal neurons freshly dissociated from the rat hippocampal CA1 region. Internal perfusion with ATP analogues such as ADP and ATP$\gamma$S prevents the decline in the GABA response. The specificity of the effect suggests the existence of intracellular ATP receptors. However, we have no idea about the mechanism of interaction between such ATP receptors and the $GABA_A$ receptor (Akaike & Shirasaki 1991).

*Hunter:* Is there a conventional ATP-binding site in the phosphorylation site region of the $InsP_3$ receptor?

*Mikoshiba:* Yes; there is a consensus binding site sequence located just downstream of the C-terminal part of the phosphorylation site.

*Hunter:* Have you mutated the binding site?

*Mikoshiba:* No, not yet. We are currently making various mutants.

*Collingridge:* Is the receptor expressed in cerebellar granule cells?

*Mikoshiba:* There is a very low level of expression.

*Collingridge:* We have evidence from calcium-imaging experiments for a $Ca^{2+}$-mobilizing receptor in these cells. Might there be a different type of receptor releasing calcium in these cells?

*Mikoshiba:* We have recently cloned cDNA encoding two new types of human $InsP_3$ receptor. Although they show fragmentary sequence identity to the cerebellar receptor, we believe that the genes they are encoded by are different from that for the cerebellar receptor. Even from a single gene different types of receptor can be produced through RNA splicing. A variety of $InsP_3$ receptors are produced.

*Changeux:* Can you make heteropolymers from the different spliced forms?

*Mikoshiba:* Probably; differential splicing could produce heterogeneity, changes in the character of the calcium channel. I believe this happens in Nature and that it will also be possible to demonstrate in an *in vitro* system.

*Changeux:* Can you confer different properties by adding one of the spliced forms?

*Mikoshiba:* We have no data yet, but combination of various types of receptors may produce various properties.

*Changeux:* In the nervous system, are the different forms only found in neurons or are they also in glial cells?

*Mikoshiba:* C6 cells, a line derived from glial cells, contain a form that has a 120-base pair deletion. This is the form found in non-nervous system tissues.

*Rasmussen:* Have you looked to see where the calcium pump in the smooth endoplasmic reticulum is in relation to the $InsP_3$ receptor?

*Mikoshiba:* It is interesting that in tissues such as the cerebellum where $Ca^{2+}$-releasing activity is high (probably because of $InsP_3$ receptor enrichment), $Ca^{2+}$ uptake (probably through the activity of the $Ca^{2+}$-ATPase) is also very high. I believe, therefore, that there is a close correspondence between calcium uptake and release mechanisms.

*Rasmussen:* Is it possible that the two are spatially separated within the smooth endoplasmic reticulum?

*Mikoshiba:* It is likely that the $InsP_3$ receptor and the $Ca^{2+}$-ATPase are enriched on the same calcium store, namely the smooth endoplasmic reticulum. We have no direct evidence for that, but immunogold staining using antibodies against the $InsP_3$ receptor and the $Ca^{2+}$-ATPase should give us the answer.

*Tsien:* In the cerebellum is the $InsP_3$ receptor mostly in the spines or in the dendritic shafts?

*Mikoshiba:* It is found everywhere where smooth endoplasmic reticulum is localized.

*Tsien:* In your transfected cells, do you know whether the $InsP_3$ receptor overlaps with the ryanodine-sensitive receptor in its ability to discharge the same $Ca^{2+}$ store?

*Mikoshiba:* We have no direct evidence about that; however, the exogenous $InsP_3$ receptor might be expressed also on the ryanodine-sensitive receptor pool.

*Tsien:* Do you mean that the distribution of the exogenous $InsP_3$ receptor is broader than that of the endogenous receptor?

*Mikoshiba:* Yes; our $Ca^{2+}$ release experiments indicated that the size of the $InsP_3$-sensitive $Ca^{2+}$ store was increased with expression of the exogenous $InsP_3$ receptor. Thus, I think that the expressed receptor is distributed more widely than the endogenous one. In fact, we find that the exogenous receptor is expressed all over the endomembrane of the L- fibroblast cell. Immunostaining should properly answer this question.

*Tsien:* That's very interesting with regard to what causes the localization of the ryanodine receptor. Was your result obtained by immunohistochemistry or did you use caffeine to deplete the $Ca^{2+}$ stores?

*Mikoshiba:* We have done only immunohistochemical studies and experiments on $InsP_3$-induced calcium release. We have not done double labelling with the antibody against the ryanodine receptor. To investigate the localization of the $InsP_3$ receptor in relation to that of the ryanodine receptor we should carry out $Ca^{2+}$ release experiments using a combination of $InsP_3$ and caffeine, as you suggested.

*Michell:* In the cerebellum there are extensive smooth membranes which you call endoplasmic reticulum and which contain a lot of the protein. In cells in which there is a small amount of the protein expressed, where do you find it to be located? I am thinking about Meldolesi's focal arrays of smooth membrane that seem to hold high concentrations of the inositol trisphosphate receptor (Meldolesi et al 1990).

*Mikoshiba:* That is a very important point. High levels of expression of a protein such as the $InsP_3$ receptor might alter the structure of a membrane or even the distribution of a structure such as the smooth endoplasmic reticulum.

*Michell:* In non-Purkinje cells where there is little of the receptor is it particularly concentrated in small areas of smooth membrane?

*Mikoshiba:* We are still analysing the detailed distribution of the receptor in non-neural tissues.

*Michell:* I suppose what I am really asking is, do you think that the membrane in which the receptor is expressed is what one might call 'typical' smooth endoplasmic reticulum, or is it some more specialized type of membrane?

*Mikoshiba:* I presume that you are referring to the 'calciosome' structure proposed by Volpe et al (1988) as a specialized calcium store. We have examined Purkinje cells in collaboration with Professor Tashiro but have not detected this structure.

*Nishizuka:* In the reconstituted receptor system what effect does phospholipid composition have on channel activity?

*Mikoshiba:* We are collaborating with Professor Kasai in this project. We do not yet have enough results to answer your question.

*Daly:* There was little difference in maximal calcium release between the cells containing the endogenous deleted form of the $InsP_3$ receptor and those with the transfected CNS form, yet your binding data indicate a large difference in the number of $InsP_3$ binding sites. Does this mean that the deleted form is coupled to calciosomes that have a lot of calcium, or is it more efficient in releasing calcium? Would you not agree that there is a lack of correlation between $InsP_3$ binding and the amount of $Ca^{2+}$ release you saw?

*Mikoshiba:* Before answering this question one needs to make sure that the $InsP_3$ receptor is expressed in a particular compartment of the smooth endoplasmic reticulum. If the cDNA-derived receptor is expressed in high amounts only in the native compartment the $V_{max}$ value would not change. Therefore, the $V_{max}$ value of $InsP_3$-induced $Ca^{2+}$ release does not necessarily correlate with the $InsP_3$ binding activity. Our data (Miyawaki et al 1990) show that the $V_{max}$ in transfected cells was about twice that in non-transfected cells, indicating that the cDNA-derived $InsP_3$ receptor is also distributed in $InsP_3$-insensitive compartments. The $EC_{50}$ value for $InsP_3$-induced $Ca^{2+}$ release in transfected cells is about 10-fold less than that in non-transfected cells. The increased sensitivity in the transfected cells roughly correlates with the increased $InsP_3$ binding activity.

*Fredholm:* I was intrigued by the suggestion that there might be some $InsP_3$ receptor in the nuclear membrane. There are experiments which suggest that $InsP_3$ might regulate calcium levels in the nucleus (Nicotera et al 1990). Have you any proof that there really are $InsP_3$ receptors on the nuclear membrane?

*Mikoshiba:* The immunogold staining method gives us a positive result at the outer nuclear membrane. We still do not know the true functional implications of this result.

*Fredholm:* In hepatocyte nuclei there is certainly pretty good evidence that $InsP_3$ regulates $Ca^{2+}$ levels (Nicotera et al 1990).

## References

Akaike N, Shirasaki T 1991 Intracellular ATP regulates GABA responses in rat dissociated hippocampal neurons. In: Imai S, Nakazawa M (eds) Role of adenosine and adenine nucleotides in the biological system. Elsevier Science Publishers, Amsterdam p 649–651

Danoff SK, Ferris CD, Donath C et al 1991 Inositol 1,4,5-trisphosphate receptors: distinct neuronal and nonneuronal forms derived by alternative splicing differ in phosphorylation. Proc Natl Acad Sci USA 88:2951–2955

Ehrlich BE, Watras J 1988 Inositol 1,4,5-trisphosphate activates a channel from smooth muscle sarcoplasmic reticulum. Nature (Lond) 336:583–586

Furuichi T, Yoshikawa S, Miyawaki A, Wada K, Maeda N, Mikoshiba K 1989 Primary structure and functional expression of the inositol 1,4,5-trisphosphate-binding protein $P_{400}$. Nature (Lond) 342:32–38

Meldolesi JB, Madeddu L, Pazzan T 1990 Intracellular $Ca^{2+}$ storage organelles in non muscle cells: heterogeneity and functional alignment. Biochim Biophys Acta 1055:130–140

Meyer T, Stryer L 1988 Molecular model for receptor-stimulated calcium spiking. Proc Natl Acad Sci USA 85:5051–5055

Meyer T, Wensel T, Stryer L 1990 Kinetics of calcium channel opening by inositol 1,4,5-trisphosphate. Biochemistry 29:32–37

Miyawaki A, Furuichi T, Maeda N, Mikoshiba K 1990 Expressed cerebellar-type inositol 1,4,5-trisphosphate receptor, $P_{400}$, has calcium release activity in a fibroblast L cell line. Neuron 5:11–18

Nicotera P, Orrenius T, Nilsson T, Berggren PO 1990 An inositol 1,4,5-trisphosphate-sensitive $Ca^{2+}$ pool in liver nuclei. Proc Natl Acad Sci USA 87:6858–6862

Volpe P, Krause K-H, Hashimoto S et al 1988 ‘Calciosome’, a cytoplasmic organelle: the inositol 1,4,5-trisphosphate sensitive $Ca^{2+}$ store of non muscle cells? Proc Natl Acad Sci USA 85:1091–1095

# Regulation of phosphoinositide and phosphatidylcholine phospholipases by G proteins

John H. Exton, Stephen J. Taylor, Jonathan S. Blank and Stephen B. Bocckino

*Howard Hughes Medical Institute, Department of Molecular Physiology and Biophysics, and Department of Pharmacology, Vanderbilt University School of Medicine, Nashville, TN 37232, USA*

*Abstract.* Two G proteins that regulate phosphoinositide phospholipase C in liver plasma membranes have been purified to homogeneity in both the heterotrimeric and dissociated forms. The heterotrimers contain a 42 kDa or 43 kDa $\alpha$ subunit and a 35 kDa $\beta$ subunit. The $\alpha$ subunits are not ADP-ribosylated by pertussis toxin and are closely related immunologically to members of the recently identified $G_q$ class of G proteins. The specific phosphoinositide phospholipase C isozyme that responds to the G proteins has been determined to be the $\beta_1$ isozyme. GTP analogues stimulate phosphatidylcholine hydrolysis in rat liver plasma membranes. The nucleotide specificity and $Mg^{2+}$ dependency of the response indicate that it is mediated by a G protein. Phosphatidic acid, diacylglycerol, choline and phosphorylcholine are the products, indicating that both phospholipase D and C activities are involved. Activation of phospholipase D is also indicated by the enhanced production of phosphatidylethanol in the presence of ethanol.

*1992 Interactions among cell signalling systems. Wiley, Chichester (Ciba Foundation Symposium 164) p 36–49*

Many hormones, neurotransmitters and related agonists stimulate the hydrolysis of phosphatidylinositol 4,5-bisphosphate, $PtdIns(4,5)P_2$, in the plasma membrane of their target cells. This produces two signalling molecules—inositol trisphosphate ($InsP_3$), which releases $Ca^{2+}$ from components of the endoplasmic reticulum, and diacylglycerol (DAG), which activates protein kinase C (Berridge 1987). In addition to releasing $Ca^{2+}$ from intracellular stores, the agonists also promote $Ca^{2+}$ influx through plasma membrane channels, but the mechanisms are unclear. There is much evidence that guanine nucleotide-binding regulatory proteins (G proteins) are involved in the activation of $PtdInsP_2$ phospholipase C by hormones and neurotransmitters, and these proteins may be involved in the regulation of the $Ca^{2+}$ channels (Exton 1988)

Recent studies have indicated that many of the agonists that stimulate $PtdInsP_2$ hydrolysis also promote phosphatidylcholine breakdown, through the activation of phospholipases of the C and D type, to yield DAG, phosphatidic acid, choline and phosphorylcholine (Exton 1990, Billah & Anthes 1990). Although the physiological significance of phosphatidylcholine breakdown is unclear, it is a major factor in the increase in phosphatidic acid and DAG and the activation of protein kinase C that are induced by many agonists in various cell types (Exton 1990, Billah & Anthes 1990). Here, we review our investigations into the molecular mechanisms involved in agonist-stimulated $PtdInsP_2$ and phosphatidylcholine breakdown.

## Regulation of phosphoinositide phospholipase by G proteins

Calcium-mobilizing agonists induce a large decrease in $PtdInsP_2$ levels within a few seconds in isolated rat hepatocytes, whereas the level of phosphatidylinositol 4-phosphate (PtdInsP) does not change significantly and that of phosphatidylinositol (PtdIns) decreases much more slowly (Augert et al 1989a). The rapid breakdown of $PtdInsP_2$ is associated with a parallel increase in $InsP_3$ (Exton 1988), consistent with the activation of a phospholipase C. There are slower changes in other inositol phosphates, representing the conversion of $InsP_3$ to inositol 1,4-bisphosphate and inositol 1,3,4,5-tetrakisphosphate, and the further metabolism of these.

Experiments with isolated rat liver plasma membranes have shown that direct addition of GTP and its stable analogues activates a $PtdInsP_2$-selective phospholipase C (Uhing et al 1986, Taylor & Exton 1987). This activation is produced by submicromolar concentrations of the GTP analogues, but not other nucleotides, requires millimolar $Mg^{2+}$ concentrations and is competitively inhibited by a stable GDP analogue. The phospholipase is also activated by $Ca^{2+}$-mobilizing agonists in the presence of low concentrations of GTP analogues. Further evidence that $Ca^{2+}$-mobilizing receptors are linked to a G protein(s) is provided by the observations that GTP and its analogues reduce the binding of agonists to these receptors (Lynch et al 1986), that $Ca^{2+}$-mobilizing agonists stimulate a low $K_m$ GTPase activity in liver plasma membranes (Fitzgerald et al 1986) and that aluminium fluoride stimulates $InsP_3$ formation and mobilization of calcium in hepatocytes (Blackmore et al 1985).

It is now evident that there are several phosphoinositide phospholipase C activities (Kriz et al 1990). These all hydrolyse PtdIns, PtdInsP and $PtdInsP_2$ in a $Ca^{2+}$-dependent manner, but $PtdInsP_2$ is the preferred substrate at low $Ca^{2+}$ concentrations. There are five major subtypes ($\alpha$–$\epsilon$) which have different molecular weights and sequences and can be distinguished immunologically. The phospholipase C isozyme responsible for growth factor-stimulated $PtdInsP_2$ hydrolysis in several tissues has been identified as the $\gamma_1$ isozyme (Kriz et al 1990).

Recent work in our laboratory has identified the G proteins that regulate $PtdInsP_2$ phospholipase C in rat and bovine liver (Taylor et al 1990, 1991). This was possible because treatment of plasma membranes with stable (i.e. non-hydrolysable) analogues of GTP produced a phospholipase activator that retained activity during purification through a series of chromatographic steps. The final purification step involved HPLC on a Mono Q column after ADP-ribosylation with pertussis toxin. This procedure resulted in separation of the phospholipase activator from contaminating $G_i$ $\alpha$ subunits. Silver-stained gels of the final preparation revealed proteins with molecular masses of 42 kDa, 43 kDa and 35 kDa (Taylor et al 1991). The 42 and 43 kDa proteins were recognized by an antiserum raised to a peptide common to many G protein $\alpha$ subunits, and the 35 kDa protein was recognized by an antiserum to a $\beta$ subunit peptide. The concentrations of the 42 and 43 kDa $\alpha$ subunits correlated with the magnitude of the phospholipase activation. In more recent work, it has been possible to resolve the two $\alpha$ subunits prepared from GTP$\gamma$S-treated membranes sufficiently to demonstrate that they apparently activate purified phospholipase equally well.

To test which phospholipase isozyme is stimulated by the activated $\alpha$ subunits, these were reconstituted with the $\beta_1$, $\gamma_1$ and $\delta_1$ isozymes purified from bovine brain according to the procedure of Ryu et al (1987a,b). The results showed unequivocally that only the $\beta_1$ isozyme was activated (Taylor et al 1991). Furthermore, only antibodies to the $\beta_1$ isozyme were able to inhibit the activation of endogenous phospholipase when the activated $\alpha$ subunits were added to liver plasma membranes.

The 42 and 43 kDa $\alpha$ subunits have been shown to be members of the newly discovered $G_q$ class of G proteins, identified as cDNAs in a mouse brain library by the polymerase chain reaction (Strathmann & Simon 1990). Polyclonal antisera (WO82, WO83) to peptides corresponding to unique regions in the deduced amino acid sequences of two members of the $G_q$ class ($\alpha_q$, $\alpha_{11}$), supplied to us by P. C. Sternweis, reacted strongly with the 42 kDa protein, whereas the 43 kDa protein was recognized only by WO83. In parallel studies, two 42 kDa G protein $\alpha$ subunits prepared from rat or bovine brain by affinity chromatography on $\beta\gamma$-agarose (Pang & Sternweis 1990) have been partially sequenced. Several tryptic peptides were found to correspond to sequences in $\alpha_q$ and $\alpha_{11}$, indicating that the brain proteins are identical to, or very similar to, these $\alpha$ subunits. Antisera WO82 and WO83 recognized the proteins, as expected. In more recent work, Sternweis and collaborators demonstrated that activation of the proteins by aluminium fluoride stimulated phosphoinositide phospholipase C, but guanine nucleotides were ineffective (Smrcka et al 1991).

The G proteins that regulate phosphoinositide phospholipase C have also been purified in the heterotrimeric ($\alpha\beta\gamma$) form (Blank et al 1991). The final purification step involved FPLC on Sephacryl S-300. A symmetrical peak of GTP$\gamma$S-dependent activation was observed when the G protein fractions were reconstituted with a pure preparation of the $\beta_1$ isozyme of phosphoinositide phospholipase. The final G protein preparation contained two 42 and 43 kDa $\alpha$ subunits (recognized by antiserum WO83) and a 35 kDa $\beta$ subunit (recognized by antiserum 113 raised against a specific peptide in this subunit). Rapid activation of the phospholipase in the presence of the G proteins was observed with GTP$\gamma$S, but relatively high concentrations of the nucleotide (1–100 μM) were needed. The activation was inhibited by a high concentration of GDP$\beta$S and by the addition of excess $\beta\gamma$ subunits. Aluminium fluoride also activated the phospholipase in the presence of the G proteins, although high concentrations were inhibitory.

In summary, we have identified two pertussis toxin-insensitive G proteins, distinguished by their $\alpha$ subunits, that regulate phosphoinositide phospholipase C. These are both members of the $G_q$ class of G proteins, and appear to have all the properties of heterotrimeric G proteins, although their affinity for GTP and its analogues is very low in the absence of agonists and relevant receptors. The selectivity of the G proteins for phosphoinositide phospholipase is extremely high, with only the $\beta_1$ isozyme being activated.

**Regulation of phosphatidylcholine phospholipases by G proteins and other mechanisms**

Evidence that $Ca^{2+}$-mobilizing agonists stimulate the breakdown of another phospholipid besides $PtdInsP_2$ came initially from chemical measurements of the accumulation of diacylglycerol in response to $Ca^{2+}$-mobilizing agonists in rat hepatocytes (Bocckino et al 1985). These measurements showed a gross discrepancy between the time courses of DAG accumulation and $InsP_3$ formation. Furthermore, fatty acid analysis of the DAG indicated that it came from another source, probably phosphatidylcholine (Bocckino et al 1985, 1987). More refined HPLC analyses of the molecular species of DAG (Augert et al 1989a) and chemical measurements of the changes in inositol phospholipids, DAG and phosphatidic acid (Augert et al 1989a,b, Bocckino et al 1985, 1987) have confirmed this conclusion. More direct proof of phosphatidylcholine breakdown was provided by experiments using cells in which phosphatidylcholine was selectively labelled using [$^3$H]alkyl-lysoglycerophosphorylcholine. In these cells $Ca^{2+}$-mobilizing agonists stimulated the production of [$^3$H]alkyl-acylglycerol (Augert et al 1989a). There was also stimulation of the formation of [$^3$H]phosphorylcholine and [$^3$H]choline from cells in which the phosphatidylcholine had been previously labelled by incubation with [$^3$H]choline for 90 min. In hepatocytes and other cell types there is now evidence that phosphatidylcholine is broken down to phosphatidic acid and DAG by phospholipase D and phospholipase C activities (Exton 1990). The activation of phospholipase D accounts for the rapid formation of phosphatidic acid observed in many agonist-stimulated cells (Exton 1990, Billah & Anthes 1990).

The ability of agonists to activate phospholipase D is also demonstrated by the formation of phosphatidylethanol when the cells are incubated with agonists in the presence of ethanol (Exton 1990, Billah & Anthes 1990). The production of phosphatidylethanol is due to the transphosphatidylating activity of phospholipase D. No other enzyme can catalyse this reaction.

Evidence for four mechanisms of hormonal stimulation of phosphatidylcholine breakdown has been presented (Exton 1990, Billah & Anthes 1990). These involve stimulation of phosphatidylcholine phospholipase C and D activities by G proteins, protein kinase C, $Ca^{2+}$ and growth factor receptors. Evidence for control of phosphatidylcholine breakdown through a G protein(s) comes from studies with isolated rat liver plasma membranes. These studies showed that addition of GTP analogues stimulates the production of DAG, phosphatidic acid, choline and phosphorylcholine (Bocckino et al 1987, Irving & Exton 1987). Because there is an associated decrease in phosphatidylcholine levels, the data are consistent with the activation of phosphatidylcholine hydrolysis by phospholipases C and D. The effect is observed with submicromolar concentrations of GTP analogues and is inhibited by GDPβS, a stable GDP analogue (Irving & Exton 1987). $P_2$-Purinergic agonists (ATP and ADP) also stimulate phosphatidylcholine breakdown in liver plasma membranes, but only in the presence of a GTP analogue (Bocckino et al 1987, Irving & Exton 1987). Neither cholera toxin nor pertussis toxin has an effect (Irving & Exton 1987), indicating that the putative G protein is not a substrate for these toxins.

GTP analogues stimulate phosphatidycholine phospholipase D activity in permeabilized pulmonary artery endothelial cells (Martin & Michaelis 1989) and fibroblasts (Diaz-Meco et al 1989), and the stimulating effect of the nucleotides is enhanced by $P_2$-purinergic and cholinergic agonists in the respective systems. These results indicate that the phospholipase can be controlled by G proteins in these tissues.

Evidence that protein kinase C controls phosphatidylcholine breakdown is provided by studies in which tumour-promoting phorbol esters and synthetic DAGs are used to promote the release of choline or phosphorylcholine from cells or to stimulate the breakdown of labelled phosphatidylcholine to DAG or phosphatidic acid (for references see Exton 1990, Billah & Anthes 1990). In some systems inhibitors of protein kinase C block the effects of agonists and phorbol esters, and down-regulation of the enzyme attenuates the response.

Because almost all the agonists that elicit phosphatidylcholine breakdown also cause $PtdInsP_2$ hydrolysis, the initial transient increase in the level of DAG arising from $PtdInsP_2$ breakdown may act as a trigger for hydrolysis of phosphatidylcholine through activation of protein kinase C. This hydrolysis would generate more DAG, providing a positive feedback to maintain elevated levels of DAG and continued protein kinase C activation. Unlike the situation in the phosphoinositide signalling system, agonist-induced phosphatidylcholine hydrolysis does not seem to be subject to negative feedback.

Studies in several cell types have indicated that an increase in cytosolic $Ca^{2+}$ concentration can activate phosphatidylcholine hydrolysis by phospholipases C and D (for references see Exton 1990, Billah & Anthes 1990). For example, the $Ca^{2+}$ inophore A23187 stimulates the conversion of [$^3$H]alkylphosphatidylcholine to [$^3$H]alkylacylglycerol or [$^3$H]alkylphosphatidic acid in hepatocytes and neutrophils, and $Ca^{2+}$ ionophores promote breakdown of phosphatidylcholine in other cells. Furthermore, the effects of agonists on DAG formation or phosphatidylcholine hydrolysis in several systems are partly dependent upon the cell $Ca^{2+}$ level.

Several growth factors and cytokines (epidermal growth factor, platelet-derived growth factor, interleukins 1 and 3, $\gamma$-interferon, $\alpha$-thrombin) stimulate phosphatidylcholine hydrolysis in fibroblasts and other cell types (for references see Exton 1990). This can result in an elevation of DAG that persists for several hours, which may be important for the stimulation of mitogenesis. The mechanism(s) by which growth factors stimulate hydrolysis of phosphatidylcholine is unknown.

In summary, it is becoming clear that many hormones, neurotransmitters and growth factors stimulate the hydrolysis of phosphatidylcholine by phospholipases C and D in a wide variety of cells, but the mechanisms involved are poorly defined and probably multiple. The physiological significance of the effect is also unknown. Phosphatidylcholine probably contributes more DAG for activation of protein kinase C at late times of agonist action than $PtdInsP_2$ does. In many systems phosphatidic acid is produced at early times of agonist action when DAG accumulation is minimal or not detectable, and some of the DAG that accumulates later is derived from phosphatidic acid through the action of phosphatidate phosphohydrolase. Many functions have been proposed for phosphatidic acid, including the stimulation of DNA synthesis, but it is possible that its physiological role has yet to be defined.

## References


Augert G, Bocckino SB, Blackmore PF, Exton JH 1989a Hormonal stimulation of diacylglycerol formation in hepatocytes. Evidence for phosphatidylcholine breakdown. J Biol Chem 264:21689–21698

Augert G, Blackmore PF, Exton JH 1989b Changes in the concentration and fatty acid composition of phosphoinositides induced by hormones in hepatocytes. J Biol Chem 264:2574–2580

Berridge MJ 1987 Inositol trisphosphate and diacylglycerol: two interacting second messengers. Annu Rev Biochem 56:159–193

Billah MM, Anthes JC 1990 The regulation and cellular functions of phosphatidylcholine hydrolysis. Biochem J 269:281–291

Blackmore PF, Bocckino SB, Waynick LE, Exton JH 1985 Role of guanine nucleotide-binding regulating protein in the hydrolysis of hepatocyte phosphatidylinositol 4,5-bisphosphate by calcium-mobilizing hormones and the control of cell calcium. Studies utilizing aluminium fluoride. J Biol Chem 260:14477–14483

Blank JL, Ross AH, Exton JH 1991 Purification and characterization of two G-proteins which activate the β1 isozyme of phosphoinositide-specific phospholipase C. Identification as members of the $G_q$ class. J Biol Chem 266:18206–18216

Bocckino SB, Blackmore PF, Exton JH 1985 Stimulation of 1,2-diacylglycerol accumulation in hepatocytes by vasopressin, epinephrine and angiotensin II. J Biol Chem 260:14201–14207

Bocckino SB, Blackmore PF, Wilson PB, Exton JH 1987 Phosphatidate accumulation in hormone-treated hepatocytes via a phospholipase D mechanism. J Biol Chem 262:15309–15315

Diaz-Meco MT, Larrodera P, Lopez-Barahona M, Cornet ME, Barreno PG, Moscat J 1989 Phospholipase C-mediated hydrolysis of phosphatidylcholine is activated by muscarinic agonists. Biochem J 263:115–120

Exton JH 1988 The roles of calcium and phosphoinositides in the mechanisms of $\alpha_1$-adrenergic and other agonists. Rev Physiol Biochem Pharmacol 111:118–224

Exton JH 1990 Signalling through phosphatidylcholine breakdown. J Biol Chem 265:1–4

Fitzgerald TJ, Uhing RJ, Exton JH 1986 Solubilization of the vasopressin receptor from rat liver membranes. J Biol Chem 261:16871–16877

Irving HR, Exton JH 1987 Phosphatidylcholine breakdown in rat liver plasma membrane. J Biol Chem 262:3440–3443

Kriz R, Lin L-L, Sultzman L et al 1990 Phospholipase C isozymes: structural and functional similarities. In: Proto-oncogenes in cell development. Wiley, Chichester (Ciba Found Symp 150) p 112–127

Lynch CJ, Prpic V, Blackmore PF, Exton JH 1986 Effect of islet-activating pertussis toxin on the binding characteristics of $Ca^{2+}$-mobilizing hormones and on agonist activation of phosphorylase in hepatocytes. Mol Pharmacol 29:196–203

Martin TW, Michaelis K 1989 $P_2$-Purinergic agonists stimulate phosphodiesteratic cleavage of phosphatidylcholine in endothelial cells. J Biol Chem 264:8847–8856

Pang I-H, Sternweis PC 1990 Purification of unique α-subunits of GTP-binding regulatory proteins (G proteins) by affinity chromatography with immobilized βγ subunits. J Biol Chem 265:18707–18712

Ryu SH, Cho KS, Lee K-Y, Suh P-G, Rhee SG 1987a Purification and characterization of two immunologically distinct phosphoinositide-specific phospholipase C from bovine brain. J Biol Chem 262:12511–12518

Ryu SH, Suh P-G, Cho KS, Lee K-Y, Rhee SG 1987b Bovine brain cytosol contains three immunologically distinct forms of inositol phospholipid-specific phospholipase C. Proc Natl Acad Sci USA 84:6649–6653

Smrcka AV, Hepler JR, Brown KO, Sternweis PC 1991 Regulation of a polyphosphoinositide-specific phospholipase C activity by purified $G_q$. Science (Wash DC) 251:804–807

Strathmann M, Simon MI 1990 G protein diversity: a distinct class of α-subunits is present in vertebrates and invertebrates. Proc Natl Acad Sci USA 87:9113–9117

Taylor SJ, Exton JH 1987 Guanine-nucleotide and hormone regulation of polyphosphoinositide phospholipase C activity of rat liver plasma membranes. Biochem J 248:791–799

Taylor SJ, Smith JA, Exton JH 1990 Purification from bovine liver membranes of a guanine nucleotide-dependent activator of phosphoinositide specific phospholipase C. Immunologic identification as a novel G-protein α-subunit. J Biol Chem 265:17150–17156

Taylor SJ, Chae HZ, Rhee SG, Exton JH 1991 Activation of the β1 isozyme of phospholipase C by purified α subunits of the $G_q$ class of G proteins. Nature (Lond) 350:516–518

Uhing RJ, Prpic V, Jiang H, Exton JH 1986 Hormone-stimulated polyphosphoinositide breakdown in rat liver plasma membranes. J Biol Chem 261:2140–2146

## DISCUSSION

*Carpenter:* Are the phosphoinositide phospholipases sensitive to pertussis toxin?

*Exton:* In the liver they are insensitive.

*Carpenter:* It has been suggested that phosphatidylinositol turnover is pertussis toxin-sensitive in hepatocytes (Johnson & Garrison 1987).

*Exton:* That is correct.

*Carpenter:* Do you know which G proteins are involved?

*Exton:* There is some evidence that the pertussis toxin-sensitive G protein involved in the activation of $PtdInsP_2$ phospholipase C is $G_o$ (Moriarty et al 1990, Padrell et al 1991), but this G protein has a limited tissue distribution and clearly cannot account for the responses seen in neutrophils, HL-60 cells, mast cells, mesangial cells and certain fibroblasts. Another possible candidate, particularly in neutrophils, HL-60 cells and mast cells is $G_{i\alpha 2}$ or $G_{i\alpha 3}$—again, the evidence is indirect and confined to a particular cell type. We are searching for the pertussis toxin-sensitive G protein(s) involved in the signal transduction mechanism.

*Fischer:* Are the 42 and 43 kDa G protein $\alpha$ subunits isoprenylated?

*Exton:* We don't yet know.

*Ui:* How much $G_q$ is there in the liver?

*Exton:* Northern analysis and Western blotting have been done (Strathmann & Simon 1990). In contrast to Pang & Sternweis (1990), we find substantial amounts of $\alpha_q$ in rat and bovine liver membranes, and we find approximately equal quantities of $\alpha_{11}$.

*Ui:* How much is there in relation to $G_i$?

*Exton:* They are less abundant than the $G_i$ proteins.

*Michell:* How widespread is the q family of G proteins?

*Exton:* The $\alpha_q$ protein can be found in all tissues, and is abundant in brain and lung. The $\alpha_{11}$ protein is probably ubiquitous—its mRNA is found in every tissue that has been examined by Northern analysis.

*Kato-Homma:* What is the molar ratio of $G_q$ to phospholipase-$\beta_1$ in the activation of the phospholipase?

*Exton:* We haven't investigated stoichiometry or kinetics in detail. Phospholipase activity increased four-fold when reconstituted with equimolar amounts of the G protein $\alpha$ subunits.

*Yang:* Do the G proteins activate phospholipase C and D equally?

*Exton:* In most tissues phosphatidylcholine phospholipase D is predominantly activated, but there are some tissues where only phospholipase C is activated.

*Yang:* Does the $G_q$ protein have $\beta$ and $\gamma$ subunits that are identical or homologous to those from other families?

*Exton:* We have not yet analysed the $\beta$ and the $\gamma$ subunits.

*Daly:* There is interest in the direct activation of G proteins by *Hymenoptera* venoms. For example, mastoparan stimulates PtdIns breakdown in many cell types through direct activation of G proteins. Did mastoparan activate your G protein?

*Exton:* Unfortunately, it did not work in our system.

*Krebs:* I know it wasn't part of your talk, but would you care to comment on the potential of phosphatidic acid as a second messenger?

*Exton:* As you know, many roles have been proposed for phosphatidic acid (for review see Exton 1990). Initially it was thought to act as an ionophore, that is, to increase calcium influx into cells. That's very questionable, because when you use freshly prepared phosphatidic acid you cannot see its ionophoretic action. Another possible role is the inhibition of adenylate cyclase, which is mediated by one of the $G_{i\alpha}$ species. Effects on protein kinases, such as phosphorylase kinase, have been reported. We suspect there is an unknown protein kinase target for phosphatidic acid. Phosphatidic acid and lysophosphatidic acid are very powerful mitogens in many fibroblast cell types.

*Parker:* Work from Moolenaar's lab. indicates that the mitogenic effect of lysophosphatidic acid may be driven at an extracellular level, perhaps by interaction with a receptor to activate G proteins— it may not in that context be acting as a second messenger.

*Exton:* We found no evidence in our cells for a surface receptor for phosphatidic acid or lysophosphatidic acid.

*Parker:* Is the $\beta_1$ form of phospholipase C expressed in hepatocytes?

*Exton:* Yes, definitely.

*Parker:* Are you sure it is $\beta_1$, not $\beta_2$ or $\beta_3$?

*Exton:* We have monoclonal and polyclonal antibodies only to the $\beta_1$, $\gamma_1$ and $\delta_1$ forms. Using these antisera we have tracked the particular isozyme in rat and bovine liver which is activated by $\alpha_q$; it is the $\beta_1$ isozyme.

*Parker:* Is it known whether those antisera will cross-react with $\beta_2$ or $\beta_3$ polypeptides?

*Exton:* To my knowledge, they do not.

*Ui:* In your reconstitution experiments did you use a mixture of GTP$\gamma$S and GTP or GTP$\gamma$S only?

*Exton:* Obviously, for studies of GTPase activity we used GTP. I mentioned only GTP$\gamma$S binding and we have shown that this is activated through the $M_1$ muscarinic receptor. You can get similar results with GTPase activity; presumably, the binding of GTP is also enhanced.

*Ui:* You used GTP$\gamma$S in all your experiments.

*Exton:* Yes, because we always use a two-step assay—activation of the G protein(s) followed by interaction with phospholipase C—and we must keep the G protein active for the second phase.

*Ui:* Perhaps your preparation contains bound GDP.

*Exton:* That is almost certainly true.

*Nishizuka:* If my memory is correct, most of the reactions of phospholipase D require agonists and can be activated by phorbol ester.

*Exton:* That is correct.

*Nishizuka:* What does that mean? Either activation of phospholipase D must be secondary to its phosphorylation, or there must be heterogeneity in phospholipase D.

*Exton:* The problem with our studies with phospholipase D is that we have not been able to purify it to homogeneity, so we are not able to do the sorts of experiments that you and I would like to do using protein kinase C. I suspect, by analogy with the $PtdInsP_2$ phospholipase C, that there will be several phospholipase D isozymes and heterogeneity of control, but that's just a prediction.

*Nishizuka:* Many people have reported the activation of phosphatidylcholine phospholipase C (see, for example, Besterman et al 1986). Is it possible in your system to distinguish between activation of phospholipase D and phospholipase C?

*Exton:* Many of the inhibitors of phosphatidate phosphohydrolase and diacylglycerol kinase do not work in the hepatocyte or in the liver plasma membrane. We would like to distinguish between phospholipase C and D. In general, it seems that phospholipase D is the predominant pathway in most cells.

*Hunter:* Is there any evidence for formation of a complex between $\alpha_q$–GTP and phospholipase C-$\beta_1$, in terms of possible recruitment to the membrane?

*Exton:* If you prepare bovine liver membranes very carefully you find the $\beta_1$ isozyme predominantly in the membrane, and not in the cytoplasm; activation therefore probably does not involve recruitment.

*Hunter:* Is that not in contrast to what one finds in the brain, where all three phospholipase C isoforms are soluble?

*Exton:* Because of protease activity it is extremely difficult to prepare brain membranes without some perturbation. In the liver, taking very good care, we find the $\beta_1$ enzyme only in the membrane, whereas the $\gamma_1$ isozyme is largely in the cytosol.

*Hunter:* Are the $\alpha_q$ and $\alpha_{11}$ subunits myristoylated?

*Exton:* There is no myristoylation signal in the sequences.

*Michell:* What are your thoughts on the relative roles of the diacylglycerols from the different pathways? I ask the question partly because of the work of Leach et al (1991), which suggests that that the inositide-derived diglyceride is a very good signal for protein kinase C, whereas the phosphatidylcholine-derived diglyceride isn't.

*Exton:* I think this is the reason why there are different protein kinase C isozymes in different cells. In the cell that they studied (the IIC9 fibroblast) the $\alpha$ form was present, but not the $\beta$ or $\gamma$ isozymes. We don't know if the $\beta$, $\gamma$, $\delta$, $\varepsilon$, $\zeta$ or $\eta$ forms can be activated by diacylglycerol coming from phosphatidylcholine. The other possibility is that this diacylglycerol has other actions, perhaps to produce arachidonic acid, or some other undiscovered function.

*Michell:* Is there anything about the specificity of known phospholipase C isozymes that would lead you to expect that sharp discrimination in activation between the different diacylglycerols?

*Exton:* There's a tremendous difference between the activation of different protein kinase C isozymes by phosphatidylserine and arachidonic acid, and also $Ca^{2+}$. It's possible that there could be such a distinction.

*Kanoh:* My general impression is that if you detect very low diacylglycerol kinase activity in homogenates or membranes, the activation of phospholipase D will predominate over diacylglycerol kinase, at least when observed at the level of intact cells. If there is a very high diacylglyerol kinase activity *in vitro*, or if you detect large amounts of a particular isozyme by immunochemistry, as we reported in lymphocytes (Kanoh et al 1990), you will see mainly diacylglyerol kinase action. Do you think that my general impression is correct?

*Exton:* Unfortunately, my knowledge is mainly of the liver. We were astonished to find that in the liver the major route of metabolism of exogenous diacylglycerol was hydrolysis by diacylglycerol lipase, and its conversion to phosphatidic acid was small.

*Daly:* It has been reported that if phosphatidic acid is produced through the phospholipase D pathway, adding ethanol should greatly reduce its formation because ethanol is a much better substrate for phospholipase D than water is.

*Exton:* The problem is that phosphatidate formation by phospholipase D is not completely shut off by ethanol.

*Daly:* That's true, but if there were significant amounts of phosphatidate still being formed then you might conclude that it was coming through the phospholipase C plus phosphorylation route, rather than directly by phospholipase D action.

*Exton:* What I am saying is that, in the liver, addition of ethanol does not switch the phospholipase D activity totally from production of phosphatidic acid to phosphatidylethanol production. There are some cells in which the switch is complete, but there are also cells in which phosphatidic acid is still produced, regardless of how much ethanol you add.

*Yang:* Is it possible that the diacylglycerol acts on protein kinase C differently depending on whether it is derived from phosphatidylcholine or from phosphatidylinositol?

*Exton:* I wish I knew! I hope that is so.

*Nishizuka:* The Golgi apparatus is frequently very rich in the $\beta_2$ isoform of protein kinase C. Sphingomyelin synthesis is active in the Golgi apparatus and it is very rich in diacylglycerol. There should be a large variety of intracellular compartments for diacylglycerol production. I shall talk about the synergistic effects of diacylglycerol and unsaturated fatty acids. Unexpectedly, in the presence of diacylglycerol unsaturated fatty acids dramatically activate protein kinase C. So, it is an attractive hypothesis that both phospholipase $A_2$ and phospholipase C, and possibly also phospholipase D, are participating in a degradation cascade to activate enzymes. We don't yet have the exact picture of events, particularly for diacylglycerol.

*Daly:* Professor Exton, you said that ATP stimulates phosphatidylcholine breakdown. Which receptor subtype is involved?

*Exton:* It's the $P_{2Y}$ subtype.

*Daly:* So 2-methylthioATP would activate phospholipase D in the membrane.

*Exton:* Yes, it does.

*Nishizuka:* Is phospholipase $A_2$ in hepatocytes activated by GTP$\gamma$S? Does phospholipase $A_2$ participate in your hepatocyte system?

*Exton:* We have never looked at that, but there are many systems in which regulation of phospholipase $A_2$ by G proteins is indicated (Silk et al 1989, Kajiyama et al 1990, Narasimhan et al 1990).

*Michell:* You said that standard calcium-mobilizing agonists are effective at activating protein kinase C through phosphatidylcholine; you gave the impression that activation of the $P_2$ purinergic receptor is particularly important for phosphatidylcholine hydrolysis. I don't understand why.

*Exton:* In our cells the $P_2$ purinergic receptor powerfully stimulates phosphatidylcholine hydrolysis directly through a G protein. Other calcium-mobilizing agonists do this only very weakly.

*Michell:* What do you think the likely role of the $P_2$ receptor is in liver cells? Where does it get its ATP from?

*Exton:* ATP is released on activation of the sympathetic nervous system. This is a classic sympathetic response—are you sympathetic to that?

*Michell:* I didn't think there was much sympathetic innervation of the liver.

*Exton:* There is (Shimazu & Amakawa 1968, Edwards & Silver 1970, Friedman 1988).

*Carpenter:* You indicated that $G_q$ proteins seem to have a low affinity for GTP. What would be the implications of that in cells activated simultaneously by $M_1$ and $M_2$ receptor agonists? Would the $M_2$ response dominate?

*Exton:* Our data are very preliminary. We think that $G_{i2}$ is the G protein which mediates cyclase inhibition. I don't know the exact quantitation between $\alpha_{i2}$ and $\alpha_q$. We also have $\alpha_{11}$, which is just as powerful a stimulant of phospholipase $\beta_1$ as $\alpha_q$. Together, $\alpha_q$ and $\alpha_{11}$ could produce a signal comparable to that mediated by $\alpha_{i2}$, but this assumes a lot.

*Carpenter:* In experiments in which the $M_2$ muscarinic receptor is overexpressed in transfected cells it begins to couple with phosphatidylinositol hydrolysis (Ashkenazi et al 1987). How do you think that occurs? Do you think that the stoichiometry of the receptors in a cell membrane affects the selective interaction with different G proteins?

*Exton:* I don't know. When you overexpress proteins in a cell many secondary events may occur.

*Carpenter:* If you add a lot of $M_2$ receptors in your *in vitro* system, can you get coupling to $G_q$?

*Exton:* The amount of $M_2$ receptor used in the reconstitution experiments was in excess of $M_1$. We cannot get coupling of the $M_2$ receptor with $G_q/G_{11}$ in this system (unpublished work).

*Parker:* Wouldn't the concentration of $G_q$ be a deciding factor?

*Exton:* We varied the amount of G protein and it always showed a high degree of selectivity for the $M_1$ receptor.

## References

Ashkenazi A, Winslow JW, Peralta EG et al 1987 An M2 muscarinic receptor subtype coupled to both adenyl cyclase and phosphoinositide turnover. Science (Wash DC) 238:672–675

Besterman JM, Duronio V, Cuatrecasas P 1986 Rapid formation of diacylglycerol from phosphatidylcholine: a pathway for generation of a second messenger. Proc Natl Acad Sci USA 83:6785–6789

Edwards AV, Silver M 1970 The glycogenolytic response to stimulation of the splanchnic nerves in adrenalectomized calves. J Physiol 211:109– 124

Exton JH 1990 Signalling through phosphatidylcholine breakdown. J Biol Chem 265:1–4

Friedman MI 1988 Hepatic nerve function. In: Arias IM, Jakoby WB, Popper H, Schachter D, Shafritz DA (eds) The liver: biology and pathobiology, 2nd edn. Raven Press, New York p 949–959

Johnson RM, Garrison JC 1987 Epidermal growth factor and angiotensin II stimulate formation of inositol 1,4,5- and inositol 1,3,4-trisphosphate in hepatocytes. Differential inhibition by pertussis toxin and phorbol 12-myristate 13-acetate. J Biol Chem 262:17285–17293

Kajiyama Y, Murayama T, Kitamura Y, Imai S-I, Nomura Y 1990 Possible involvement of different GTP-binding proteins in noradrenaline- and thrombin-stimulated release of arachidonic acid in rabbit platelets. Biochem J 270:69–75

Kanoh H, Yamada K, Sakane F 1990 Diacylglycerol kinase: a key modulator of signal transduction? Trends Biochem Sci 15:47–50

Leach KL, Ruff VA, Wright TM, Pessin MS, Raben MS 1991 Dissociation of protein kinase C activation and *sn*-1,2-diacylglycerol formation: comparison of phosphatidylinositol-derived and phosphatidylchloine-derived glycerides in α-thrombin-stimulated fibroblasts. J Biol Chem 266:3215–3221

Moriarty TM, Padrell E, Carty DJ, Omri G, Landau EM, Iyengar R 1990 $G_o$ protein as signal transducer in the pertussis toxin-sensitive phosphatidylinositol pathway. Nature (Lond) 343:79–82

Narasimhan V, Holowka D, Baird B 1990 A guanine nucleotide-binding protein participates in IgE receptor-mediated activation of endogenous and reconstituted phospholipase $A_2$ in a permeabilized cell system. J Biol Chem 265:1459–1464

Padrell E, Carty DJ, Moriarty TM, Hildebrandt JD, Landau EM, Iyengar R 1991 Two forms of the bovine brain $G_o$ that stimulate the inositol trisphosphpate-mediated $Cl^-$ currents in *Xenopus* oocytes: distinct guanine nucleotide binding properties. J Biol Chem 266:9771–9777

Pang I-H, Sternweis PC 1990 Purification of unique α-subunits of GTP-binding regulatory proteins (G proteins) by affinity chromatography with immortalized βγ subunits. J Biol Chem 265:18707–18712

Shimazu T, Amakawa A 1968 Regulation of glycogen metabolism in liver by the autonomic nervous system. II. Neural control of glycogenolytic enzymes. Biochim Biophys Acta 165:335–348

Silk ST, Clejan S, Witkom K 1989 Evidence of GTP-binding protein regulation of phospholipase $A_2$ activity in isolated human platelet membranes. J Biol Chem 264:21466–21469

Strathmann M, Simon MI 1990 G protein diversity: a distinct class of $\alpha$-subunits is present in vertebrates and invertebrates. Proc Natl Acad Sci USA 87:9113–9117

# The signal-induced phospholipid degradation cascade and protein kinase C activation

Yoshinori Asaoka*, Kimihisa Yoshida*, Masahiro Oka*, Tetsutaro Shinomura†, Hiroyuki Mishima†, Shinji Matsushima† and Yasutomi Nishizuka*†

**Biosignal Research Center, Kobe University, Kobe 657 and †Department of Biochemistry, Kobe University School of Medicine, Kobe 650, Japan*

*Abstract.* Acting in synergy with diacylglycerol, unsaturated free fatty acids such as arachidonic, oleic, linoleic, linolenic and docosahexaenoic acids dramatically activate some members of the protein kinase C family at the basal level of $Ca^{2+}$ concentration. It is plausible that phospholipase C and phospholipase $A_2$, and possibly phospholipase D as well, are involved in the activation of protein kinase C. Presumably, this enzyme activation is integrated into the signal-induced membrane phospholipid degradation cascade, prolonging the activation of protein kinase C. The sustained activity of this enzyme appears to be of importance for long-term cellular responses such as development of neuronal plasticity and gene activation.

*1992 Interactions among cell signalling systems. Wiley, Chichester (Ciba Foundation Symposium 164) p 50–65*

An early report from this laboratory has shown that diacylglycerol (DAG) derived from inositol phospholipids activates protein kinase C (PKC) in the presence of $Ca^{2+}$ and phosphatidylserine (PtdSer) (Kishimoto et al 1980). Although the hydrolysis of phosphatidylinositol and its polyphosphates was once thought to be the sole mechanism leading to the activation of PKC, there now appear to be several additional routes to provide the DAG that is needed for PKC activation. For instance, phosphatidylcholine has been proposed as another source of DAG (see Exton 1990; Exton et al 1992, this volume). Recent studies indicate that PKC is activated further by the simultaneous addition of unsaturated free fatty acids (FFAs) and DAG, particularly at low $Ca^{2+}$ concentrations (Shinomura et al 1991). The responses of several PKC subspecies to these lipids differ slightly from one another, and various metabolites produced in the signal-induced membrane phospholipid degradation cascade may play roles in sustaining the activation of PKC. Here, we briefly describe such synergistic action of unsaturated FFAs and DAG on the activation of PKC, and discuss its possible implications for cellular regulation, such as in the development of synaptic plasticity and gene activation.

## The synergistic action of unsaturated free fatty acids and diacylglycerol

Several subspecies of PKC isolated from mammalian tissues absolutely require PtdSer and $Ca^{2+}$ for their actions, and DAG greatly increases the apparent affinity of the enzyme for $Ca^{2+}$, thereby rendering it active over a micromolar $Ca^{2+}$ concentration range (Kishimoto et al 1980, Nishizuka 1988). The experiments shown in Fig. 1 confirm this early observation, and show that in the presence of DAG and PtdSer the addition of arachidonic acid further increases the affinity of PKC for $Ca^{2+}$, and greatly enhances the reaction velocity, particularly at near basal $Ca^{2+}$ concentrations. These results were obtained with subspecies of PKC isolated from rat brain. The synergistic action of arachidonic acid and DAG was most marked with the $\alpha$ and $\beta$ subspecies, which have been found to be present in many of the tissues and cell types so far tested. The $\gamma$ subspecies was activated slightly by arachidonic acid in the absence of DAG and PtdSer at relatively high concentrations of $Ca^{2+}$, as described previously (Sekiguchi et al 1987). This subspecies is expressed only in central nervous tissues after birth (Hashimoto et al 1988).

The synergistic effect of FFA and DAG on activation is also observed with many other naturally occurring *cis*-unsaturated FFAs, including oleic, linoleic, linolenic and docosahexaenoic acids, as shown in Fig. 2. Saturated FFAs, including palmitic and stearic acids, are inactive. Elaidic acid is also inactive with all PKC subspecies tested.

The results described above were obtained with calf thymus H1 histone as a model substrate. The DAG, PtdSer and $Ca^{2+}$ required for PKC activation vary significantly with the phosphate acceptor protein used (Bazzi & Nelsestuen 1987, Nishizuka 1988, Buday & Faragó 1990). The synergistic action of unsaturated FFAs and DAG, however, can be observed with various other model phosphate acceptor proteins, such as myelin basic protein (Shinomura et al 1991), as well as with physiological substrate proteins, such as growth-associated protein 43 (GAP-43, also known as F1 protein, B50 protein and neuromodulin, see below), although the kinetics of this synergistic action vary slightly according to the phosphate acceptor protein used. It may be important to note that with all substrates so far examined several subspecies of PKC exhibit nearly full activity at low $Ca^{2+}$ concentrations in the presence of both DAG and unsaturated FFAs.

## An implication for synaptic processes

The activation of PKC has been proposed to play a key role in the development of long-term potentiation (LTP) in the hippocampus (for reviews, see Alkon &

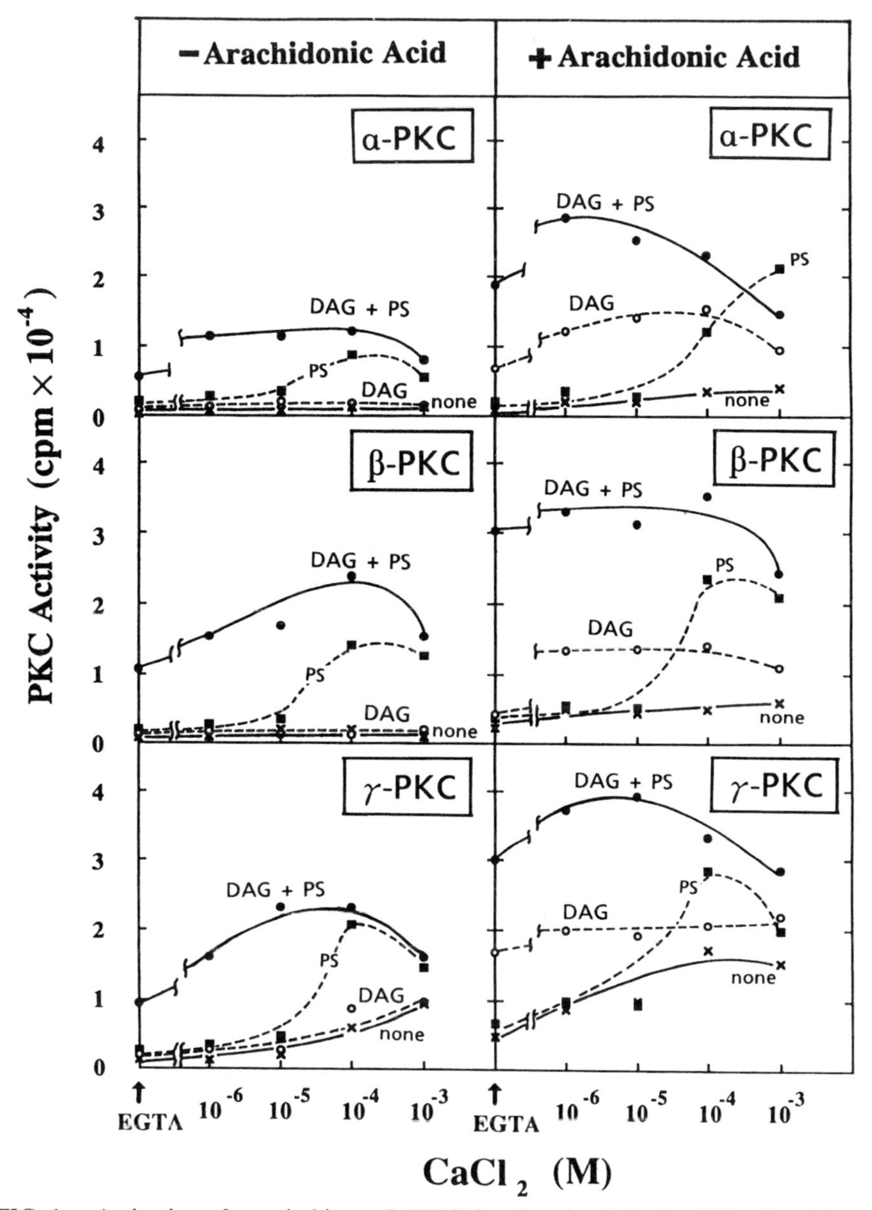


FIG. 1. Activation of protein kinase C (PKC) by phosphatidylserine (PS), diacylglycerol (DAG) and arachidonic acid at various concentrations of $Ca^{2+}$. The activity of each PKC subspecies (isolated from rat brain) was assayed with H1 histone as a model substrate under the conditions described by Shinomura et al (1991), except that various concentrations of $CaCl_2$ were added in the absence (*left panel*) or presence (*right panel*) of 50 μM arachidonic acid. EGTA (5 mM) instead of $CaCl_2$ was added to the reaction mixture where indicated by an arrow.

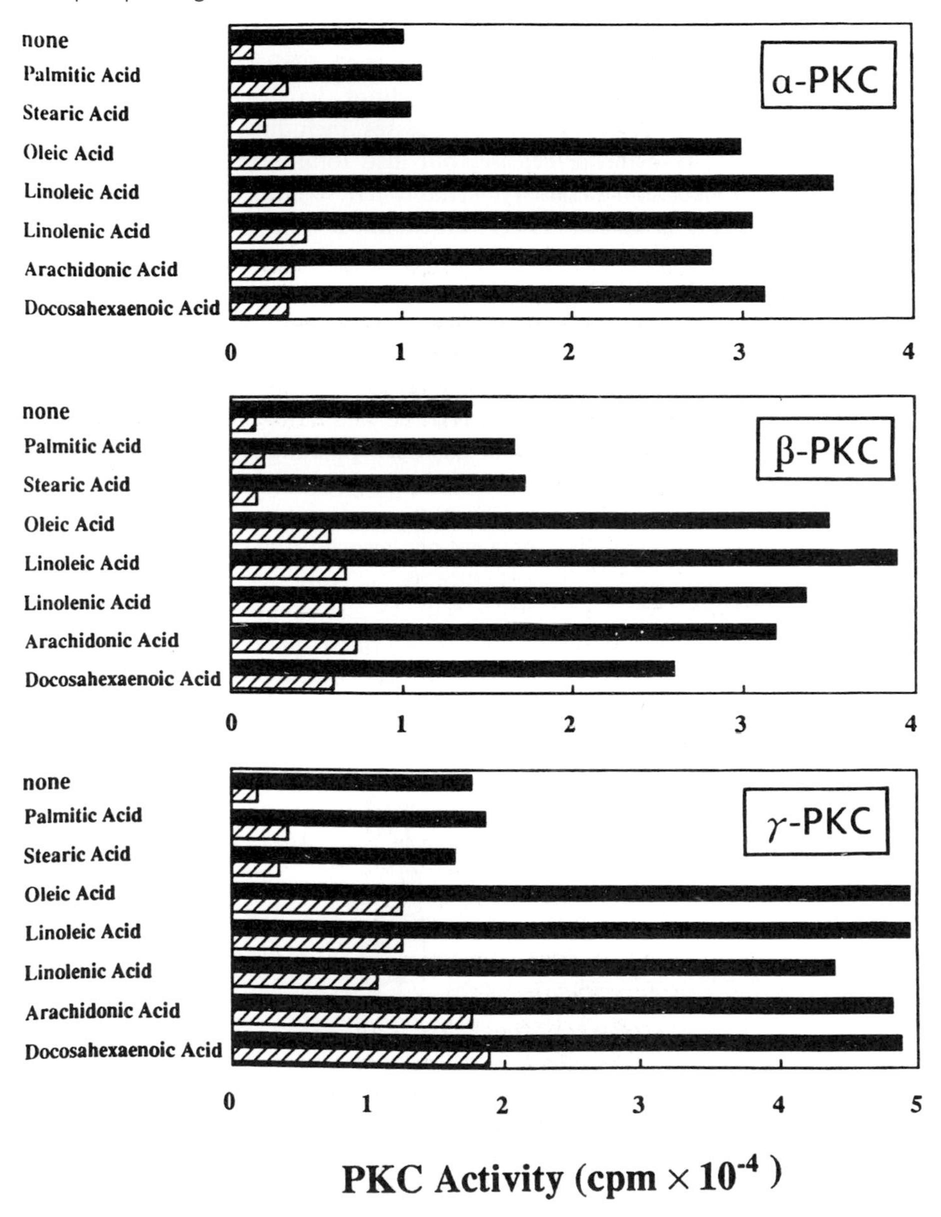


FIG. 2. Activation of PKC by various free fatty acids in the presence of PtdSer, DAG and $Ca^{2+}$. Each subspecies of PKC was assayed with H1 histone as a model substrate under standard conditions as described by Shinomura et al (1991) except that various fatty acids (50 μM each) were added as indicated. Black bars, assayed in the presence of PtdSer and DAG; hatched bars, assayed in the presence of PtdSer.

Rasmussen 1988, Gispen 1988, Linden & Routtenberg 1989, Tsien et al 1990). One of the major target proteins of PKC, GAP-43, which is tightly associated with the membrane of presynaptic nerve endings, has been proposed to play a role in the control of diverse synaptic processes, including not only immediate responses such as transmitter release, but also long-term responses such as development of neuronal plasticity, synaptogenesis and regeneration (for a review see Coggins & Zwiers 1991). Biochemical and immunocytochemical analyses have shown that in the adult rat hippocampus the $\alpha$ and $\beta$ subspecies are present in both pre- and postsynaptic regions, whereas the $\gamma$ subspecies is expressed only in the postsynaptic region, as illustrated schematically in Fig. 3 (Kose et al 1990, Shearman et al 1991).

It has been proposed that unsaturated FFAs may be involved in the maintenance of LTP, probably through the activation of PKC (Linden & Routtenberg 1989). Arachidonic acid has been postulated to be a retrograde messenger in this process from the postsynaptic cell to the presynaptic terminal (Bliss et al 1989). Although the biochemical mechanism of LTP and the role of PKC in synaptic transmission have not yet been fully clarified, the synergistic action of unsaturated FFAs and DAG on PKC activation reported here suggests a potential role of FFAs in the activation of the enzyme, particularly the $\alpha$ and $\beta$ subspecies, which are abundant both in the presynaptic terminal and in the postsynaptic cell. It has been reported, in fact, that in the hippocampus on stimulation with serotonin arachidonic acid is released by the activation of

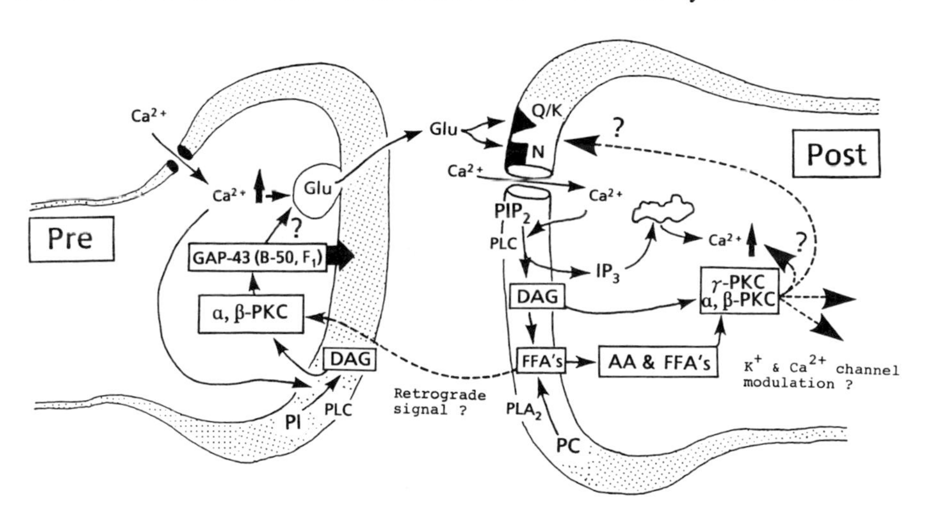


FIG. 3. A schematic representation of the localization and possible mode of activation of PKC subspecies in the presynaptic and postsynaptic regions of the adult rat hippocampus. N, *N*-methyl-D-aspartate receptor; Q/K, quisqualate/kainate receptor; PLC, phospholipase C; $PLA_2$, phospholipase $A_2$; PI, phosphatidylinositol; $IP_3$, inositol trisphosphate; AA, arachidonic acid; PC, phosphatidylcholine; GAP-43, growth-associated protein-43.

phospholipase $A_2$ (Felder et al 1990). Docosahexaenoic acid is known to be a major constituent of the phospholipids of the central nervous system, especially during early postnatal development and synaptogenesis (Scott & Bazan 1989). This fatty acid, when combined with DAG, is a potent enhancer of PKC activity, and can activate the $\gamma$ subspecies to some extent by itself in the absence of DAG. This fatty acid may have some functions in neuronal processes.

## An implication for gene activation

Another possible implication is that the synergistic action of unsaturated FFAs and DAG may be involved in the sustained activation of PKC that is apparently needed for cell growth and differentiation. Phosphatidylcholine is broken down to produce DAG at a relatively late phase in cellular responses (for a review see Exton 1990). The experiments shown in Fig. 4 confirmed our previous finding (Berry et al 1990) that in the presence of ionomycin repeated doses of a membrane-permeable diacylglycerol, dioctanoylglycerol, mimic a single dose of tumour-promoting phorbol ester, inducing T cell activation, as measured

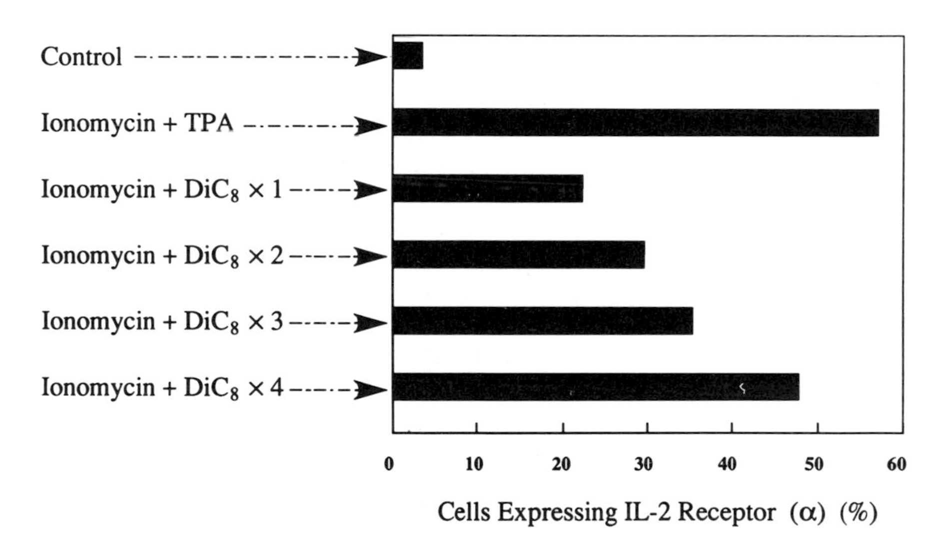


FIG. 4. Effect of repeated additions of dioctanyoylglycerol on IL-2 $\alpha$-receptor expression. Accessory cell-depleted T cells ($5 \times 10^5$ cells/ml) from human venous blood were cultured in Roswell Park Memorial Institute medium containing 5% autologous serum. Cells were stimulated with either dioctanoylglycerol or TPA (12-*O*-tetradecanoylphorbol-13-acetate) in the presence of ionomycin. Dioctanoylglycerol was added repeatedly (1–4 times, each time 25 μM) in the presence of 0.25 μM ionomycin as indicated. A single dose of phorbol ester (10 nM) and ionomycin (1 μM) were also added where specified. Cells were harvested 16 hours after the first stimulation, and IL-2 $\alpha$-receptors were stained by the direct immunofluorescence staining method. Fluorescence was quantified using a flow cytometer. $DiC_8$, dioctanoylglycerol.

by interleukin 2 (IL-2) $\alpha$-receptor expression. Increasing the duration of PKC activation by treating T cells with repeated doses of dioctanoylglycerol potentiates the level of IL-2 $\alpha$-receptor expression in a manner that is dependent on the frequency of the DAG additions. A recent biochemical analysis with radioactive dioctanoylglycerol has indicated that a major portion of such exogenously added DAG is metabolized within T cells by DAG lipase very rapidly, within several minutes to one hour, depending upon the cell density, whereas the phorbol ester is hardly metabolized and is retained in the membrane for a prolonged period of time, irrespective of cell density. Similarly, thymidine incorporation into resting T cells can be greatly enhanced by multiple doses of dioctanoylglycerol, just as it can by a single dose of phorbol ester. Down-regulation of PKC does not take place during the experiment, although multiple doses of dioctanoylglycerol and a single dose of phorbol ester cause a similar redistribution of PKC subspecies (Berry et al 1989). It is plausible, therefore, that prolonged PKC activation, as well as elevation of the intracellular $Ca^{2+}$ concentration, is required for the initiation of IL-2 $\alpha$-receptor expression and IL-2 secretion.

On the other hand, it is known that an antigenic signal provided by anti-Ti/CD3 antibodies causes only a transient rise of $Ca^{2+}$ and DAG within the T cell. The hypothesis that several accessory signals function, at least partially, by prolonging PKC activation is supported by several recent studies (for a review see Berry & Nishizuka 1990). Although it is still premature to discuss the precise relationship between the accessory signals and the sustained activation of PKC, it is attractive to investigate whether the accessory signals generate second messengers, such as DAG and unsaturated FFAs, to sustain the activation of PKC long enough to initiate T cell activation, even after the $Ca^{2+}$ concentration has returned to its basal level. Szamel et al (1989) have observed that linoleic and arachidonic acids may potentiate the IL-2 synthesis that is initiated by the combination of a membrane-permeable DAG and ionomycin.

## A hypothesis

Most probably, the intracellular events involved in signal-induced cellular responses, in both the short-term and the long-term, are the result of interactions between a number of signal transduction pathways. One of these pathways is obviously PKC activation, which is apparently involved in many physiological processes. Initially, we thought that the receptor-mediated hydrolysis of inositol phospholipids was the sole route of production of DAG for PKC activation. Subsequently, however, receptor-mediated and voltage-dependent $Ca^{2+}$ channel opening was shown to produce DAG through an activation of phospholipase C that is perhaps secondary to the increase in the intracellular $Ca^{2+}$ concentration. Phospholipase C that is reactive against phosphatidylcholine has also been described (Besterman et al 1986). In recent years attention has been paid to the receptor-mediated activation of phospholipase $A_2$ (Axelrod et al 1988), as well as to phospholipase D (Exton 1990, Exton et al 1992). The results of the test-tube experiments described here seem to suggest that the signal-induced release of unsaturated FFAs from several phospholipids may also, acting in synergy with DAG also derived from several phospholipids, participate in the activation of PKC, even when the $Ca^{2+}$ concentration remains at the basal level. Thus, the mechanism of activation of various PKC subspecies may vary greatly with cell type, extracellular signal, and presumably with the duration after cell stimulation. It is possible that the activation of PKC is an integral part of the signal-induced degradation cascade of membrane phospholipids, which is catalysed by phospholipase C, phospholipase $A_2$ and perhaps also by phospholipase D, as shown schematically in Fig. 5. Analysis of interactions among various pathways that activate these phospholipases may provide a clue to help us understand further the mechanisms of cellular responses, particularly longer term ones.

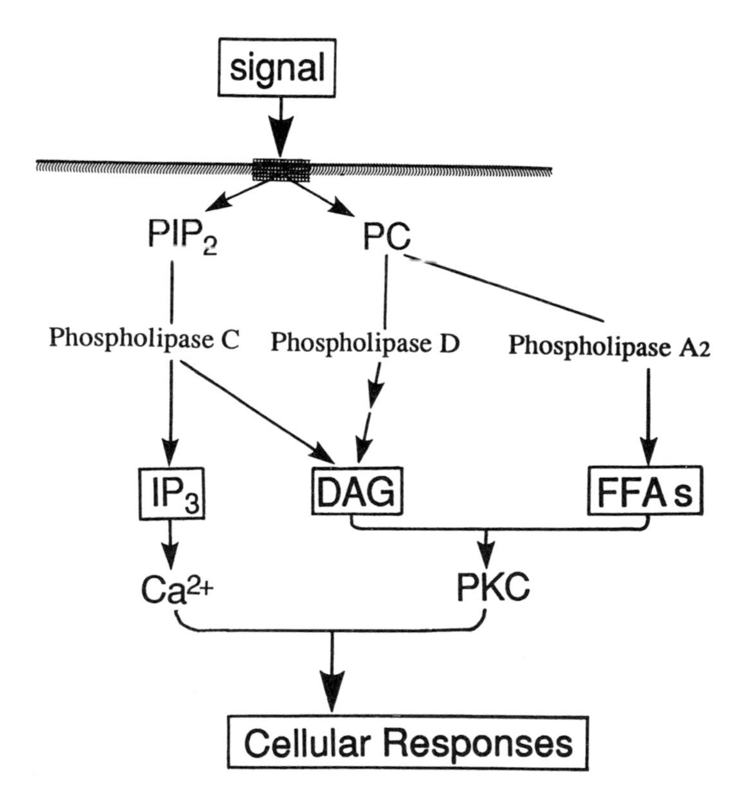


FIG. 5. A schematic representation of the induction of cellular responses by the signal-induced degradation cascade of membrane phospholipids. $PIP_2$, phosphatidylinositol bisphosphate; $IP_3$, inositol trisphosphate; PC, phosphatidylcholine; DAG, diacylglycerol; FFAs, free fatty acids; PKC, protein kinase C.

*Acknowledgements*

The skillful secretarial assistance of Mrs S. Nishiyama, Miss Y. Kimura and Miss Y. Yamaguchi is cordially acknowledged. This work was supported in part by research grants from the Special Research Fund of the Ministry of Education, Science and Culture, Japan; Muscular Dystrophy Association, USA; Juvenile Diabetes Foundation International, USA; Yamanouchi Foundation for Research on Metabolic Disorders; Sankyo Foundation of Life Science; Merck Sharp & Dohme Research Laboratories; Biotechnology Laboratories of Takeda Chemical Industries; and the New Lead Research Laboratories of Sankyo Company.

## References

Alkon DL, Rasmussen H 1988 A spatial–temporal model of cell activation. Science (Wash DC) 239:998–1005

Axelrod J, Burch RM, Jelsema CL 1988 Receptor-mediated activation of phospholipase $A_2$ via GTP-binding proteins: arachidonic acid and its metabolites as second messengers. Trends Neurosci 11:117–123

Bazzi MD, Nelsestuen GL 1987 Role of substrate in imparting calcium and phospholipid requirements to protein kinase C activation. Biochemistry 26:1974–1982

Berry N, Nishizuka Y 1990 Protein kinase C and T cell activation. Eur J Biochem 189:205–214

Berry N, Ase K, Kikkawa U, Kishimoto A, Nishizuka Y 1989 Human T cell activation by phorbol esters and diacylglycerol analogues. J Immunol 143:1407–1413

Berry N, Ase K, Kishimoto A, Nishizuka Y 1990 Activation of resting human T cells requires prolonged stimulation of protein kinase C. Proc Natl Acad Sci USA 87:2294–2298

Besterman JM, Duronio V, Cuatrecasas P 1986 Rapid formation of diacylglycerol from phosphatidylcholine: a pathway for generation of a second messenger. Proc Natl Acad Sci USA 83:6785–6789

Bliss TVP, Clements MP, Errington ML, Lynch MA, Williams JH 1989 Presynaptic mechanisms underlying the maintenance of long-term potentiation in the hippocampus. In: Ito M, Nishizuka Y (eds) Brain signal transduction and memory. Academic Press, Tokyo p 185–196

Buday L, Faragó A 1990 Dual effect of arachidonic acid on protein kinase C isoenzymes isolated from rabbit thymus cells. FEBS (Fed Eur Biochem Soc) Lett 276: 223–226

Coggins PJ, Zwiers H 1991 B-50 (GAP-43): biochemistry and functional neurochemistry of a neuron-specific phospho-protein. J Neurochem 56:1095–1106

Exton JH 1990 Signaling through phosphatidylcholine breakdown. J Biol Chem 265:1–4

Exton JH, Taylor SJ, Blank JS, Bocckino SB 1992 Regulation of phosphoinositide and phosphatidylcholine phospholipases by G proteins. Wiley, Chichester (Ciba Found Symp 164) p 36–49

Felder CC, Kanterman RY, Ma AL, Axelrod J 1990 Serotonin stimulates phospholipase $A_2$ and the release of arachidonic acid in hippocampal neurons by a type 2 serotonin receptor that is independent of inositolphospholipid hydrolysis. Proc Natl Acad Sci USA 87:2187–2191

Gispen WH 1988 Transmembrane signal transduction and ACTH-induced excessive grooming in the rat. Ann NY Acad Sci 525:141–149

Hashimoto T, Ase K, Sawamura S et al 1988 Postnatal development of a brain-specific subspecies of protein kinase C in rat. J Neurosci 8:1678–1683

Kishimoto A, Takai Y, Mori T, Kikkawa U, Nishizuka Y 1980 Activation of calcium and phospholipid-dependent protein kinase by diacylglycerol, its possible relation to phosphatidylinositol turnover. J Biol Chem 255:2273–2276
Kose A, Ito A, Saito N, Tanaka C 1990 Electron microscopic localization of γ- and βII-subspecies of protein kinase C in rat hippocampus. Brain Res 518:209–217
Linden DJ, Routtenberg A 1989 The role of protein kinase C in long-term potentiation: a testable model. Brain Res Rev 14:279–296
Nishizuka Y 1988 The molecular heterogeneity of protein kinase C and its implications for cellular regulation. Nature (Lond) 334:661–665
Scott BL, Bazan NG 1989 Membrane docosahexaenoate is supplied to the developing brain and retina by the liver. Proc Natl Acad Sci USA 86:2903–2907
Sekiguchi K, Tsukuda M, Ogita K, Kikkawa U, Nishizuka Y 1987 Three distinct forms of rat brain protein kinase C: differential response to unsaturated fatty acids. Biochem Biophys Res Commun 145:797–802
Shearman MS, Shinomura T, Oda T, Nishizuka Y 1991 Synaptosomal protein kinase C subspecies: A. Dynamic changes in the hippocampus and cerebellar cortex concomitant with synaptogenesis. J Neurochem 56:1255–1262
Shinomura T, Asaoka Y, Oka M, Yoshida K, Nishizuka Y 1991 Synergistic action of diacylglycerol and unsaturated fatty acid for protein kinase C activation, its possible implications. Proc Natl Acad Sci USA 88:5149–5153
Szamel M, Rehermann B, Krebs B, Kurrle R, Resch K 1989 Activation signals in human lymphocytes. J Immunol 143:2806–2813
Tsien RW, Schulman H, Malinow R 1990 Peptide inhibitors of PKC and CaMK block induction but not expression of long-term potentiation. In: Nishizuka Y, Endo M, Tanaka C (eds) The biology and medicine of signal transduction. Raven Press, New York. (Adv Second Messenger Phosphoprotein Res 24) p 101–107

## DISCUSSION

*Tsien:* Have you tried to test whether a brief calcium transient is sufficient to trigger persistent kinase activation, by raising the calcium concentration and then lowering it by the immediate addition of EGTA?

*Nishizuka:* This is a very good experiment, which we have not yet done. These are recent observations made *in vitro* and we are planning to test this possibility with cell systems such as T lymphocyte activation, HL-60 differentiation and platelet activation.

*Exton:* We have done experiments very similar to yours in hepatocytes (unpublished). When we added dioctanoylglycerol it was extremely rapidly metabolized by the cells. Repetitive addition was needed to reproduce the effect of TPA. Unless dioctanoylglycerol is metabolized differently from endogenous diacylglycerol produced by phospholipid breakdown, this result indicates that there must be continuous production of diacylglycerol and continuous breakdown of the phospholipid to sustain the signal.

*Rasmussen:* I don't agree. Jorge and Monica Eloun-Christiason in my laboratory have done similar experiments in fibroblasts (NIH-3T3 cells) using dioctanoylglycerol. This synthetic diglyceride disappears very rapidly, but higher chain length fatty acid diacylglycerols, either mixed stearate/arachidonate or dioleate, disappear much more slowly. Most of the octanoate actually disappears from the plasma membrane and ends up inside the cell, whereas longer chain fatty acids seem to stay in the membrane longer.

Also, when adrenal glomerulosa cells are activated by angiotensin II an increase in diacylglycerol concentration is seen. After the hormone has been removed the time course of disappearance of the diacylglycerol is very slow. If we label the diacylglycerol pool with either arachidonate or myristate we find that the arachidonate-labelled diacylglycerol appears first and reaches a steady-state level, whereas myristate-labelled diacylglycerol levels continue to increase with time. When we take the hormone away the arachindonate-labelled diacylglycerol disappears rapidly but the myristate-labelled diacylglycerol takes a long time to return to normal levels (Bollag et al 1991).

*Nishizuka:* I think it is possible that the chain length of the fatty acyl moiety significantly affects the half length of diacylglycerol. Dioctanoylglycerol is metabolized much more quickly by non-specific esterases than by diacylglycerol lipase or kinase. Is there any myristate in the natural phospholipid?

*Rasmussen:* Yes; there is myristate in both phosphatidylcholine and phosphatidylinositol.

*Fisher:* Gary Nelsestuen has shown that after translocation of protein kinase C to the membrane the active conformation could be retained. In your system, I suppose that this activation (of PKC by unsaturated fatty acids) would allow protein kinase C to remain in the cytosolic fraction, without being translocated—is that correct?

*Nishizuka:* Translocation is very difficult to demonstrate with natural signals. Phorbol esters are probably an extreme case. We have no idea about the real mechanism of activation of protein kinase C; obviously, simple ligand–ligand interactions are not involved. Also, I think there are a wide variety of states of activation of protein kinase C, and small changes in lipid bilayer structure might be important in keeping the enzyme active.

*Michell:* I wonder whether we are all being too simplistic when talking about diacylglycerol in cells. As you and I have discussed many times over the years, diacylglycerol is a perfectly normal constituent of unstimulated cells as a metabolic intermediate in lipid synthesis, where it obviously doesn't activate protein kinase C. It is made in the plasma membrane in response to stimuli, where it obviously does stimulate protein kinase C. In some of the experiments that have been done recently on the long-term generation of diacylglycerol, such as those by Leach et al (1991), there is clearly a stimulation of diacylglycerol accumulation that occurs without the activation of protein kinase C. So, it seems that the various pools of diacylglycerol that are made under different circumstances have the potential to behave quite differently, either because they are different molecular species or, more probably, because they are at different cellular sites and/or in different physical contexts in which protein kinase C doesn't see them similarly. Thinking about Howard Rasmussen's comment about turnover rate, I wonder whether the different rates of diacylglycerol turnover in pools may be a result of different diacylglycerol pools being made in different places and therefore being accessible to different metabolic fates. In old studies of red cells it was perfectly clear that diacylglycerol can flip between the leaflets of a lipid bilayer quickly; red cells are nice and simple in that they have only a plasma membrane, but if you tickle them a little with external phospholipase C then the diacylglycerol always disappears from the membrane quickly, either through diacylglycerol lipase action or by switching leaflets and then being phosphorylated. If the diacylglycerol is made at an intracellular site by diacylglycerol kinase in some way that is secondary to the actions of the second messengers generated by stimulation, perhaps it doesn't activate protein kinase C.

*Nishizuka:* We have very little information available about the intracellular topography of diacylglycerol. Normally, the rate of turnover of diacylglycerol is very high. The $\alpha$, $\beta$ and $\gamma$ subspecies of protein kinase C are located differently within the cell. The second group of subspecies—$\delta$, $\varepsilon$, $\zeta$ and $\eta$—probably have a location that differs from that of the first group; these four species are less sensitive to calcium and diacylglycerol. We really don't know what the signal is for the second subgroup of protein kinase C subspecies, but they are obviously activated by phospholipid. There are still plenty of problems. There must be very fine topographical arrangements of both diacylglycerol and the enzymes.

*Michell:* Do you have any evidence that phosphatidates have any influence on any of the protein kinase C subspecies?

*Nishizuka:* In the presence of a high concentration of calcium they can activate protein kinase C, but under physiological conditions, at submicromolar calcium concentrations, they have no effect, by themselves at least.

*Parker:* Bob Michell just mentioned the work by Leach et al (1991) which suggested that there is no activation of protein kinase C in the second phase of diacylglycerol production. I think the work that's been done in that context is limited and I don't think it has clearly defined whether any of the PKCs are activated in that second phase. The best example of activation is in the GH4C1 cell, where TRH causes a second phase response. Here 'second phase' translocation of ε-PKC, which is expressed in those cells, is not detected; however, what you do see is a down-regulation of ε-PKC that is typical of that normally associated with chronic activation (S. Kiley, unpublished work). The implication is that the ε form may be activated in this second phase response.

*Daly:* With agents that cause calcium influx and PtdIns breakdown, such as maitotoxin, you do see translocation of protein kinase C in cultured cells (Gusovsky et al 1989). Thus, though receptor-induced PtdIns breakdown doesn't always cause translocation, maitotoxin does in all cells examined, apparently by triggering PtdIns breakdown and generating diacylglycerols.

*Fischer:* I like the idea that unsaturated fatty acids sustain or prolong a transient calcium signal. This is analogous to what happens with

$Ca^{2+}$/calmodulin-dependent protein kinase II and other calcium-dependent protein kinases: a calcium transient activates the enzyme, then the enzyme undergoes autophosphorylation. After autophosphorylation, the enzyme loses its dependence on calcium. The calcium signal can then disappear and the enzyme remains active, just as Professor Nishizuka has shown.

*Rasmussen:* Before we get carried away, I should point out that the evidence that the autophosphorylated enzyme remains present in the long term is not very good.

*Fischer:* Once phosphorylase kinase has undergone autophosphorylation it loses its calcium dependency, and the same is true of $Ca^{2+}$/calmodulin-dependent protein kinase II.

*Rasmussen:* But how long does it remain active in its non-calcium-dependent form?

*Nairn:* The results suggest that it remains active for seconds to minutes, depending on the tissue preparation. For example, in synaptosomes the transient activation lasts for seconds (Gorelick et al 1988), whereas in hippocampal slices it possibly lasts longer (Ocorr & Schulman 1991).

*Rasmussen:* So this cannot be important in any long term responses.

*Fischer:* Repeated calcium transients could allow the maintenance of a steady level of kinase activity.

*Rasmussen:* That's not what Professor Nishizuka was talking about.

*Fischer:* Perhaps not, but nevertheless the signal is being prolonged and there must be a dependence on phosphatases to return the system to its original state.

*Rasmussen:* I agree with that, but $Ca^{2+}$/calmodulin-dependent protein kinase II has been championed as a memory molecule and there is no good evidence that it serves such a role.

*Fischer:* It would be a reasonable mechanism though. One should be able to measure what's happening within the cell. One can measure a calcium signal fairly well but it's difficult to measure the intracellular activity of an enzyme. Changes in conformation are greatly affected by metabolites or effectors. In our work we can see that phosphorylase and phosphorylase phosphatase are regulated by effectors that transiently maintain the active conformation. There ought to be a logical reason why phosphorylation should render an enzyme independent of an original signal. That is a marvellous way for the cell to prolong a transient signal, to maintain the 'memory' of this signal in the form of an active conformation of an enzyme.

*Tsien:* This is a matter of degree. I think that Angus Nairn and Howard Rasmussen would say that $Ca^{2+}$/calmodulin-dependent protein kinase II autophosphorylation might be helpful in taking a calcium signal that lasts two seconds and prolonging it to two minutes. That's not memory of the kind you are thinking about, but two minutes is a lot longer than two seconds and there are a lot of interesting things that might happen in the brain and in other cells as a result of that degree of prolongation.

Professor Nishizuka, if I understood Fig. 1 correctly, diacylglycerol is absolutely required for protein kinase C activity; if you lower the diacylglycerol concentration to zero, it doesn't matter whether you have arachidonic acid *or* calcium, or arachidonic acid *and* calcium—the enzyme is not active. To get sustained protein kinase C activity, the diacylglycerol needs to be delivered to the cell and its concentration needs to remain raised.

*Nishizuka:* That is correct.

*Tsien:* Thus, the problem of where the maintained signal comes from has been pushed back one step; it's not obvious what would maintain the high intracellular diacylglycerol concentration.

*Parker:* In the activated PKC complex the diacylglycerol is (or might be) in a form in which it is no longer seen by the enzymes that would normally metabolize it. The real question is how long that complex lasts—what's the stability of that complex? The diacylglycerol–PKC complex may persist longer than the measurable free diacylglycerol.

*Tsien:* You could imagine that the cofactor is taken up by the molecule and becomes invisible to the degradative enzymes. That can be tested experimentally. Is there evidence that the affinity of protein kinase C for some cofactor such as diacylglycerol changes from a low or moderate affinity to a very high affinity?

*Parker:* No, but from the wealth of studies on translocation it's clear that the rate of dissociation of the complex is relatively slow.

*Yang:* Free fatty acids, especially many unsaturated fatty acids, would be derived from phosphatidylcholine hydrolysis by phospholipase $A_2$. Phospholipase $A_2$ will hydrolyse many other phospholipids. Membranes from different tissues have different ratios of phospholipids, so phospholipase $A_2$ might cause different effects.

*Nishizuka:* Phosphatidylcholine is the main source of unsaturated fatty acids. As far as I know, phosphatidylethanolamine does not act as a major source.

*Kanoh:* That is puzzling. Phosphatidylethanolamine has been found to be the major target of phospholipase $A_2$ in most *in vitro* studies reported.

*Rasmussen:* There is a report that in mesangial cells phosphatidylethanolamine can serve as a source of diacylglycerol (Kester et al 1989).

*Kanoh:* Professor Nishizuka, in your experiments with the repetitive additions of dioctanoylglycerol I gather that you could not detect accumulation of phosphatidic acid of shorter chain length fatty acids.

*Nishizuka:* We saw octanoic acid only.

*Kanoh:* I am uncertain why that should be so. If you had incubated the cells for a much shorter time, for 2–5 min, I am sure you would have detected accumulation of phosphatidic acid of shorter fatty acid chain length.

*Nishizuka:* As far as dioctanoylglycerol is concerned, we could not find any accumulation of the phosphatidic acid. If any was present, it was a very small quantity. This is in T cells—I don't know about other cell systems.

*Kanoh:* I have done similar experiments in Jurkat cells and saw a brief accumulation of phosphatidic acid composed of short chain fatty acids from dihexanoyl-, dioctanoyl- and didecanoylglycerol in 2–5 min incubations.

*Exton:* We also saw only octanoic acid—we never have seen any other metabolites, but we started only with dioctanoylglycerol.

*Kanoh:* In cells labelled with $^{32}P$ the action of diacylglycerol kinase can become quite marked. In that case, you can see a brief accumulation of phosphatidic acid of $C_6$, $C_8$, or $C_{10}$ chain length from exogenous permeant diacylglycerols.

*Exton:* You must have a different cell!

*Kanoh:* With $^{32}P$-labelled cells can you follow mainly the action of diacylglycerol kinase. If you use $^{14}C$-labelled fatty acids to follow diacylglycerol metabolism then you can follow the net outcome of the actions of diacylglycerol kinase and lipase.

*Krebs:* In comparison with the natural signal phorbol esters prolong the activation of protein kinase C, but the extent of rapid activation is not necessarily any greater with phorbol ester, is it?

*Nishizuka:* This may be true, but it is complicated because phorbol esters frequently cause unphysiological artifactual events. Perhaps activation of the enzyme by phorbol esters is an extreme case for the cell, and the process might be irreversible. In most cells examined phorbol esters deplete protein kinase C to various extents. The rate of this depletion or down-regulation differs between PKC subspecies. This is probably because of the differing susceptibility of the subspecies to calpain, which would initiate proteolysis of the enzyme at the variable region, $V_5$, connecting the regulatory and catalytic domains. The amino acid sequence of this region varies greatly between subtypes.

The key question is the difference between natural signals and phorbol esters; the answer is not simple and it depends on cell type, the nature of the signal, concentration and time after stimulation. I suppose depletion of the enzyme by phorbol esters is a kind of self-protection and may be an extreme situation. Usually, phorbol esters cannot deplete protein kinase C completely from the cell and the enzyme reappears soon. We have not observed depletion of PKC in response to natural signals. Of course, the balance between calpain activity and activation mechanisms is another complication.

*Krebs:* I sometimes wonder if the extent of protein kinase C activation that one gets with phorbol ester is so great that there are times at which it is actually utilizing protein kinase A substrates, for example. *In vitro*, the specificities are not tremendously different, and in many cases the same residues are phosphorylated by the two enzymes. If you over-activate protein kinase C you may lose some of its specificity.

## References

Bollag WB, Barrett PQ, Isales CM, Rasmussen H 1991 Angiotensin II-induced changes in diacylglycerol levels and their potential role in modulating the steroidogenic response. Endocrinology 128:231–241

Gorelick FS, Wang JKT, Lai Y, Nairn AC, Greengard P 1988 Autophosphorylation and activation of $Ca^{2+}$-calmodulin-dependent protein kinase II in intact nerve terminals. J Biol Chem 263:17209–17212

Gusovsky F, Yasumoto T, Daly JW 1989 Calcium-dependent effects of maitotoxin on phosphoinositide breakdown and on cyclic AMP accumulation in PC12 and NCB-20 cells. Mol Pharmacol 36:44–53

Kester M, Simonson MS, Mené P, Sedor JR 1989 Interleukin-1 generates transmembrane signals from phospholipids through novel pathways in cultured rat mesangial cells. J Clin Invest 83:718–723

Leach KL, Ruff VA, Wright TM, Pessin MS, Raben DM 1991 Dissociation of protein kinase C activation and *sn*-1,2-diacylglycerol formation: comparison of phosphatidylinositol-derived and phosphatidylcholine-derived diglycerides in $\alpha$-thrombin-stimulated fibroblasts. J Biol Chem 266:3215–3221

Ocorr KA, Schulman H 1991 Activation of multifunctional $Ca^{2+}$/calmodulin-dependent protein kinase in hippocampal slices. Neuron 6:907–914

# The acetylcholine receptor: a model of an allosteric membrane protein mediating intercellular communication

Jean-Pierre Changeux, Anne Devillers-Thiéry, Jean-Luc Galzi and Frédéric Revah

*Institut Pasteur, Neurobiologie Moléculaire, Bâtiment des Biotechnologies, 25, rue du Dr Roux, 75015 Paris, France*

*Abstract.* Over the past 20 years the nicotinic acetylcholine receptor has become the prototype of a superfamily of ligand-gated ion channels. As a single macromolecular entity of $M_r$ about 300 000, the receptor protein mediates, altogether, the activation and the desensitization of the associated ion channel and the regulation of these processes by extracellular and intracellular signals. The notion is discussed that the acetylcholine receptor is a membrane-bound allosteric protein which possesses several categories of specific sites for neurotransmitters and for regulatory ligands, and undergoes conformational transitions which link these diverse sites together. At this elementary molecular level, interactions between signalling pathways may be mediated by membrane-bound allosteric receptors and/or by other categories of cytoplasmic allosteric proteins.

*1992 Interactions among cell signalling systems. Wiley, Chichester (Ciba Foundation Symposium 164) p 66–97*

The title of this symposium, 'Interactions among cell signalling systems' refers to two distinct notions: first, that cells produce and propagate communication signals, and second, that these signalling systems may interact. The level of organization at which such communications and their interactions take place is that of the cell; but the signals may either be *intra*cellular, and thus be present in single cells (or a homogeneous population of cells), or *inter*cellular, and thus involve groups or networks of cells, as in the nervous system.

In the early 1960s, from studies on the regulation of intracellular metabolism, mainly in bacteria (but also in eukaryotes), the concept arose that chemical substances present in the cell may serve not only as substrates entering into metabolic pathways as building blocks or energy sources for the cell, but may also act as regulatory signals. In the absence of any chemical transformation, such compounds regulate metabolic pathways at the level of critical molecular targets by the simple virtue of reversible binding to specific complementary sites. This notion was first discussed in the case of the 'induction' of enzyme biosynthesis in bacteria (or enzymic adaptation). This elementary regulatory interaction became directly accessible to experimentation *in vitro* with enzymes such as L-threonine deaminase (Changeux 1961, 1965) or aspartate transcarbamylase (Gerhart & Pardee 1962). These two enzymes are the first in the pathways of L-isoleucine and pyrimidine biosynthesis, respectively, and they are selectively inhibited, by feedback, by the end-product of their specific pathway.

The first simple scheme suggested to account for the observed feedback inhibition was based on classical enzymology: the idea was that the regulatory ligand would bind to the same site as the substrate of the enzyme (or to an overlapping site) so that there would be mutual exclusion by *steric* hindrance. Yet, two series of observations led to a rejection of this scheme. First, these enzymes display unusual kinetic properties, such as deviation from strict competitive inhibition by the regulatory ligand, and apparent cooperative behaviour of both substrate and regulatory signal; second, various chemical or physical treatments uncouple the inhibition by the regulatory ligands while keeping the enzymes active (Changeux 1961, Gerhart & Pardee 1962). It was thus postulated that the regulatory process involves two distinct sites (Changeux 1961)—the active site, responsible for the biological activity, and another category of site, referred to as *allosteric*, with a structure complementary to that of the regulatory ligand which it binds specifically and reversibly (Monod et al 1963). The regulatory interaction would be indirect and take place between at least two stereospecifically different, non-overlapping binding sites. It would then be mediated by a discrete reversible 'alteration of the molecular structure of the protein, or allosteric transition, which modifies the properties of the active site, changing one or several of the kinetic parameters which characterize the biological activity of the protein' (Monod et al 1963).

In a subsequent step, mainly to account for cooperative effects between identical ligand-binding sites frequently encountered in regulatory molecules, allosteric proteins were viewed as 'closed microcrystals' or *oligomers*, made up of a finite number of identical subunits (or protomers), and, as a consequence, possessing at least one axis of symmetry (Monod et al 1965). Allosteric interactions between identical (homotropic) and between different (heterotropic) binding sites were then postulated to result from 'concerted', all-or-none, transitions between two discrete states ($T \rightleftharpoons R$) with different ligand-binding properties and biological activities. To relate the cooperative binding to a cooperative structural change, it was assumed that the transition preserved the symmetry of the oligomer and that the two states exist, in reversible equilibrium, prior to the interaction with ligands (Monod et al 1965). By convention, the R, 'relaxed', state was postulated to be the biologically active conformation, and the T, 'constrained', state the inactive one. The transduction process was thus assigned to a particular flexibility of the quaternary organization of the protein molecule. The ligands which preferentially stabilize the R state would then behave as activators or 'agonists', and those stabilizing the T state as inhibitors or 'antagonists'. Also, the occupancy of the binding sites by the ligand, Y, or *binding function*, was expected to differ from the fraction of the molecules in the active conformation, R, or *state function*.

Over the past 25 years, several predictions of this model have been validated at the atomic level with a few regulatory proteins, mostly by X-ray crystallography (for review see Perutz 1989). Evidence has been provided for interaction between stereospecific sites often more than 30 Å apart and for discrete changes of quaternary structure which preserve the symmetry of the molecule, with only a few exceptions. Still, an issue which sometimes remains debated is the relative importance of: (1) the global conformational transitions which affect the quaternary structure of the oligomeric molecule and are selectively stabilized by the binding of the ligands to their site, and (2) local modifications of protein structure 'induced' by binding of the ligand to the regulatory site, which are directly dependent on the actual structure of the ligand (Koshland et al 1966). One straightforward experimental distinction between the two processes is that in (1) the state function might differ from the binding function, whereas in (2) the two functions would in principle coincide. Experiments initially carried out under equilibrium conditions with aspartate transcarbamylase and phosphorylase *b*, and X-ray diffraction studies on haemoglobin and on several regulatory enzymes (Perutz 1989), unambiguously showed that, in these systems, the binding function clearly differs from the state function. Thus, 'the change in geometry at one site is not produced by *direct* propagation of structural changes induced by the binding of ligand at another site. Only very small structural changes occur on ligand binding to a subunit unless the quaternary structure changes' (Baldwin & Chothia 1979). But, again, in some particular cases there may be exceptions to this rule.

Initially, the concept of allosteric interaction, and, by extension, that of allosteric proteins, was designed to account for the effect of metabolites on critical reactions of intermediary metabolism. The target enzymes were 'receptors' for metabolic effectors, but strictly *intracellular* ones. These chemical signals included 'second messengers' such as cyclic AMP and the covalent modifications (for example, phosphorylation–dephosphorylation) that these second messengers elicit (consider, for example, phosphorylase *b*). The notion was soon extended to *extracellular* signals (steroid hormones, thyroxine) and to non-enzymic processes such as repression of gene transcription (Monod et al 1963). Yet, neither membrane receptors for neurotransmitters nor electrogenesis were mentioned at the time. The possibility of the involvement of allosteric interactions in neurotransmission was subsequently suggested (Changeux 1965, 1966) and elaborated (see Changeux 1990 for review) to take into account: (1) the contribution of neurotransmitters as chemical signals for *inter*cellular communication in the nervous system; (2) the integration of such receptors into biological membranes; and (3) the eventual control by neurotransmitters of transmembrane channels (or 'ionophores') engaged in electrogenic responses.

Signal transduction at the cell surface was thus viewed as a reversible transmembrane allosteric transition of an integral protein complex which includes the ion channel (or an enzymically active component facing the cytoplasmic side of the membrane). Such views have since been well documented, with the addition of the concept that the efficacy of the signal transduction step mediated might itself be subject to *higher order regulations*, implicating the participation of additional classes of allosteric sites and conformational transitions (see subsequent chapters and Changeux 1990). Such additional allosteric sites and transitions may contribute to interactions among signalling pathways. Allosteric receptors would mediate, altogether, signal transduction and its regulation. Accordingly, the interactions among cell signalling systems would ultimately be mediated by allosteric proteins (or receptors) located at 'nodal points' of the signalling networks. The much-studied nicotinic acetylcholine receptor offers an example of how such imbricated or 'nested' allosteric properties can be implemented through protein structure.

## The acetylcholine receptor as a prototype of ligand-gated ion channels

Chemical communication between nerve cells and with other categories of cells takes place through a specialized structure—the synapse—and most of our knowledge about the physiology of chemical transmission derives from studies of the motor end-plate in vertebrates where the neurotransmitter is acetylcholine. In this system the intercellular space between the cytoplasmic membranes of the motor nerve ending and of the muscle fibre is only a few $\mu m^3$, and the invasion of the motor ending by an electrical impulse causes the local concentration of neurotransmitters to rise from about $10^{-9}$ M to $3 \times 10^{-4}$ M in a fraction of millisecond. In a few tenths of a millisecond, the acetylcholine molecules cross the synaptic cleft and reach the postsynaptic membrane where they cause the collective fast opening or 'activation' of cationic channels when they bind, most probably only once, to the closely packed receptor molecules. The layer of acetylcholine receptors thus decodes the chemical impulse into an electrical one, under non-equilibrium conditions, over a millisecond timescale. Furthermore, exposure to steady low concentrations of acetylcholine (one to two orders of magnitude lower than its concentration in the cleft during transmission) causes, on a 0.1 second-to-minutes timescale, a reversible decline (desensitization) of the conductance response to acetylcholine (Katz & Thesleff 1957). This consists of two kinetically distinct processes, a rapid phase occurring at subsecond times and a slow process taking place within seconds or minutes (see Changeux 1990 for review). As discussed below, a variety of effectors may differentially regulate the rate of acetylcholine receptor desensitization. Such regulation may then contribute to potential interactions between signalling pathways. The acetylcholine receptor thus constitutes an excellent model system to unravel such mechanisms.

The membrane component engaged in the regulation of ion permeability by acetylcholine was identified (Changeux et al 1970, Changeux 1990) through the combined use of fish electric organ and snake venom $\alpha$-toxins as selective ligands. $\alpha$-Bungarotoxin was shown to block the permeability response to cholinergic agonists of the single electrocyte and of excitable microsacs from *Electrophorus* and to inhibit the binding of the radioactively labelled nicotinic agonist decamethonium to a protein present in deoxycholate extracts of the excitable membranes (Changeux et al 1970). The protein purified from *Electrophorus electricus* (Olsen et al 1972) and *Torpedo* (Schmidt & Raftery 1972) electric tissue had an apparent relative molecular mass of 300 000 and was composed of smaller subunits, with stoichiometry $\alpha_2\beta\gamma\delta$ (for review see Karlin 1991, Changeux 1990).

The question then arose as to whether the $\alpha_2\beta\gamma\delta$ molecule was solely responsible for the electrogenic action of acetylcholine. Recovery of acetylcholine-regulated ion fluxes after reincorporation into artificial lipid microsacs of detergent extracts of highly purified receptor-rich membranes (Hazelbauer & Changeux 1974) and of purified receptor protein supplemented with a complex lipid mixture (Epstein & Racker 1978; for review see Popot et al 1981), and the observation of acetylcholine-gated single channels in recordings from the pure protein reconstituted into planar lipid bilayers (Nelson et al 1980), finally demonstrated this. The $\alpha_2\beta\gamma\delta$ oligomer carries the acetylcholine-binding sites, contains the ion channel and possesses all the structural elements required for the activation and desensitization of the ionic response to agonists over a physiological concentration range and time scale. In support of this conclusion, injection of the four mRNAs from transcribed cDNAs encoding each of the subunits yielded acetylcholine-gated ion channels and $\alpha$-toxin-binding protein (Mishina et al 1984). The $\alpha_2\beta\gamma\delta$ oligomer thus suffices for the physiological nicotinic response to acetylcholine in muscle and electric organ.

However, the apparently asymmetrical quaternary structure of the receptor molecule would seem to contradict the notion that the receptor, being an allosteric protein, should be symmetrical. Several observations showed that, in fact, the receptor possesses definite, though unconventional, properties of symmetry: (1) electron microscopy of the membrane-bound and purified receptor protein revealed ring-like particles (Cartaud et al 1973) with five subunits regularly distributed around a central pit, the two $\alpha$ subunits not being adjacent (Unwin et al 1988); (2) affinity labelling of the high affinity site for non-competitive channel blockers resulted in the agonist-dependent incorporation of label into the non-$\alpha$ subunits, thus demonstrating a functional role for these subunits (Oswald & Changeux 1981); (3) amino acid sequence analysis of the purified subunits (Devillers-Thiéry et al 1979, Raftery et al 1980) disclosed striking sequence identities between the subunits (Raftery et al 1980), subsequently confirmed by the determination of the complete cDNA coding sequence (for review see Noda et al 1983a,b). The receptor oligomer thus displays a cryptic symmetry, with a five-fold rotational axis perpendicular to the plane of the membrane.

As a consequence, the receptor also displays a transverse polarity across the membrane, as manifested by the exclusive sensitivity of its synaptic face to acetylcholine, by the presence of carbohydrate residues exposed to the synaptic cleft, by the phosphorylation sites which face the cytoplasm and by the differential sensitivity of its two faces to proteolytic enzymes and specific antibodies (references in Changeux 1990). Such transmembrane disposition makes signal transduction across the membrane possible. It also renders the transducing molecule sensitive to regulatory signals from the two sides of the membrane.

The cloning and complete sequencing of the four subunits of the *Torpedo* receptor (Noda et al 1982, 1983a,b, Claudio et al 1983, Devillers-Thiéry et al 1983), followed by the quantitative analysis of the distribution of hydrophilic and hydrophobic amino acids along the aligned sequences of the four subunits, initiated direct investigations into the transmembrane organization of the receptor.

First, the four homologous sequences were subdivided into a large $NH_2$-terminal hydrophilic domain (with, in the $\alpha$ subunit, two characteristic pairs of cysteine residues, 128–142 and 192–193), a compact hydrophobic domain split into three segments of 19–27 uncharged amino acids (named MI, MII and MIII), a small hydrophilic domain and a hydrophobic C-terminal segment about 20 amino acids long (MIV). Several structural models have been suggested on the basis of this. The most plausible, but still largely hypothetical, model (Claudio et al 1983, Devillers-Thiéry et al 1983, Noda et al 1983a,b) has the four hydrophobic stretches MI to MIV as transmembrane $\alpha$-helices, with the large and small hydrophilic domains exposed to the synaptic cleft and cytoplasm, respectively. This organization is supported by the demonstration that: (1) MIV is exposed to the lipid bilayer (Giraudat et al 1985), (2) the C-terminus faces the synaptic cleft (McCrea et al 1987, Karlin 1991), (3) the small hydrophilic domain can be phosphorylated from the cytoplasm (Huganir & Greengard 1990) and (4) amino acids from the large hydrophilic domain constitute the acetylcholine-binding site.

A direct test of this model is the identification of the amino acids that constitute the cholinergic ligand-binding sites and the ion channel.

## The recognition site for cholinergic nicotinic ligands

The $\alpha_2\beta\gamma\delta$ oligomer carries two main binding sites for nicotinic ligands (review Reynolds & Karlin 1978, Galzi et al 1991a); these sites also bind snake venom $\alpha$-toxins. The two $\alpha$ subunits are not adjacent, yet they interact in a positively cooperative manner (Changeux 1990, Pedersen & Cohen 1990a,b, Galzi et al 1991a). Their interaction is thus indirect, or allosteric. Even though the two $\alpha$ subunits are encoded by a single gene (in both *Torpedo* and mouse), these two sites nevertheless differ in their ligand-binding properties (for snake $\alpha$-toxins, *d*-tubocurarine, several affinity-labelling reagents, and the coral toxin lophotoxin). The cooperative homotropic interactions mediated by the receptor protein take place between sites which are *not* strictly equivalent. These differences are most probably due to the contribution of the non-$\alpha$ subunits to the two acetylcholine-binding areas (see Oswald & Changeux 1982, Pedersen & Cohen 1990a, Blount & Merlie 1989).

Early affinity labelling experiments of the reduced receptor with a maleimide reagent led to the identification of Cys-192 and possibly Cys-193 in or near the acetylcholine-binding sites in the $\alpha$ subunit (Kao et al 1984). The use of another reagent, *p*(*N*,*N*-dimethylamino)benzenediazonium fluoroborate (DDF), which acts as a competitive antagonist in the dark and can be photoactivated by energy transfer from the protein (Langenbuch-Cachat et al 1988), revealed a more complex organization of the cholinergic ligand-binding area (Dennis et al 1988, Galzi et al 1990, 1991b) (Fig. 1). DDF labels amino acids from three different regions of the large N-terminal domain: $\alpha$-Tyr-93 and $\alpha$-Trp-149, and a larger set which includes $\alpha$-Cys-192 and 193 and $\alpha$-Tyr-190, an amino acid which also reacts with a derivative of the coral toxin lophotoxin (Abramson et al 1989). None of the side chains labelled are those of the aspartic and glutamic acid residues often assumed to contribute, with their carboxylate anion, to a 'negative subsite' of the acetylcholine-binding site. However, two (or possibly four) of the labelled residues are tyrosines, which together exhibit the electronegative character required to complement the quaternary ammonium of the ligand. Tyrosine residues in proteins are commonly located at positions exposed to solvents, in particular in antibody-combining sites (Chothia et al 1989), and their presence in the acetylcholine-binding area may reflect a rather general feature of ligand recognition sites in proteins (Galzi et al 1990). Site-directed mutagenesis of these tyrosines, indeed, alters the sensitivity of the mouse receptor reconstituted in oocytes (Tomaselli et al 1991, Galzi et al 1991c). None of the identified amino acids is present in muscle non-$\alpha$ subunits, but they are all conserved in all $\alpha$ subunits from *Torpedo* to humans and from muscle to brain (except $\alpha_5$) (review Galzi et al 1991a), and may determine acetylcholine-binding specificity. In contrast, neighbouring amino acids in the sequence of the various $\alpha$ subunits vary significantly and this contributes to the observed diversity of pharmaco-logical properties among the several nicotinic acetylcholine receptors, in particular those from different brain regions.

Another source of binding diversity may be the contribution of non-$\alpha$ subunits to the acetylcholine-binding sites. UV irradiation of complexes of the receptor with snake $\alpha$-toxins (Oswald & Changeux 1982), DDF (Lagenbuch-Cachat et al 1988) or *d*-tubocurarine (Pedersen & Cohen 1990a) results in the incorporation

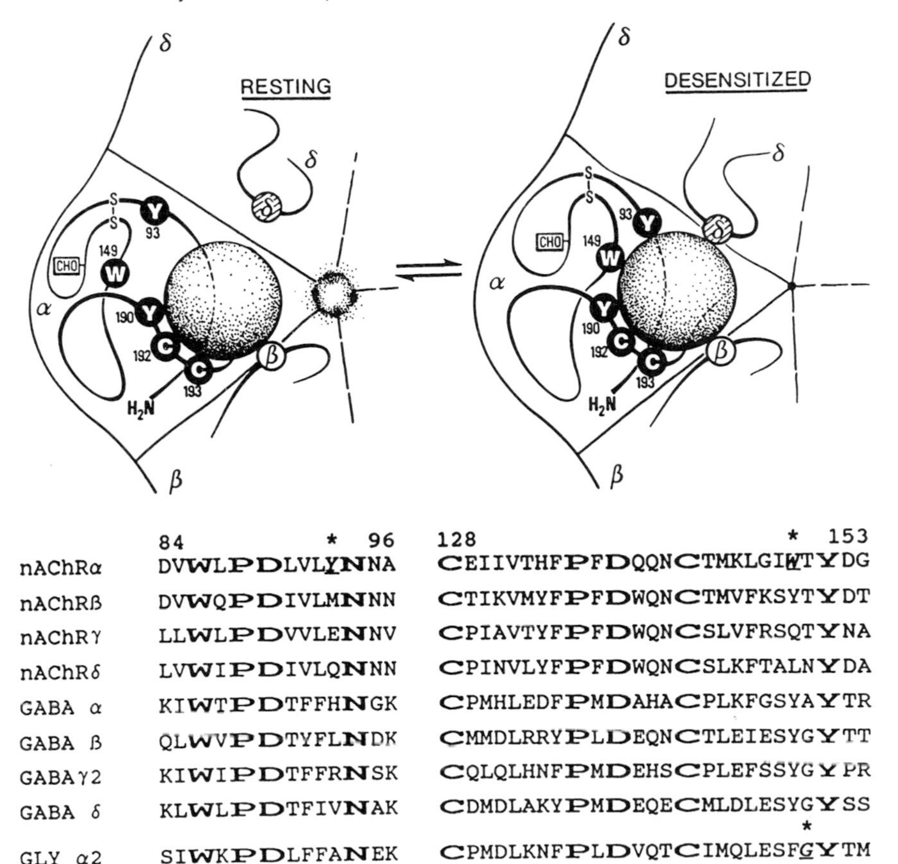


FIG. 1. Model of the cholinergic ligand-binding domain on the $\alpha$ subunit of the acetylcholine receptor from *Torpedo marmorata*, and of its transitions upon desensitization. The affinity label DDF is represented as a shaded sphere, and the amino acids labelled by DDF are circled in black. The contribution of the non-$\alpha$ subunits is indicated by white circles. Underneath are shown aligned sequences from various nicotinic acetylcholine receptor (nAChR) subunits, GABA receptor subunits and a glycine receptor (GLY) subunit. (From Galzi et al 1991b).

of label into the $\gamma$ and $\delta$ subunits (in addition to the $\alpha$ subunit). Also, expression of different pairs of mouse muscle $\alpha$ and non-$\alpha$ subunits in fibroblasts showed that the $\gamma$ and $\delta$ (but not the $\beta$) subunits associate with the $\alpha$ subunit into complexes that have different affinities for *d*-tubocurarine (Blount & Merlie 1989). Accordingly, *d*-tubocurarine would bind with high affinity to the $\alpha$–$\gamma$ pair and with low affinity to the $\alpha$–$\delta$ pair, most probably at the boundary between subunits, as occurs in other regulatory proteins (see Perutz 1989). A striking diversity of binding of nicotinic ligands and toxins as a function of subunit composition is also seen with the neuronal receptor (review Galzi et al 1991a). For instance, binding to the $\alpha_3$–$\beta_2$ combination, but not to $\alpha_2$–$\beta_2$, is blocked by neuronal ($\varkappa$) bungarotoxin; the $\alpha_7$ subunit is sensitive to $\alpha$-bungarotoxin, but not to neuronal bungarotoxin. Furthermore, the functional and pharmacological properties of the multiple combinations of subunits obtained vary significantly. Because the distribution of the diverse subunits also varies between different brain areas, such variability may result in a diversity of the physiological responses of the postsynaptic cells and thus may have important behavioural consequences. These studies offer a way to develop drugs targeted to defined circuits within the brain.

## Identification and structure of the ion channel

Reconstitution experiments unambiguously showed that the $\alpha_2\beta\gamma\delta$ oligomer contains the ion channel. The identification of the amino acids which directly contribute to its function relied on two different, but complementary, approaches. First, affinity labelling with pharmacological agents known to block ion transport at a site distinct from the cholinergic ligand-binding site was used (Giraudat et al 1986, Hucho et al 1986). These compounds, referred to as non-competitive blockers, were shown to bind to a high affinity site, present as one copy per $\alpha_2\beta\gamma\delta$ oligomer, to which the four different subunits contribute (Oswald & Changeux 1981, Heidmann et al 1983a). It was thus proposed that this site lies on the axis of pseudo-symmetry of the receptor oligomer, within the ion channel (Heidmann et al 1983b).

*In vitro* experimental evidence supports the view that this site does lie *within* the ion channel: (1) permeant cations competitively inhibit the binding of the non-competitive blockers ethidium or chlorpromazine to their high affinity site (Herz et al 1989; F. Revah, unpublished results); (2) rapid photolabelling experiments with chlorpromazine (Heidmann & Changeux 1984, 1986) show that its rate of covalent attachment to the high affinity site increases $10^2$–$10^3$-fold on mixing with acetylcholine, within the time course expected for channel opening; under these conditions, the apparent dissociation constant for acetylcholine is about 30 μM, a concentration which is close to that required for the activation of the ion channel, and this effect is blocked by the competitive antagonists *d*-tubocurarine or the $\alpha$-toxins (Heidmann & Changeux 1984); (3) the rate of chlorpromazine incorporation is reduced when the receptor is pre-exposed to a constant concentration of acetylcholine, showing kinetics close to those measured for the rapid desensitization of the ion flux response (Heidmann & Changeux 1986). In other words, chlorpromazine has more direct access to, and upon UV irradiation covalently reacts with, its high affinity site when the ion channel opens.

The amino acid labelled by chlorpromazine under equilibrium conditions in the presence of agonists was initially identified on the $\delta$ subunit of *T. marmorata* as $\delta$-Ser-262 (Giraudat et al 1986). This amino acid belongs to the hydrophobic

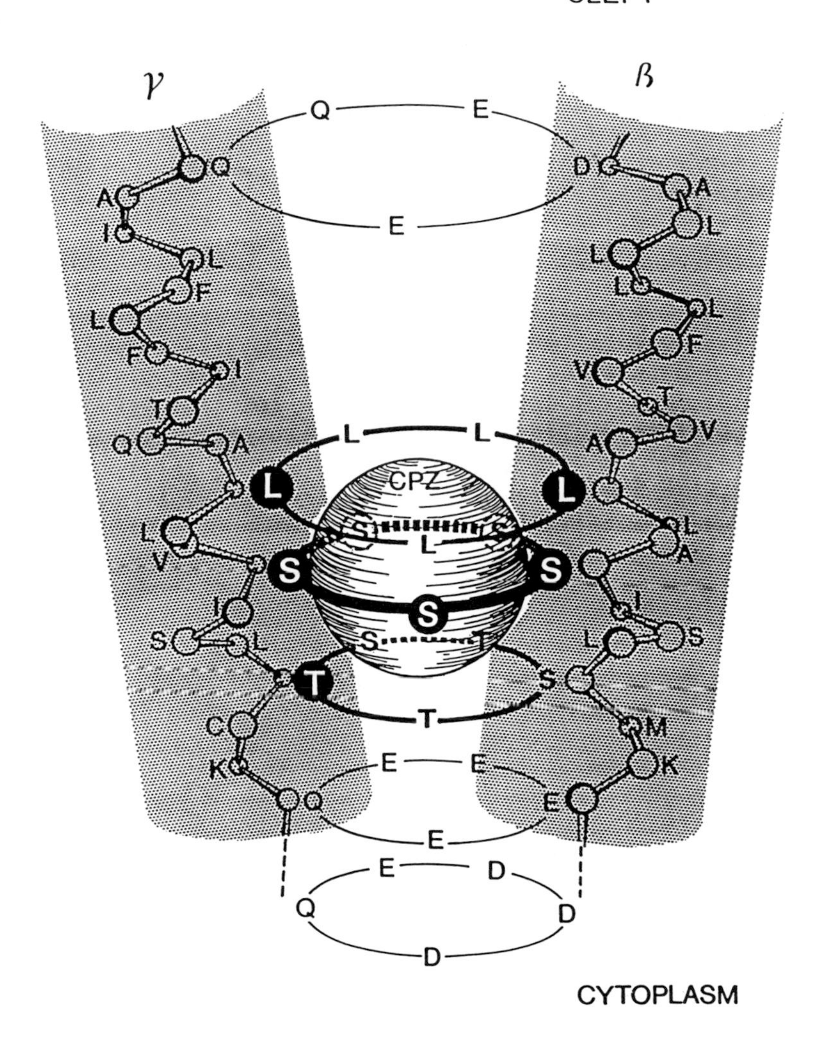


FIG. 2. Model of the ion channel from *Torpedo marmorata* acetylcholine receptor. The shaded sphere represents chlorpromazine. The indicated rings of amino acids are, from the synaptic cleft to the cytoplasm: (1) the 'outer charged ring' (Q,E,D,E,Q) corresponding to $\alpha$-Glu-261 and homologous residues in other subunits (mutated by Imoto et al 1988); (2) the three rings of amino acids labelled by chlorpromazine (filled circles) (Giraudat et al 1986, 1987, 1989; Revah et al 1990) and triphenylmethylphosphonium (Hucho et al 1986); the serine (S) and threonine (T) rings have been mutated by Leonard et al (1988) and Villarroel et al (1991); and (3) the 'intermediate' and 'inner' charged rings mutated by Imoto et al (1988) and Konno et al (1991) (from Revah et al 1990).

segment MII; thus, for the first time MII was implicated as a possible component of the ion channel (Fig. 2). Subsequent work with the other subunits disclosed a ring of homologous serine residues, $\alpha$-248, $\beta$-254, $\gamma$-257 and $\delta$-262, which were all labelled by chlorpromazine (Giraudat et al 1986, 1987, 1989, Revah et al 1990), a finding consistent with the proposal that the high affinity site for non-competitive blockers lies on the axis of pseudo-symmetry of the $\alpha_2\beta\gamma\delta$ oligomer. In addition, chlorpromazine was found to label the flanking amino acids Leu-257 on the $\beta$ subunit (Giraudat et al 1987) and Thr-253 and Leu-260 (Revah et al 1990) on the $\gamma$ subunit; these residues are located 3–4 amino acids above (leucine ring) or below (threonine ring) the serine ring (see Fig. 2), thus supporting an $\alpha$-helical organization of the MII segments. Triphenylmethylphosphonium was also reported to label serine residues in the serine ring (Oberthür et al 1986, Hucho et al 1986). In *T. californica*, meproadifen mustard labelled Glu-262 on the $\alpha$ subunit. This residue lies at the end of MII, close to MIII (Pedersen & Cohen 1990b).

The putative contribution of segment MII to the ion channel has subsequently received support from a complementary series of electrophysiological experiments carried out with functional acetylcholine receptor channels expressed after injection of *T. californica* mRNAs into *Xenopus* oocytes (Imoto et al 1986, 1988). Point mutations were introduced into the *Torpedo* receptor subunit mRNAs to alter the three rings of negatively charged (or polar) residues $\alpha$-Asp-238, $\alpha$-Glu-241 and $\alpha$-Glu-262 and homologous residues located on either side of the MII segment. Under conditions of low divalent ion concentrations, these mutations were accompanied by changes of channel conductance for monovalent cations. Despite the fact that under more physiological conditions several of these mutations might not be accompanied by significant changes of channel properties (Imoto et al 1986), these studies brought useful information about the topology of the channel. A sidedness of the effects of $Mg^{2+}$ was noticed, and from this the anionic ring between MII and MIII ($\alpha$-Glu-262 and homologous residues [outer charged ring, see Fig. 2]) was located on the extracellular side, and the two other rings between MI and MII [$\alpha$-Asp-238 and homologous residues (inner charged ring) and $\alpha$-Glu-241 and homologous residues (intermediate charged ring)], were located on the cytoplasmic side. As suggested by all models of transmembrane folding of the subunits, the segment MII thus spans the membrane. Moreover, the results of these mutagenesis experiments strikingly confirm the proposed location of chlorpromazine- and triphenylmethylphosphonium-binding sites, which would, indeed, be 'framed' by these anionic rings *within* the membrane (Giraudat et al 1987, Hucho et al 1986).

In parallel studies, the contribution of the amino acids labelled by chlorpromazine was investigated by site-directed mutagenesis. The physiological importance of the $\delta$-262 serine ring (Giraudat et al 1986, 1989) was demonstrated using mouse subunit mRNAs (Leonard et al 1988, Charnet et al 1990). Mutation to alanine of several serine residues from this ring caused both a selective decrease in outward single-channel potassium currents, and a decrease in residence time (and thus in the affinity) of the channel blocker QX-222 (Leonard et al 1988). Similar mutagenesis experiments were carried out with the threonine ring (composed of residues homologous to β-Thr-253, see Fig. 2) from mouse muscle receptor, and slightly reduced conductance was seen with some combinations of mutant subunits. With the receptor from *T. californica* (Villaroel et al 1991), substitution of α-Thr-264 with valine also caused a slight decrease of the conductance for $K^+$ (about 20%), whereas its replacement by alanine or glycine increased it (by about 10%).

The serine ring is conserved through all neuronal nicotinic acetylcholine receptor subunits and in frog and chick kainate-type glutamate receptors (for review see Betz 1990), and is replaced by a threonine ring in glycine and GABA (α-aminobutyric acid) receptors. Similarly, the threonine ring is conserved in acetylcholine receptor subunit sequences, but not in those of GABA and glycine receptors. The hydrophilic character of these two rings might contribute to the exchange of water molecules surrounding the transported ion and thus both rings may 'probe' the size of the partially dehydrated ion (whatever its charge). Moreover, their mild electronegative character may, by trapping the water molecules and reducing the size of the permeant ion, facilitate (or even 'catalyse') ion translocation through the channel (instead of creating barriers, as rings of negative charges might do, Changeux 1990).

The inner and outer charged rings identified by Imoto et al (1988) are conserved (as aspartic acid, glutamic acid and/or glutamine) at equivalent positions in neuronal nicotinic receptor subunits; but, in the case of the GABA and glycine receptors, where the transported ion is negatively charged, the inner ring remains negatively charged and the outer one is negative or neutral (Betz 1990). The function of MII in GABA and glycine receptors has not been tested, thus the actual contribution of the inner and outer charged rings to ion transport under physiological conditions in any of these ligand-gated ion channels remains uncertain. Changing the amino acids in the intermediate charged ring has a strong effect on conductance (Imoto et al 1988), and has been reported to modify relative permeabilities slightly (Konno et al 1991, Imoto et al 1991). The mutations studied generally affect the conductance for $Rb^+$ and $Cs^+$ ions, without changing the permeability preference for the physiological ions, $K^+ > Na^+ > Li^+$. This negatively charged ring is conserved in all neuronal nicotinic receptors but is absent from the GABA and glycine receptor MII segments (Betz 1990). Provided that the data obtained in the absence of divalent cations are valid under physiological conditions, this ring may be responsible for probing the nature of the transported ion and, in particular, its charge. The chlorpromazine-labelled leucine ring is conserved in all known subunits of acetylcholine, glycine and GABA receptors and has recently been shown to lock the ion channel in its desensitized state (Revah et al 1991).

**Allosteric transitions of the acetylcholine receptor**

The distance between the agonist-binding sites and the high affinity site for non-competitive channel blockers has been evaluated by energy transfer with a fluorescent agonist and ethidium bromide; it was found to be 21–40 Å (Herz et al 1989). This is in the range of the distances between haem groups in haemoglobin and those between regulatory and catalytic sites in regulatory enzymes (Perutz 1989). The positive interaction between the acetylcholine-binding sites and the high affinity site for non-competitive blockers thus takes place between topographically distinct (and distant) sites. In this respect, it is a typical allosteric interaction, in the original sense of the term (Monod et al 1963).

The fast conformational transition which accounts for the rapid opening or 'activation' of the ion channel by acetylcholine has not been resolved at the atomic level. Carbamylcholine causes decreased labelling of all subunits by a hydrophobic 'non-specific' probe (White & Cohen 1988). Rapid labelling experiments with quinacrine azide reveal further that amino acids from the N-terminal end of MI are labelled under conditions where the channel opens (Karlin 1991). If MII $\alpha$-helices were closely packed in the open state of the channel, non-competitive blockers as large as chlorpromazine, triphenylmethylphosphonium or quinacrine would not be able to enter the channel. Thus, the possibility has been considered (review Changeux 1990, Pullman 1991) that, in the open-channel state, the MII helices are no longer closely packed and that the gap is filled by MI helices with which quinacrine azide would react. Also, mutations of Cys-230 in MI from the $\gamma$ subunit (but not from the $\alpha$ subunit) alter a single gating rate constant that governs the duration of channel opening (Lo et al 1990, 1991). A contribution of MI to the propagation of the conformational transition which links the active sites and the ion channel thus appears plausible, but still requires demonstration. Also still hypothetical, strictly, are the mechanisms for the overall 'expansion' of the protein's quaternary structure during the opening transition, which might involve a concerted 'bloom' or 'twist' of the five transmembrane subunits and may affect either their entire transmembrane structures or only the upper synaptic moiety (see Changeux 1990).

In addition to showing the fast transition of activation, muscle, electroplaque and brain nicotinic receptors respond to prolonged exposure to acetylcholine and cholinergic ligands, within seconds to minutes, with a two-step reversible decline of the conductance response to acetylcholine. Such desensitization persists after reconstitution of the purified protein; thus, it is an intrinsic property of the $\alpha_2\beta\gamma\delta$ oligomer. Parallel measurements, on the same membranes, of ion fluxes and agonist binding (Neubig et al 1982, Heidmann et al 1983b) gave a direct demonstration of the theoretical scheme of Katz & Thesleff (1957), which proposed that the opening of the channel is associated with low affinity binding ($K_d$ 50 μM) and its desensitization with high affinity states ($K_d$ 1 μM and

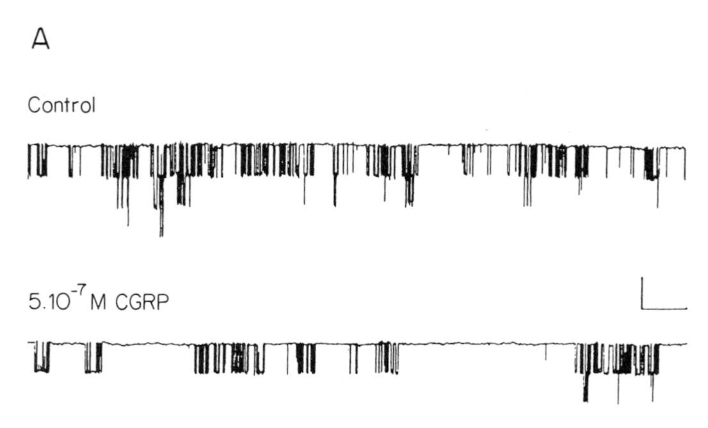


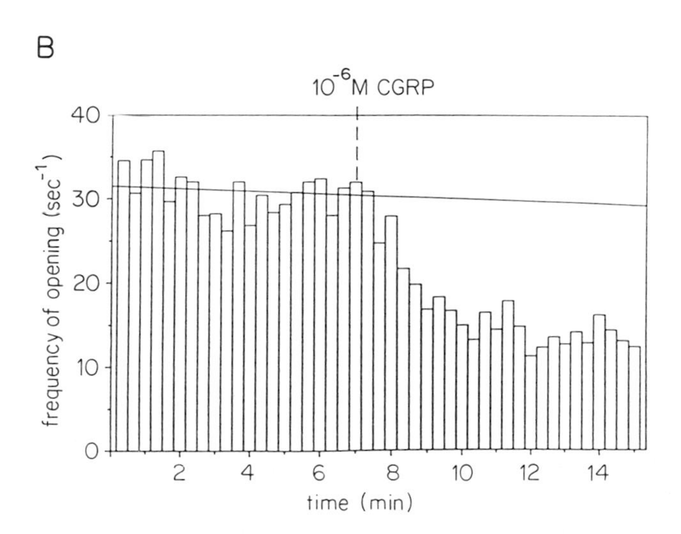


FIG. 3. Calcitonin gene-related peptide (CGRP) accelerates desensitization of the nicotinic receptor in cultured Sol-8 mouse muscle cells under conditions of cell-attached patch-clamp recordings (from Mulle et al 1988). External application of CGRP causes a decrease of single channel opening frequency in cell-attached patch recordings. (A) Recordings obtained before (upper trace) and after (lower trace) application of 500 nM CGRP. (B) Effect of CGRP over time.

3–5 μM, respectively, for the rapid and slow desensitized states). In agreement with the two-state model of allosteric transition, a significant fraction of the sites (about 20%) spontaneously exhibit high affinity in the absence of agonists. The distinction between the binding function and the state function, a characteristic prediction of the allosteric scheme, is thus verified. However, in

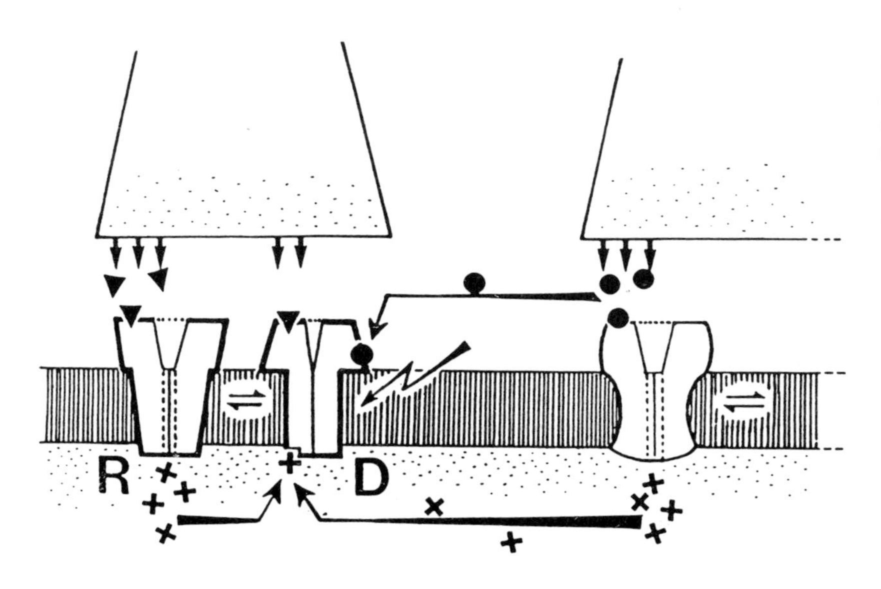
R
D

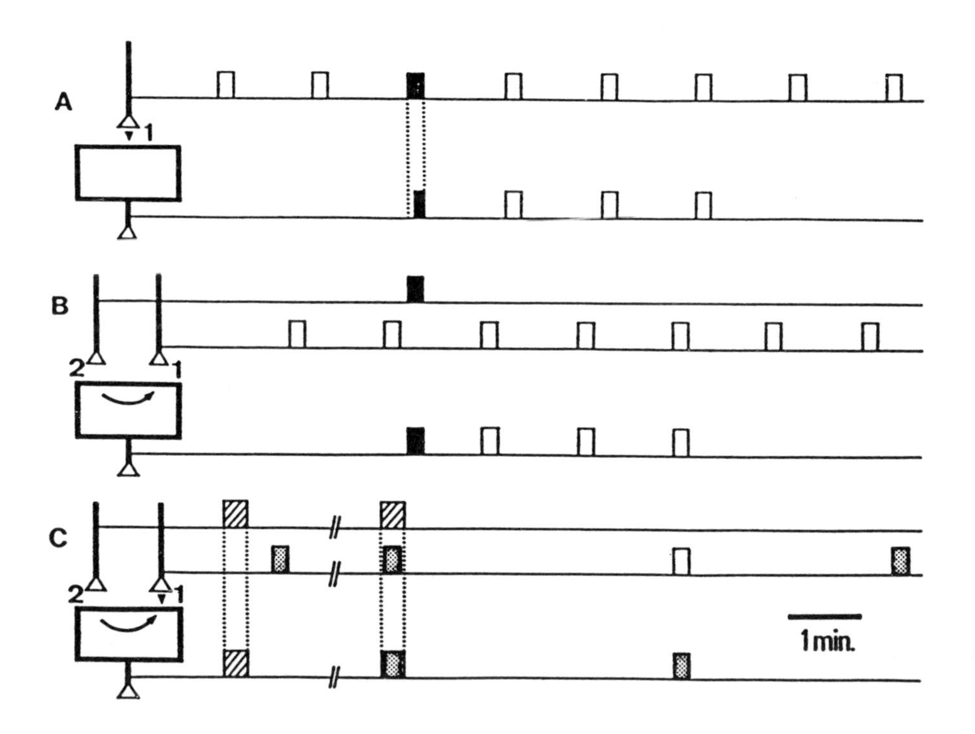
A
1
B
2
1
C
2
1
1 min.

contrast to classical allosteric proteins, the receptor undergoes *multiple* conformational transitions, with different rates of interconversion between states, which also possess different binding properties.

The structural changes which occur on transition from the resting to the desensitized 'equilibrium' state were probed at the level of the cholinergic ligand-binding site with DDF under rapid mixing conditions (Galzi et al 1991b) (Fig. 1). The allosteric effector meproadifen was used to stabilize the high affinity desensitized state. Equilibration with meproadifen caused a major increase in the rapid DDF labelling of the $\alpha$ subunit and, to a lesser extent, an enhanced incorporation of DDF in the $\delta$ subunit from the low affinity *d*-tubocurarine site, and a decreased one in the $\gamma$ subunit from the high affinity *d*-tubocurarine site. These experiments thus reveal changes in the contribution of the $\alpha$ *and* non-$\alpha$ subunits to cholinergic ligand binding in the course of the desensitization transition. These observations are consistent with cryoelectron microscopy experiments showing that after equilibration of membrane-bound receptor with carbamylcholine, the quaternary structure rearrangements detected involve mainly the $\gamma$ and $\delta$ subunits (Unwin et al 1988).

Also, in agreement with a tighter binding of cholinergic ligands to the desensitized receptor, differential labelling of three peptide loops of the $\alpha$ subunit was detected: whereas amino acids $\alpha$-Tyr-190, $\alpha$-Cys-192 and $\alpha$-Cys-193 were labelled in a roughly identical manner in resting and desensitized conformations, the labelling of $\alpha$-Tyr-93 and $\alpha$-Trp-149 increased up to six-fold relative to the desensitized state (Galzi et al 1991b). Thus, the desensitization transition affects both the quaternary structure of the receptor oligomer and the tertiary folding of at least some of its subunits.

Comparison of the deduced amino acid sequences of all cloned nicotinic cholinergic, GABA and glycine receptor subunits reveals, in addition to a similar distribution of hydrophilic and hydrophobic amino acids, the occurrence of several strictly conserved 'canonical' amino acids. For example, the loop which

---

FIG. 4. Model of allosteric regulation of synapse efficacy at the postsynaptic level. Computer simulation of homosynaptic regulation, heterosynaptic regulation and of a 'chemical Hebb synapse' (with an obligatory timing relationship) using the constants determined with the acetylcholine receptor from *T. marmorata* (from Heidmann & Changeux 1982). *Top*: The ratio of the desensitized (D) to active (R) conformations determines the efficacy of the synapse. This ratio, and/or the rate of interconversion between the two states, can be regulated by the neurotransmitter or a coexisting messenger from the synapse (triangles) or from a neighbouring synapse (filled circles) or an internal chemical signal (crosses). *Bottom*: Computer simulations of (A) homosynaptic potentiation, (B) heterosynaptic regulation without a requirement for a timing relationship, (C) heterosynaptic regulation including an obligatory timing relationship ('classical conditioning' or 'Hebb synapse'). The constants determined from the acetylcholine receptor from *T. marmorata* were used in the simulation. Hatched bars, square pulse; unfilled bars, 1 Hz; stippled bars, 10 Hz; filled bars, 50 Hz.

contains the DDF-labelled Tyr-93 from the acetylcholine receptor $\alpha$ subunit contains the canonical amino acids Trp-86, Pro-88, Asp-89 and Asn-94. Also, another DDF-labelled amino acid, $\alpha$-Trp-149, is located in the vicinity of the canonical $\alpha$-Cys-128–Cys-142 loop and $\alpha$-Tyr-151. Furthermore, in the same region of the glycine receptor $\alpha$ subunit, mutation at position 167 (equivalent to acetylcholine receptor $\alpha$-150) strikingly modifies the binding of the competitive antagonist strychnine (see Betz 1990). Thus, in contrast to the loop of the $\alpha$ subunit containing Tyr-190, Cys-192 and Cys-193, which has no equivalent in the GABA and glycine receptors of the CNS, those peptide $\alpha$ subunit loops containing Tyr-93 and Trp-149 may adopt common main chain backbone structures in all these ligand-gated ion channels, as observed, for instance, in immunoglobulin hypervariable regions (Chothia et al 1989). Discrete variations of key amino acid side chains would confer ligand-binding specificity and/or ion selectivity, the three-loop structure plausibly playing a role in the allosteric transitions which may be common to all members of the superfamily of ligand-gated ion channels.

A variety of effectors which bind to additional allosteric sites distinct from the acetylcholine-binding site and from the ion channel (and the associated sites for channel blockers) may differentially bind to some of the multiple conformations accessible to the receptor protein, and thus indirectly regulate the permeability response to acetylcholine. For example, $Ca^{2+}$, electrical potential, reversible allosteric effectors such as non-competitive blockers binding to their low affinity sites, and components of the lipid phase (such as cholesterol) regulate the rate of desensitization (for review see Changeux 1990). Also, phosphorylation of the cytoplasmic loop at particular sites accelerates desensitization *in vitro* in reconstituted receptor-rich membranes (Huganir & Greengard 1990).

Accelerated desensitization of the acetylcholine receptor (Mulle et al 1988) has been observed in cultured muscle cells derived from soleus muscle after treatment with calcitonin gene-related peptide (CGRP) (Fig. 3). This neuropeptide coexists with acetylcholine in spinal cord motor neurons (for review see Villar et al 1989) and can be released by nerve cells. It is plausible that the progressive enhancement of the rapid-decay phase of receptor desensitization by CGRP (Mulle et al 1988) involves cyclic AMP as a second messenger and cyclic AMP-dependent phosphorylation of the receptor (Miles et al 1989). CGRP is thus an excellent candidate for the first messenger of neural origin to regulate postsynaptic receptor efficacy by enhancing its desensitization via phosphorylation. Covalent modifications may thus contribute to the regulation of the receptor's response over a prolonged time scale in addition to, or as an alternative to, the reversible binding of allosteric ligands.

Moreover, the transmembrane polarity of the receptor molecule allows allosteric effectors to cooperate 'through' the membrane, even when their sites are located on its two opposite faces (Revah et al 1987). The convergence of their effects within a given time window may then result in a Hebbian regulation of postsynaptic receptor efficacy in the direction of a 'depression' but also in the direction of a 'facilitation' (Heidmann & Changeux 1982, Changeux 1990) (Fig. 4). Such a mechanism would differ from, and possibly complement, that suggested for the *N*-methyl-D-aspartate-type glutamate receptor, which is based on the direct effect of electrical fields on ion channel block by $Mg^{2+}$ ions (Wigström & Gustaffson 1985). This mechanism provides a potential means for interaction between signalling systems with the additional constraint, essential in learning theory, of a precise timing relationship between the converging signals.

In this respect, it may be noted that at the neuromuscular junction the equilibrium between activatable and desensitized states is spontaneously shifted in favour of the activatable state, as one would expect in view of the function of the neuromuscular junction as an excitatory synapse that transmits nerve signals with high fidelity and little, if any, modulation. In contrast, the balance may be spontaneously shifted for central nicotinic cholinergic, GABA and NMDA receptors in favour of refractory states. Indeed, the benzodiazepines (and/or the corresponding endogenous ligand) and glycine activate, in an allosteric manner, the GABA and glutamate (NMDA) receptors, respectively. Being spontaneously stabilized in a 'silent refractory state', these brain receptors become more susceptible to positive regulation than if they were preferentially stabilized in an activatable conformation, as in the case of peripheral synapses.

Several of the amino acid-gated ion channels, such as the GABA and glycine receptors and the various species of glutamate receptor, display multiple conductance states, and direct transitions between some of these states have been reported to occur (Cull-Candy & Usowicz 1989). Moreover, in the case of the non-NMDA-type (AMPA) glutamate receptor, different agonists, such as kainate and quisqualate, selectively activate some of these conductance levels (Ascher & Nowak 1988), suggesting that these receptors may exist in several conformations with different channel-open states, but also with differing pharmacological specificity. Also, the response to some of the agonists desensitizes (quisqualate or AMPA) whereas that to other agonists (kainate or domoate) does not. Striking differences in binding specificity have also been reported with the nicotinic receptor between resting, active and desensitized states (Grünhagen & Changeux 1976, Cohen & Strnad 1987). Thus, it appears plausible that the multiple-states model proposed for the acetylcholine receptor (Heidmann & Changeux 1982, Neubig et al 1982) may be extended to fit the properties of the other members of the superfamily of ligand-gated ion channels, provided that several (not only one) of these multiple states are conducting. Site-directed mutagenesis and biochemical studies on these receptors may test this hypothesis (Revah et al 1991, Bertrand et al 1992).

## Conclusion

The acetylcholine receptor from electric organ and motor end-plate has become a useful model of ligand-gated ion channels because of the large amount of information available on its structure, molecular biology and functional properties, but also because several of its critical properties may be shared with other members of the superfamily of ligand-gated ion channels (see Galzi et al 1991b, Changeux 1990).

The acetylcholine receptor $\alpha_2\beta\gamma\delta$ oligomer contains all the structural elements responsible for the conversion of the chemical signal in the cleft into an electrical response of the postsynaptic membrane in the physiological millisecond time scale and range of neurotransmitter concentration. In addition, the receptor protein undergoes slow transitions that control the amplitude of the ionic response and are regulated by a distinct population of sites. Such sites may bind ligands from (or be sensitive to second messengers produced by) different signalling systems. Through these sites and slow transitions the acetylcholine receptors may thus mediate interactions between cell signalling pathways and be viewed as a prototype of membrane-bound regulatory proteins engaged in such functions.

In several instances, the nicotinic receptor protein possesses properties expected of typical allosteric proteins: (1) it is an oligomer carrying several categories of topographically distinct sites, linked together by both hetero- and homotropic interactions; (2) acetylcholine causes all-or-none opening and closing of the ion channel with an intrinsic conductance that is independent of the nature of the agonist, and both the open and desensitized states may spontaneously exist in the absence of ligand; (3) the receptor protein undergoes conformational transitions which affect both the tertiary and quaternary structure of the molecule.

Conversely, the receptor protein displays distinctive features atypical of globular allosteric proteins that may be associated with its transmembrane disposition: (1) it possesses an unusual quaternary structure with a single axis of pseudo-symmetry; (2) the $\alpha_2\beta\gamma\delta$ oligomer possesses two binding sites for acetylcholine which differ in their binding properties and thus are not strictly equivalent, but which nevertheless interact in a positively cooperative manner; (3) the biologically active site is an ionic channel, tentatively located along the rotational axis of pseudo-symmetry, that also contains a site (or sites) for pharmacologically active ligands distinct from those acting on the acetylcholine-binding sites; (4) the receptor protein has access to multiple discrete states with different binding properties, activities and rates of interconversion.

These general principles of functional architecture may be extended to other members of the superfamily of neurotransmitter-gated ion channels for chemo–electrical transduction and its short-term regulation by physiological allosteric effectors. The properties of these proteins differ from those of other superfamilies of membrane receptors linked with G proteins or with protein kinase activities. However, in these high affinity ‘slow’ metabotropic receptors allosteric interactions are also expected to play a critical role in transmembrane transduction and its regulation. The process of indirect interaction between topographically distinct sites mediated by conformational transitions of an oligomeric molecule may thus provide a general mechanism for signal transduction and its regulation.

*Acknowledgements*

This work has been supported by research grants from the Association Française contre les Myopathies, the Collège de France, the Centre National de la Recherche Scientifique, the Ministère de la Recherche et de la Technologie, Institut National de la Santé et de la Recherche Médicale (contract 872-004) and the Direction des Recherches Etudes et Techniques (contract 87-211).

## References

Abramson SN, Li Y, Culver P, Taylor P 1989 An analog of lophotoxin reacts covalently with Tyr 190 in the α-subunit of the nicotinic acetylcholine receptor. J Biol Chem 264:12666–12672

Ascher P, Nowak L 1988 Quisqualate and kainate activated channels in mouse central neurons in culture. J Physiol (Lond) 399:227–245

Baldwin J, Chothia C 1979 Haemoglobin: the structural changes related to ligand binding and its allosteric mechanism. J Mol Biol 129:175–220

Bertrand D, Revah F, Galzi JL et al 1992 Unconventional pharmacology of a neuronal nicotinic receptor mutated in the channel domain. Proc Natl Acad Sci USA, in press

Betz H 1990 Homology and analogy in transmembrane channel design: lessons from synaptic membrane proteins. Biochemistry 29:3591–3599

Blount P, Merlie JP 1989 Molecular basis of the two nonequivalent ligand binding sites of the muscle nicotinic acetylcholine receptor. Neuron 3:349–357

Cartaud J, Benedetti L, Cohen JB, Meunier JC, Changeux JP 1973 Presence of a lattice structure in membrane fragments rich in nicotinic receptor protein from the electric organ of *Torpedo marmorata*. FEBS (Fed Eur Biochem Soc) Lett 33:109–113

Changeux JP 1961 The feedback control mechanism of biosynthetic L-threonine deaminase by L-isoleucine. Cold Spring Harbor Symp Quant Biol 26:313–318

Changeux JP 1965 Sur les propriétés allostériques de la l-thréonine désaminase de biosynthese. VI. Discussion générale. Bull Soc Chim Biol 47:281–300

Changeux JP 1966 Responses of acetylcholinesterase from *Torpedo marmorata* to salts and curarizing drugs. Mol Pharmacol 2:369–392

Changeux JP 1990 Functional architecture and dynamics of the nicotinic acetylcholine receptor: an allosteric ligand-gated ion channel. Fidia Res Found Neurosci Award Lect 4:21–168

Changeux JP, Kasai M, Lee CY 1970 The use of a snake venom toxin to characterize the cholinergic receptor protein. Proc Natl Acad Sci USA 67:1241–1247

Charnet P, Labarca C, Leonard PJ et al 1990 An open-channel blocker interacts with adjacent turns of α-helices in the nicotinic acetylcholine receptor. Neuron 2:87–95

Chothia C, Lesk AM, Tramontano A et al 1989 Conformations of immunoglobulin hypervariable regions. Nature (Lond) 342:877–883

Claudio T, Ballivet M, Patrick J, Heinemann S 1983 Nucleotide and deduced amino acid sequences of *Torpedo californica* acetylcholine receptor gamma-subunit. Proc Natl Acad Sci USA 80:1111–1115

Cohen JB, Strnad 1987 Permeability control and desensitization by nicotinic acetylcholine receptors. In: Konijn TM et al (eds) Molecular mechanisms of desensitization to signal molecules. Springer-Verlag Berlin Heidelberg (NATO ASI (Adv Sci Inst) Ser H Cell Biol Vol 6) p 257–273

Cull-Candy SG, Usowicz MM 1989 On the multiple-conductance single channels activated by excitatory amino acids in large cerebellar neurons of the rat. J Physiol (Lond) 415:555–582

Dennis M, Giraudat J, Kotzyba-Hibert F et al 1988 Amino acids of the *Torpedo marmorata* acetylcholine receptor $\alpha$ subunit labeled by a photoaffinity ligand for the acetylcholine binding site. Biochemistry 27:2346–2357

Devillers-Thiéry A, Changeux JP, Paroutaud P, Strosberg AD 1979 The amino-terminal sequence of the 40.000 molecular weight subunit of the acetylcholine receptor protein from *Torpedo marmorata*. FEBS (Fed Eur Biochem Soc) Lett 104:99–105

Devillers-Thiéry A, Giraudat J, Bentaboulet M, Changeux JP 1983 Complete mRNA coding sequence of the acetylcholine binding alpha subunit of *Torpedo marmorata* acetylcholine receptor: a model for the transmembrane organization of the polypeptide chain. Proc Natl Acad Sci USA 80:2067–2071

Epstein M, Racker E 1978 Reconstitution of carbamylcholine-dependent sodium ion flux and desensitization of the acetylcholine receptor from *Torpedo californica*. J Biol Chem 253:6660–6662

Galzi JL, Revah F, Black D, Goeldner M, Hirth C, Changeux JP 1990 Identification of a novel amino acid $\alpha$ Tyr 93 within the active site of the acetylcholine receptor by photoaffinity labeling: additional evidence for a three-loop model of the acetylcholine binding site. J Biol Chem 265:10430–10437

Galzi JL, Revah F, Bessis A, Changeux JP 1991a Functional architecture of the nicotinic acetylcholine receptor: from the electric organ to brain. Annu Rev Pharmacol Toxicol 31:37–72

Galzi JL, Revah F, Bouet F et al 1991b Allosteric transitions of the acetylcholine receptor probed at the amino acid level with a photolabile cholinergic ligand. Proc Natl Acad Sci USA 88:5051–5055

Galzi JL, Bertrand D, Devillers-Thiéry A, Revah F, Bertrand S, Changeux JP 1991c Functional significance of aromatic amino acids from three peptide loops of the $\alpha_7$ neuronal nicotinic receptor site investigated by site directed mutagenesis. FEBS (Fed Eur Biochem Soc) Lett 294:198–202

Gerhart JC, Pardee AB 1962 The enzymology of control by feedback inhibition. J Biol Chem 237:891–896

Giraudat J, Montecucco C, Bisson R, Changeux JP 1985 Transmembrane topology of acetylcholine receptor subunits probed with photoreactive phospholipids. Biochemistry 24:3121–3127

Giraudat J, Dennis M, Heidmann T, Chang JY, Changeux JP 1986 Structure of the high affinity binding site for noncompetitive blockers of the acetylcholine receptor: serine-262 of the $\delta$ subunit is labeled by $^3$H chlorpromazine. Proc Natl Acad Sci USA 83:2719–2723

Giraudat J, Dennis M, Heidmann T, Haumont PT, Lederer F, Changeux JP 1987 Structure of the high-affinity binding site for noncompetitive blockers of the acetylcholine receptor: $^3$H chlorpromazine labels homologous residues in the beta and delta chains. Biochemistry 26:2410–2418

Giraudat J, Galzi JL, Revah F, Haumont PY, Lederer F, Changeux JP 1989 The noncompetitive blocker [$^3$H]chlorpromazine labels segment M2 but not segment MI of the nicotinic acetylcholine receptor $\alpha$-subunits. FEBS (Fed Eur Biochem Soc) Lett 253:190–198

Grünhagen HH, Changeux JP 1976 Studies on the electrogenic action of acetylcholine with *Torpedo marmorata* electric organ. Quinacrine: a fluorescent probe for the

conformational transitions of the cholinergic receptor protein in its membrane bound state. J Mol Biol 106:497–516
Hazelbauer G, Changeux JP 1974 Reconstitution of a chemically excitable membrane. Proc Natl Acad Sci USA 71:1479–1483
Heidmann T, Changeux JP 1982 A molecular model for the regulation of synapse efficacy at the postsynaptic level. CR Acad Sci Ser III Sci Vie 295:665–670
Heidmann T, Changeux JP 1984 Time-resolved photolabeling by the noncompetitive blocker chlorpromazine of the acetylcholine receptor in its transiently open and closed ion channel conformations. Proc Natl Acad Sci USA 81:1897–1901
Heidmann T, Changeux JP 1986 Characterization of the transient agonist-triggered state of the acetylcholine receptor rapidly labeled by the noncompetitive blocker [$^3$H]chlorpromazine: additional evidence for the open channel conformation. Biochemistry 25:6109–6113
Heidmann T, Oswald RE, Changeux JP 1983a Multiple sites of action for non competitive blockers on acetylcholine receptor rich membrane fragments from *Torpedo marmorata*. Biochemistry 22:3112–3127
Heidmann T, Bernhardt J, Neumann E, Changeux JP 1983b Rapid kinetics of agonist binding and permeability response analysed in parallel on acetylcholine receptor rich membranes from *Torpedo marmorata*. Biochemistry 22:5452–5459
Herz JM, Johnson DA, Taylor P 1989 Distance between the agonist and noncompetitive inhibitor sites on the nicotinic acetylcholine receptor. J Biol Chem 264:12439–12448
Hucho FL, Oberthür W, Lottspeich F 1986 The ion channel of the nicotinic acetylcholine receptor is formed by the homologous helices M II of the receptor subunits. FEBS (Fed Eur Biochem Soc) Lett 205:137–142
Huganir RL, Greengard P 1990 Regulation of neurotransmitter receptor desensitization by protein phosphorylation. Neuron 5:555–567
Imoto K, Methfessel C, Sakmann B et al 1986 Location of a δ-subunit region determining ion transport through the acetylcholine receptor channel. Nature (Lond) 324:670–674
Imoto K, Busch C, Sakmann B et al 1988 Rings of negatively charged amino acids determine the acetylcholine receptor channel conductance. Nature (Lond) 335:645–648
Imoto K, Kohno T, Wakai J, Wang F, Nishino M, Numa S 1991 A ring of uncharged polar amino acids as a component of channel construction in the nicotinic acetylcholine receptor. FEBS (Fed Eur Biochem Soc) Lett 289:193–200
Kao P, Dwork A, Kaldany R et al 1984 Identification of the α-subunit half-cystine specifically labeled by an affinity reagent for the acetylcholine receptor binding site. J Biol Chem 259:11662–11665
Karlin A 1991 Explorations of the nicotinic acetylcholine receptor. Harvey Lect 85:71–107
Katz B, Thesleff S 1957 A study of the 'desensitization' produced by acetylcholine at the motor end-plate. J Physiol (Lond) 138:63–80
Konno T, Busch C, Von Kitzing E et al 1991 Rings of anionic amino acids as structural determinants of ion selectivity in the acetylcholine receptor channel. Proc R Soc London B Biol Sci 244:69–79
Koshland D, Nemethy G, Filmer D 1966 Comparison of experimental binding data and theoretical models in proteins containing subunits. Biochemistry 5:365–385
Lagenbuch-Cachat J, Bon C, Mulle C, Goeldner M, Hirth C, Changeux JP 1988 Photoaffinity labeling by aryldiazonium derivatives of *Torpedo marmorata* acetylcholine receptor. Biochemistry 27:2337–2345
Leonard RJ, Labarca CG, Charnet P, Davidson N, Lester HA 1988 Evidence that the M2 membrane-spanning region lines the ion channel pore of the nicotinic receptor. Science (Wash DC) 242:1578–1581

Lo DC, Pinkham JL, Stevens CF 1990 Influence of the $\gamma$ subunit and expression system on acetylcholine receptor gating. Neuron 5:857–866

Lo DC, Pinkham JL, Stevens CF 1991 Role of a key cysteine residue in the gating of the acetylcholine receptor. Neuron 6:31–40

McCrea PD, Popot JL, Engelman DM 1987 Transmembrane topography of the nicotinic acetylcholine receptor delta-subunit. EMBO (Eur Mol Biol Organ) J 6:3619–3626

Miles K, Greengard P, Huganir RL 1989 Calcitonin gene-related peptide regulates phosphorylation of the nicotinic acetylcholine receptor in rat myotubes. Neuron 2:1517–1524

Mishina M, Kurosaki T, Tobimatsu T et al 1984 Expression of functional acetylcholine receptor from cloned cDNAs. Nature (Lond) 307:604–608

Monod J, Changeux JP, Jacob F 1963 Allosteric proteins and cellular control systems. J Mol Biol 6:306–329

Monod J, Wyman J, Changeux JP 1965 On the nature of allosteric transitions: a plausible model. J Mol Biol 12:88–118

Mulle C, Benoit P, Pinset C, Roa M, Changeux JP 1988 Calcitonin gene-related peptide enhances the rate of desensitization of the nicotinic acetylcholine receptor in cultured mouse muscle cells. Proc Natl Acad Sci USA 85:5728–5732

Nelson N, Anholt R, Lindstrom J, Montal M 1980 Reconstitution of purified acetylcholine receptor with functional ion channels in planar lipid bilayers. Proc Natl Acad Sci USA 77:3057–3061

Neubig RR, Boyd ND, Cohen JB 1982 Conformations of *Torpedo* acetylcholine receptor associated with ion transport and desensitization. Biochemistry 21:3460–3467

Noda M, Takahashi H, Tanabe T et al 1982 Primary structure of alpha-subunit precursor of *Torpedo californica* acetylcholine receptor deduced from cDNA sequence. Nature (Lond) 299:793–797

Noda M, Takahashi H, Tanabe T et al 1983a Structural homology of *Torpedo californica* acetylcholine receptor subunits. Nature (Lond) 302:528–532

Noda M, Takahashi H, Tanabe T et al 1983b Primary structures of beta and delta-subunit precursors of *Torpedo californica* acetylcholine receptor deduced from cDNA sequences. Nature (Lond) 301:251–255

Oberthür W, Muhn P, Baumann H, Lottspeich F, Wittmann-Liebold B, Hucho F 1986 The reaction site of a noncompetitive anatagonist in the $\delta$-subunit of the nicotinic acetylcholine receptor. EMBO (Eur Mol Biol Organ) J 5:1815–1819

Olsen RW, Meunier JC, Changeux JP 1972 Progress in purification of the cholinergic receptor protein from *Electrophorus electricus* by affinity chromatography. FEBS (Fed Eur Biochem Soc) Lett 28:96–100

Oswald R, Changeux JP 1981 Ultraviolet light-induced labeling by noncompetitive blockers of the acetylcholine receptor from *Torpedo marmorata*. Proc Natl Acad Sci USA 78:3925–3929

Oswald RE, Changeux JP 1982 Crosslinking of $\alpha$-bungarotoxin to the acetylcholine receptor from *Torpedo marmorata* by ultraviolet light irradiation. FEBS (Fed Eur Biochem Soc) Lett 139:225–229

Pedersen SE, Cohen JB 1990a d-Tubocurarine binding sites are located at $\alpha$–$\gamma$ and $\alpha$–$\delta$ subunit interfaces of the nicotinic acetylcholine receptor. Proc Natl Acad Sci USA 87:2785–2789

Pedersen SE, Cohen JB 1990b [$^3$H]-Mepraodifen mustard reacts with glu-262 of the nicotinic acetylcholine receptor (AChR) $\alpha$-subunit. Biophys J 57:A126 (abstr)

Perutz MF 1989 Mechanisms of cooperativity and allosteric regulation in proteins. Q Rev Biophys 22:139–236

Popot JL, Cartaud J, Changeux JP 1981 Reconstitution of a functional acetylcholine

receptor: incorporation into artificial lipid vesicles and pharmacology of the agonist-controlled permeability changes. Eur J Biochem 118:203–214
Pullman A 1991 Contribution of theoretical chemistry to the study of ion transport through membranes. Chem Rev 91:793–812
Raftery MA, Hunkapiller M, Strader CD, Hood LE 1980. Acetylcholine receptor: complex of homologous subunits. Science (Wash DC) 208:1454–1457
Revah F, Mulle C, Audhya T, Goldstein G, Changeux JP 1987 Calcium dependent effect of the thymic polypeptide, thymopoietin, on the desensitization of the acetylcholine nicotinic receptor. Proc Natl Acad Sci USA 84:3477–3481
Revah F, Galzi JL, Giraudat J, Haumont PY, Lederer F, Changeux JP 1990 The noncompetitive blocker [$^3$H]chlorpromazine labels three amino acids of the acetylcholine receptor $\gamma$ subunit: implications for the $\alpha$ helical organization of the MII segments and the structure of the ion channel. Proc Natl Acad Sci USA 87:4675–4679
Revah F, Bertrand D, Galzi JL et al 1991 Mutations in the channel domain alter desensitization of a neuronal nicotinic receptor. Nature (Lond) 353:846–849
Reynolds JA, Karlin A 1978 Molecular weight in detergent solution of acetylcholine receptor from *Torpedo californica*. Biochemistry 17:2035–2038
Schmidt TJ, Raftery MA 1972 Use of affinity chromatography for acetylcholine receptor purification. Biochem Biophys Res Commun 49:572–578
Tomaselli GF, McLaughlin JT, Jurman M, Hawrot E, Yellen G 1991 Site-directed mutagenesis alters agonist sensitivity of the nicotinic acetylcholine receptor. Biophys J 59:A33 (abstr)
Unwin N, Toyoshima C, Kubalek E 1988 Arrangement of the acetylcholine receptor subunits in the resting and desensitized states, determined by cryoelectron microscopy of crystallized *Torpedo* postsynaptic membranes. J Cell Biol 107:1123–1138
Villar MJ, Roa M, Huchet M et al 1989 Immunoreactive calcitonin gene-related peptide, vaso-active intestinal polypeptide and somatostatin: distribution in developing chicken spinal cord motoneurons and role in regulation of muscle acetylcholine receptor synthesis. Eur J Neurosci 1:269–287
Villarroel A, Herlitze S, Koenen M, Sakmann B 1991 Location of a threonine residue in the $\alpha$-subunit M2 transmembrane segment that determines the ion flow through the acetylcholine receptor channel. Proc R Soc Lond B Biol Sci 243:69–74
White BH, Cohen JB 1988 Photolabeling of membrane-bound *Torpedo* nicotinic acetylcholine receptor with the hydrophobic probe 3-trifluoromethyl-3-(m-[$^{125}$]iodophenyl)diazirine. Biochemistry 27:8741–8751
Wigström H, Gustafsson B, Huang YY, Abraham WC 1985 Hippocampal long-lasting potentiation is induced by pairing single afferent volleys with intracellularly injected depolarising current pulses. Acta Physiol Scand 126:317–319

## DISCUSSION

*Fredholm:* My question is of a quasi-philosophical nature. What is the alternative to an allosteric mechanism?

*Changeux:* Your question is not quasi-philosophical—you are raising a genuine scientific issue. The concept of allosteric interaction was first proposed to explain the inhibition of the catalytic reactions of threonine deaminase and aspartate transcarbamylase by the end product of the pathways in which they are the first enzymes. Two mechanisms were considered at the time (Changeux 1961, Monod et al 1963): (1) a 'classical' *steric effect*, where substrate and regulatory ligand bind at the same site, or at overlapping sites, and directly interact with each other such that there is mutual exclusion by steric hindrance; (2) an alternative mechanism would be for the two classes of ligands to bind at sites which are topographically distant from each other and therefore interact in an *indirect* or 'allosteric' manner. The effect is mediated by some kind of conformational change that we also referred to as an 'allosteric' transition. The term allosteric thus primarily refers to sites with different stereospecificities, and its meaning was extended to define a particular class of globular proteins that are flexible enough to mediate interactions between topographically distinct sites. Such a mechanism also differs from activation of a chemical reaction by binding of a cofactor to a site different from that to which the substrate binds but close enough to allow the substrate and cofactor to interact directly.

*Fredholm:* So what you mean by an allosteric mechanism is that there is a ligand-induced conformational change which affects another ligand or another protein.

*Changeux:* Yes. This is the main alternative to the classical steric effect. However, I would be careful about using the word 'induced'. Some of the conformational changes have been shown to take place spontaneously in the absence of ligand. This is the case for the opening of the ion channel (Jackson 1984) and for the high affinity desensitized state (Heidmann & Changeux 1979). Under these conditions, the ligand would not 'induce' the conformational transition but selectively stabilize a state to which it binds preferentially and which exists before the interaction.

*Daly:* You indicated that phosphorylation has effects on desensitization; is it primarily cyclic AMP-dependent phosphorylation that is involved?

*Changeux:* The regulation of acetylcholine receptor desensitization by CGRP has not been completely analysed. It is known that CGRP stimulates adenylate cyclase, which results in an increased level of cyclic AMP in the cell. It is also known that protein kinase A can phosphorylate the receptor and can accelerate desensitization.

*Daly:* Is there a protein kinase A phosphorylation sequence?

*Changeux:* Yes, on the $\gamma$ and $\delta$ subunits (see Huganir & Greengard 1990).

*Akaike:* There are reports that the acetylcholine receptor desensitization process in the frog end-plate is affected by extracellular calcium concentration (Magazanik & Vyskocil 1970, Scubon-Mulieri & Parsons 1977, Chesnut 1983). What is your opinion on this?

*Changeux:* We have shown with *Torpedo* acetylcholine receptor-rich membrane fragments and the fluorescent agonist DNS-$C_6$-choline, [1-(5-dimethylaminonaphthalene)-sulphonamido]*n*-hexanoic acid-β-(*N*-trimethylammonium bromide) ethyl ester, that addition of calcium ions to the membrane suspension accelerates desensitization *in vitro*. Miledi et al (1980) have also shown that injection of calcium ions into skeletal muscle fibres accelerates desensitization *in vivo*. Calcium seems to behave as a potent regulatory ligand of the nicotinic receptor. We also know that calcium ions can enter the ion channel in addition to sodium and potassium. This is also true for the neuronal nicotinic receptor, and the entry of calcium ions through its channel in the nerve cell may trigger important intracellular regulatory mechanisms.

*Tsien:* Are there any consensus sites in the sequence for calmodulin-dependent kinases? Do calmodulin inhibitors or peptide inhibitors of $Ca^{2+}$/calmodulin-dependent protein kinase II block the calcium effect?

*Changeux:* I don't think that has been looked at.

*Daly:* Is it extracellular or intracellular calcium that has these effects?

*Changeux:* The *in vivo* experiments have been done by intracellular injections of $Ca^{2+}$, the *in vitro* ones by mixing a membrane suspension with a $Ca^{2+}$ solution.

*Nishizuka:* Ion channels are in the lipid bilayer. Is there any indication that lipid composition affects channel functioning?

*Changeux:* Purified synaptic membranes from *Torpedo marmorata* electric organ have a lipid composition characterized by a high cholesterol to phospholipid ratio, a relative abundance of ethanolamine and serine phosphoglycerides and an exceptionally high content of long-chain polyunsaturated docosahexaenoic acid residues (Popot et al 1978). Moreover, surface-pressure area measurements with the purified receptor in monolayers of lipids show that the receptor exhibits a preferential affinity for the lipids that are most characteristic of subsynaptic membranes (cholesterol, long-chain fatty acids etc.) (Popot et al 1978). Reconstitution experiments have shown that a minimum of 45 lipid molecules per receptor oligomer are required for stabilization of a fully functional state (Jones et al 1988). Cholesterol affects the allosteric transitions and channel gating events at the level of 5–10 sites that are not accessible to phospholipids (Jones & McNamee 1988). Such effects could be re-evaluated now by site-directed mutagenesis. Finally, several different types of molecules that affect the lipid phase (such as general anaesthetics and phospholipases) strikingly modify the rate of desensitization (see Changeux 1981 for review).

*Fischer:* There are a lot of tyrosine residues in the sequences of the acetylcholine receptor subunits that could be targets for tyrosine kinases. Are these sites real or only on paper?

*Nairn:* Rick Huganir has recent results on this; three of the subunits (β, γ and δ) are phosphorylated on tyrosine. This phosphorylation is possibly associated with the clustering of the receptor.

*Carpenter:* Doesn't the work show that tyrosine phosphorylation affects the rate of desensitization?

*Nairn:* Initially it was shown that tyrosine phosphorylation increases desensitization in a similar way to phosphorylation by cAMP-dependent protein kinase (Hopfield et al 1988). There are additional studies which suggest that the clustering of the receptor is associated with increased tyrosine phosphorylation (Qu et al 1990, Wallace et al 1991).

*Changeux:* What has been shown is that the clustered receptor is phosphorylated.

*Nairn:* The non-clustered receptors are not phosphorylated on tyrosine and the clustered receptors are.

*Changeux:* Whether phosphorylation is the clustering agent is another question. The receptor could be phosphorylated *after* clustering. The mechanism is still unclear but the finding is interesting.

*Nairn:* In his most recent work Rick Huganir has been trying to identify the tyrosine kinase involved and to show how it might be activated during the clustering process. He has some results suggesting that agrin induces phosphorylation of the acetylcholine receptor and that this has a causative role in receptor clustering (Wallace et al 1991).

*Changeux:* Froehner et al (1990) and others have shown that injection of mRNA for the 43 kDa membrane-associated protein into *Xenopus* oocytes causes clustering of the nicotinic acetylcholine receptor.

*Mikoshiba:* Your model of synapse regulation in relation to the timing of ligand binding in the formation of neural networks is attractive. Do you think that the cytoskeletal protein might provide the connection between information from ligand binding and synapse formation?

*Changeux:* I do not want to be too speculative this time. However, this is an idea that we have used in our attempts to model neural networks (Fig. 4). I think it is not sufficiently appreciated that the acetylcholine receptor is an oligomeric protein that displays positive cooperativity and undergoes allosteric transitions between discrete conformational states. These properties may be exploited to model a chemical Hebb synapse with a postsynaptic receptor sensitive to the timing relationships of ligand binding (Heidmann & Changeux 1982). Such features have been documented for regulatory enzymes. Moreover, stopped-flow experiments with acetylcholine receptor-rich membranes reveal that this could also be the case for the acetylcholine receptor and possibly for other ligand-gated ion channels. One could even make a Hebb synapse *in vitro*, in the mixing chamber of a stopped-flow apparatus! For biochemists this is a rather trivial mechanism, yet the scheme can be exploited in physiology to account for the requirement of pre- and postsynaptic signal convergence within a defined time window for both potentiation and desensitization. Such mechanisms might be more general than that described for the NMDA receptor; for this receptor only voltage-dependent 'steric' block of the ion channel by magnesium is taken into account. Allosteric transitions are found with many ligand-gated ion channels and provide a simple and efficient way to regulate the efficiency of a chemical synapse.

The second part of your question concerned the cytoskeleton. One may say that phosphorylation of the receptor regulates its interaction with the 43 kDa protein (or vice versa). Indeed, recent evidence reveals that the 43 kDa protein can be phosphorylated on serine(s) by protein kinase A (Hill et al 1991) and preliminary observations suggest that the attachment of the 43 kDa protein to the receptor might be regulated by phosphorylation (J. A. Hill, personal communication). On this basis, Yeramian and I have constructed a model for long-term regulation of synaptic efficacy (Yeramian & Changeux 1986), but this is pure speculation at this stage.

*Tsien:* The idea of Hebbian conditioning goes back to William James in the last century. We are talking about it mostly in a neurobiological context, but, in fact, the term could be applied to the situation that Bob Michell described (p 3–5), in which a series of signalling systems have to be initiated in the right order and with the right conjunction—when they happen in the correct order, something exciting happens. What is the best studied example where one can show, for example, that tyrosine phosphorylation must precede cyclic AMP-dependent phosphorylation to produce a certain final result?

*Changeux:* That is a different question. In your example it is the order of events, rather than the time coincidence within a narrow window, that is important.

*Rasmussen:* Would glycogen synthase be a good example?

*Fischer:* That is very complicated. There are seven or eight phosphorylation sites that are targets of as many different protein kinases.

*Hunter:* But there are some contingent events in the phosphorylation of glycogen synthase. For example, casein kinase II has to phosphorylate glycogen synthase before glycogen synthase kinase 3.

*Fischer:* For the molecules that are hit by both tyrosine and serine/threonine kinases this is an important question about which we know little. There are several instances of 'second-site phosphorylation', where phosphorylation at one site is obligatory for phosphorylation at two or three other sites; this has been well documented for glycogen synthase (Roach 1990).

*Parker:* But there's no dynamic change in the casein kinase II site.

*Fischer:* That's correct. I was thinking more of enzymes such as $p34^{cdc2}$ or MAP kinase that are phosphorylated on both serine and tyrosine, where removal of either phosphate inactivates the enzyme. It is unclear what precedes what, or what allows a given event to occur at the proper time.

*Changeux:* You are talking of a timing relationship—one signal has to come before the other. Such a relationship may not result exclusively from an elementary mechanism at the receptor level. The topology of synapses may also have to be taken into account. For example, we have described a 'triad' of synapses whose topology and functional properties imply an order of activation of the synapses for the modulation of efficiency of the end synapse (Dehaene et al 1987).

*Tsien:* Neurobiologists seem to agree about the timing between voltage changes and ligand binding needed to produce LTP. I was trying to turn the discussion from a neurobiological situation to a more biochemical setting where I thought there might be more common ground. I was thinking about situations where not only temporal order, but also the temporal proximity of two stimuli, is important. The biochemical systems that have been talked about start with conjunction—two things have to be phosphorylated—but they also raise questions about possible requirements for temporal order—first A has to happen, and then B happens, and only if A precedes B do you get the right outcome. I thought Jean-Pierre Changeux was referring to biochemical systems in which not only is it important that A should precede B, but also it is necessary for A to precede B by only a short amount of time, or a certain key amount of time.

*Changeux:* The point I wished to make was that a single macromolecule can possess such properties, yet the nervous system has a complex organization and, as I have said, the topology of synaptic connections may also contribute to the way stimuli are ordered.

Non-receptor regulatory proteins, such as protein kinases, are genuine allosteric proteins—they possess regulatory and catalytic subunits, and are sensitive to allosteric effectors. It would be interesting to test for timing relationships at the level of protein kinases.

*Daly:* There are two reports of an ATP channel and a nicotinic channel co-existing in a single cell, where they might provide an opportunity for a common response to two signals arriving at different times (Nakazawa et al 1991, Igusa 1988). One of those systems is the pheochromocytoma cell, in which the action of the nicotinic channel triggers release of ATP and noradrenaline. The ATP would then reactivate a channel. Indeed, the nicotinic and ATP channels had such similar properties that the authors of these two papers have proposed that the nicotinic channel and the ATP channel in pheochromocytoma PC12 cells and in frog myocytes are the same channel.

*Changeux:* There are clear pharmacological differences between the nicotinic receptors from muscle and electric organ and those from the brain and other classes of cells. There are differences in the amino acid sequences of the $\alpha$ subunits but the association with the non-$\alpha$ subunits may markedly affect the specificity of the agonist-binding sites. The pharmacology of the receptor is changed by associating, for instance, $\alpha_4$ with $\beta_2$ or with $\beta_4$. In the case of the GABA receptor, some subunits confer sensitivity to benzodiazepines. One speculation is that the acetylcholine receptor becomes sensitive to ATP when one particular subunit is associated with a given $\alpha$ subunit.

*Daly:* One of these systems with both ATP- and nicotinic-sensitive channels was a frog myocyte, which I assume would have the same nicotinic channel as the mammalian muscle and the *Torpedo* electroplax, and the other was a pheochromocytoma PC12 cell, which would have a ganglionic-type nicotinic channel, as do chromaffin cells.

*Changeux:* Another possibility is that there are three completely distinct receptors regulated by different ligands but with channels of similar pharmacological properties. For example, histrionicotoxin can block not only the acetylcholine receptor channel but also the NMDA receptor channel.

*Daly:* Histrionicotoxin did not block the ATP channel in pheochromocytoma cells, at least not in our work.

**TABLE 1** (*Su*) **Effect of carbamylcholine treatment on ligand binding to $M_2$ muscarinic cholinergic receptors, production of cyclic AMP and inositol trisphosphate and protein kinase C activity in CCL-137 cells**

| | *Ligand binding*[a] | | | | | | |
|---|---|---|---|---|---|---|---|
| | *[$^3$H] QNB* | | *[$^3$H] NMS* | | *cyclic AMP*[b] *(pmol/$10^6$ cells)* | *$InsP_3$*[c] *produced (c.p.m.)* | *PKC*[d] *(c.p.m. $\times 10^{-3}$/ $10^6$ cells/5 min)* |
| | $B_{max}$ | $K_d$ | $B_{max}$ | $K_d$ | | | |
| Control | 726.0[e] (±25.9) | 0.14 (±0.02) | 688.0 (±21.8) | 0.19 (±0.06) | 2.19 (±0.21) | 279 (±42) | 42 (±4) |
| Carbamylcholine 1 min | — | — | — | — | 1.22* (±0.12) | 264 (±38) | 44 (±6) |
| Carbamylcholine 60 min | 514.0* (±18.3) | 0.15 (±0.02) | 488.5* (±16.0) | 0.15 (±0.09) | 4.44* (±0.35) | 523** (±54) | 365** (±12) |

[a]QNB was incubated with cells for 15 min at 37 °C, and NMS for 40 min at 37 °C. Non-specific binding was assessed in the presence of 1 μM atropine.
[b]Production of cyclic AMP was measured by radioimmunoassay (kit from Amersham International plc).
[c]Cells were cultured with [$^3$H]inositol (4 μCi/ml) for 12 hours. 20 mM LiCl was added and the cells were recultured for 20 min before the addition of carbamylcholine.
[d]Protein kinase C activity in cells was assessed by measurement of incorporation of $^{32}$P (from 12 mM [$^{32}$P]ATP) into histone and comparison with standard curves.
[e]All values in the table are mean ± S.D., $n = 5$.
QNB, quinuclidinylbenzilate.
NMS, *N*-methylscopolamine.
*Values significantly different from control, $P < 0.001$.
**Values significantly different from control cells and from cells treated for 1 min with carbamylcholine, $P < 0.001$.

*Changeux:* ATP may be an allosteric ligand for the classical nicotinic receptor, but that remains to be shown.

*Su:* We have been investigating the effects of the activator carbamylcholine on the $M_2$ muscarinic cholinergic receptor of the CCL-137 cell, a human embryonic lung fibroblast cell. Our results are shown in Table 1. The effects of carbamylcholine are quite different after short-term (1 min) and long-term (60 min) treatment. Long-term treatment with the agonist causes desensitization, as shown by the decrease in maximum binding capacity ($B_{max}$) for both [$^3$H]*N*-methylscopolamine (NMS) and [$^3$H]3-quinuclidinylbenzilate (QNB). The affinity of binding did not change. The adenylate cyclase system is negatively coupled to the $M_2$ receptor, but after the long-term treatment with carbamylcholine the adenylate cyclase system is stimulated, as shown by the increased production of cAMP. Short-term carbamylcholine treatment has no effect on the phosphatidylinositol system; after long-term treatment the PtdIns system is stimulated; $InsP_3$ production and protein kinase C activation are observed. We have done further work in our attempts to reveal the mechanism of these effects and have found that phospholipase $A_2$ plays a role.

Professor Changeux's paper on the structure–function relationships of the acetylcholine receptor, should be helpful to us in our further studies.

*Changeux:* Your work concerns the muscarinic cholinergic receptor, whereas our work is on the nicotinic receptor. The mechanism of desensitization of the ligand-gated ion channel receptors seems, at a glance, different from that of receptors linked to G proteins. Phosphorylation is required for desensitization of G protein-linked receptors, whereas there is no requirement for phosphorylation of the nicotinic receptor for desensitization. The pure receptor incorporated into lipid bilayers can desensitize without phosphorylation, but phosphorylation accelerates the process (Huganir & Greengard 1990). I think the mechanisms involved in the ligand-gated ion channels that we are working with are different: for example, the transitions between states are much faster. If you want to capture the binding states which correspond to the active and desensitized states you have to use rapid mixing conditions.

## References

Changeux JP 1961 The feedback control mechanisms of biosynthetic L-threonine deaminase by L-isoleucine. Cold Spring Harbor Symp Quant Biol 26:313–318

Changeux JP 1981 The acetylcholine receptor—an 'allosteric' membrane protein. Harvey Lect 75:85–254

Chesnut TJ 1983 Two-component desensitization at the neuromuscular junction of the frog. J Physiol (Lond) 336:229–241

Dehaene S, Changeux JP, Nadal JP 1987 Neural networks that learn temporal sequences by selection. Proc Natl Acad Sci USA 84:2727–2731

Froehner SC, Luetje CW, Scotland PB, Patrick J 1990 The postsynaptic 43K protein clusters muscle nicotinic acetylcholine receptors in *Xenopus* oocytes. Neuron 5:403–410

Heidmann T, Changeux JP 1979 Fast kinetic studies on the allosteric interactions between acetylcholine receptor and local anesthetic binding sites. Eur J Biochem 94:281–296
Heidmann T, Changeux JP 1982 A molecular model for the regulation of synapse efficacy at the postsynaptic level. CR Acad Sci Ser III Sci Vie 295:665–670
Hill JA, Ngheim HO, Changeux JP 1991 Serine specific phosphorylation of nicotinic receptor associated 43K protein. Biochemistry 30:5579–5585
Hopfield JF, Tank JW, Greengard P, Huganir RL 1988 Functional modulation of the nicotinic acetylcholine receptor by tyrosine phosphorylation. Nature (Lond) 336:677–680
Huganir RL, Greengard P 1990 Regulation of neurotransmitter desensitization by protein phosphorylation. Neuron 5:555–567
Igusa Y 1988 Adenosine 5′-triphosphate activates acetylcholine receptor channels in cultures of *Xenopus* myotomal muscle cells. J Physiol (Lond) 405:169–185
Jackson MB 1984 Spontaneous openings of the acetylcholine receptor channel. Proc Natl Acad Sci USA 81:3901–3904
Jones OT, McNamee MG 1988 Annular and nonannular binding sites for cholesterol associated with the nicotinic acetylcholine receptor. Biochemistry 27:2364–2374
Jones OT, Eubanks JH, McNamee MG 1988 A minimum number of lipids are required to support the functional properties of the nicotinic acetylcholine receptor. Biochemistry 27:3733–3742
Magazanik LG, Vyskocil F 1970 Dependence of acetylcholine desensitization on the membrane potential of frog muscle fibre and on the ionic changes in the medium J Physiol (Lond) 210:507–518
Miledi R, Parker I, Schalow G 1980 Transmitter induced calcium entry across the post-synaptic membrane at frog end-plates measured using arsenazo III. J Physiol (Lond) 300:197–212
Monod J, Changeux JP, Jacob F 1963 Allosteric proteins and cellular control systems. J Mol Biol 6:306–329
Nakazawa K, Fujimori K, Takanaka A, Inoue K 1991 Comparison of adenosine triphosphate-activated and nicotine-activated inward currents in rat pheochromocytoma cells. J Physiol (Lond) 434:647–660
Popot JL, Demel RA, Sobel A, Vandeene LL, Changeux JP 1978 Interaction of acetylcholine (nicotinic) receptor protein from *Torpedo marmorata* electric organ with monolayers of pure lipids. Eur J Biochem 85:27–42
Qu Z, Moritz E, Huganir RL 1990 Regulation of tyrosine phosphorylation of the nicotinic acetylcholine receptor at the rat neuromuscular junction. Neuron 2:367–378
Roach PJ 1990 Control of glycogen synthetase by hierarchical protein phosphorylation. FASEB (Fed Am Soc Exp Biol) J 4:2961–2968
Scubon-Mulieri B, Parsons RL 1977 Desensitization and recovery at the frog neuromuscular junction. J Gen Physiol 69:431–447
Wallace BG, Qu Z, Huganir RL 1991 Agrin induces phosphorylation of the nicotinic acetylcholine receptor. Neuron 6:869–878
Yeramian E, Changeux JP 1986 A model for long-term changes in synaptic efficacy based upon the interaction of the acetylcholine receptor with the subsynaptic protein of 43,000 daltons. CR Acad Sci Ser III Sci Vie 302:609–616

# $Ca^{2+}$-cyclic AMP interactions in sustained cellular responses

Howard Rasmussen, Carlos Isales, Shridar Ganesan, Roberto Calle and Walter Zawalich

*Division of Endocrinology, Department of Internal Medicine, Yale University School of Medicine, PO Box 3333, New Haven, CT 06510-8056, USA*

*Abstract.* As early as 1970 it was apparent that the cyclic AMP (cAMP) and $Ca^{2+}$ messenger systems often interact to regulate cellular responses. Work over the past 20 years has greatly expanded our knowledge of these interactions, and has shown that these signalling systems interact in complex ways to regulate sustained cellular responses such as aldosterone secretion, smooth muscle contraction and insulin secretion. The latter system is considered in detail because it illustrates several types of interactions, both positive and negative, which help to determine the normal response of β-cells to physiological stimuli, and how abnormalities in secretory patterns can develop as a consequence of the prolonged stimulation of a messenger system.

*1992 Interactions among cell signalling systems. Wiley, Chichester (Ciba Foundation Symposium 164) p 98–112*

When $Ca^{2+}$ and cyclic AMP (cAMP) were discovered to function as intracellular messengers they were considered to act more or less independently as intracellular signals in excitable (Douglas & Rubin 1961, Katz & Miledi 1969) and non-excitable cells (Sutherland 1961), respectively. However, it soon became apparent that in many cellular responses $Ca^{2+}$ and cAMP act as synarchic messengers in regulating cell function (Rasmussen 1970, 1981). Further, as our understanding of the behaviour of these messenger systems has become more complete, the types of interactions known to occur between their components have increased (Barrett et al 1989, Brami et al 1987, Limbird 1988, Pfeilschifter 1989, Rasmussen et al 1990a, Zawalich & Rasmussen 1990). A summary of these interactions is shown in Table 1. It is not our intent to discuss each of these interactions, but rather to use this summary as an introduction to the interactions between the cAMP and $Ca^{2+}$ messenger systems in the regulation of sustained cellular responses in three types of cells—adrenal glomerulosa cells, pancreatic β-cells and smooth muscle cells.

## Characteristics of sustained cellular responses

Our studies on the secretion of aldosterone from adrenal glomerulosa cells (Barrett et al 1989), the secretion of insulin from β-cells in the endocrine pancreas

**TABLE 1 Interaction between the cyclic AMP and calcium messenger systems**

| | *Interaction at the level of signal transduction mechanisms* | | *Interaction at the level of protein kinase activation* |
|---|---|---|---|
| Effects of the calcium messenger system on cAMP | ↑Calcium influx→↑Adenylate cyclase activation<br>↑Calcium influx→↑cAMP phosphodiesterase | ↑cAMP turnover | ↑Activation of protein kinase C→↑Adenylate cyclase activation<br>Calmodulin kinase II has activation sites for both cAMP and calcium |
| | ↑PtdIns activating agonist binding→↑Adenylate cyclase activation | | |
| Effects of the cAMP messenger system on calcium | $G_s$ protein activation→↑cAMP→↑Calcium influx (L-type channels)<br>↑cAMP→↑Intracellular calcium from internal pools<br>$G_i$ protein activation→↓cAMP/↓Calcium influx | | ↑Protein kinase A→↓Phospholipase C activation |
| Interaction at the level of protein phosphorylation | The two protein kinases may act on separate substrates but both affect the same response<br>The two protein kinases may act in sequence to phosphorylate substrate proteins and thus modify each other's response<br>The two proteins may phosphorylate the same substrate protein | | |

(Rasmussen et al 1990b, Zawalich & Rasmussen 1990) and the tonic contraction of smooth muscle (Rasmussen et al 1987, 1990a) have shown that in each case a sustained cellular response can be induced by the continued presence of an agonist that activates the $Ca^{2+}$ messenger system. Furthermore, the activity of the $Ca^{2+}$-phosphatidylinositol system is modulated in either a positive (aldosterone and insulin secretion) or a negative (smooth muscle contraction) fashion by the activity of the cAMP messenger system. Studies of the regulatory events that underlie these sustained responses show that simple second messenger models of cell activation are inadequate to account for the regulatory properties of these systems. Rather than a single signal or set of signals acting constantly throughout the period of hormonal action, cell activation leads to a sequence of changes in metabolism of $Ca^{2+}$, inositol phosphate and diacylglycerol, and in protein kinase activation; that is, there is a temporal progression of signalling events. In each case, the cellular response during the first hour or two can be characterized as occurring in two phases—an initial phase and a sustained phase (Barrett et al 1989, Rasmussen et al 1987, Zawalich & Rasmussen 1990).

During the initial phase the hydrolysis of phosphatidylinositol 4,5-bisphosphate ($PtdInsP_2$) by PtdIns-specific phospholipase C (PtdIns-PLC) leads to the generation of inositol 1,4,5-trisphosphate [$Ins(1,4,5)P_3$] and diacylglycerol (DAG). The $Ins(1,4,5)P_3$ acts to release $Ca^{2+}$ from an intracellular pool, causing a transient rise in the $Ca^{2+}$ concentration in the cell cytosol, $[Ca^{2+}]_c$. This rise in $[Ca^{2+}]_c$ activates calmodulin-dependent protein kinases, resulting in the phosphorylation of a subset of cellular proteins. The change in function of these phosphorylated proteins is thought to initiate the cellular response. Also during this phase DAG is produced as a result of $PtdInsP_2$ hydrolysis, and this plasma membrane-associated DAG, along with the transient rise in $[Ca^{2+}]_c$, causes protein kinase C (PKC) to associate with the plasma membrane.

During the sustained phase there is a continued production of DAG as a result of the activities of PtdIns-PLC, a phosphatidylcholine-specific phospholipase D and possibly a phosphatidylcholine-specific PLC (Billah & Anthes 1990, Bollag et al 1991, Exton 1990, Exton et al 1992 [this volume], Peter-Reisch et al 1988). Hence, rather than one species of DAG being produced, there are multiple forms which appear at different times during sustained activation of the cell. A sustained increase in $Ca^{2+}$ influx across the plasma membrane also occurs, which leads to a compensatory increase in energy-dependent $Ca^{2+}$ efflux via a $Ca^{2+}$-transporting ATPase or a $Ca^{2+}$ pump. The increase in $Ca^{2+}$ cycling across the plasma membrane results in a rise in the $Ca^{2+}$ concentration in a restricted cellular domain on the endoplasmic face of the plasma membrane, $[Ca^{2+}]_{sm}$. The $[Ca^{2+}]_{sm}$ regulates the activity of the plasma membrane-associated PKC, which in turn catalyses the phosphorylation of a second subset of proteins—those thought to be responsible for mediating the sustained phase of the cellular response (Barrett et al 1986, Takuwa et al 1988a).

These data and this model emphasize that in addition to there being temporally distinct phases of cell activation there are spatially restricted messengers regulating these phases.

It is noteworthy that in none of the three systems that we have studied is an increase in plasma membrane calcium influx alone sufficient to induce a cellular response. This kind of evidence implies that some $Ca^{2+}$-sensitive, plasma membrane-associated transducer must be present for the spatially restricted $Ca^{2+}$ signal to act. In all three cases PKC is an important, if not the most important, transducer. In addition, in adrenal glomerulosa cells in the actions of both adrenocorticotropic hormone (ACTH) and high extracellular $K^+$ concentrations, adenylate cyclase is an important $Ca^{2+}$-sensitive transducer (Kojima et al 1985a,b).

Of equal interest is the observation that a rise in the cAMP concentration in each of these cell types leads to an inhibition of PtdIns-PLC activity. However, the consequences of this inhibition, in terms of cellular response, are quite different in different cell types because cAMP also acts at other sites in the intracellular signalling pathways. These sites are unique to each cell type. This aspect of cAMP action is particularly striking if one compares the interactions of the $Ca^{2+}$ and cAMP messenger systems in smooth muscle (Rasmussen et al 1990a) and in β-cells (Zawalich & Rasmussen 1990). In the former, activation of the cAMP messenger system leads to an inhibition of PtdIns-PLC activity and of the cellular response, even though cAMP also causes an influx of $Ca^{2+}$ across the plasma membrane (Takuwa et al 1988b). In contrast, although in the β-cell a rise in cAMP also causes an inhibition of PtdIns-PLC and a stimulation of $Ca^{2+}$ influx, just as in the muscle cell, cAMP causes an enhanced insulin secretory response in the β-cell. This difference in the way the cAMP messenger system modulates events in the calcium messenger system in these two cell types depends on a difference in the way cAMP influences the late events in the cells. In the β-cell, a rise in cAMP content leads to an enhanced response in both the first and second phases of insulin secretion, either by enhancing the activity of protein kinases specific to each phase of the response or by inhibiting the activity of phosphoprotein phosphatases. In smooth muscle cells, cAMP has the opposite effect, either inhibition of kinases or activation of phosphoprotein phosphatases.

Another control feature of interest is that the generation of $Ins(1,4,5)P_3$ and DAG can occur in response to an appropriate agonist under conditions where there is little or no cellular response—that is, there is no necessary correspondence between signal generation and signal read-out. This is most dramatically illustrated by angiotensin II-induced aldosterone secretion and the combined effects of acetylcholine and glucagon-like peptide 1 on the secretion of insulin. A critical factor in the control of aldosterone secretion is the extracellular $K^+$ concentration. A rise in $K^+$ concentration from 3.5 to 6.0 mM causes a depolarization of the plasma membrane, the opening of voltage-dependent

(T-type) $Ca^{2+}$ channels (Cohen et al 1988) and stimulation of aldosterone secretion. Within this range of $K^+$ concentrations the addition of angiotensin II causes a further increase in aldosterone production. A fall of $K^+$ concentration from 3.5 to 2.0 mM hyperpolarizes the membrane. Stimulation of such cells with angiotensin II causes no sustained change in the rate of aldosterone secretion, even though phosphatidylinositol hydrolysis is stimulated by angiotensin II. An aldosterone secretory response does not occur because angiotensin II does not cause an increase in $Ca^{2+}$ influx in cells incubated in 2.0 mM $K^+$. This arrangement can be viewed as a physiological fail-safe mechanism to prevent further aldosterone production regardless of the stimulus when plasma $K^+$ concentration is low, because a further fall in $K^+$ concentration, due to aldosterone's action on the renal tubule, could cause the death of the animal.

A similar relationship exists between glucose and insulin secretion. At a normal preprandial blood glucose concentration (4 mM), a combination of one agonist that activates PtdIns-PLC, such as acetylcholine, and another that activates adenylate cyclase, such as glucagon-like pepetide 1, leads to the generation of the Ins(1,4,5)$P_3$, DAG and cAMP signals, but little or no increase in insulin secretion (Zawalich & Rasmussen 1990). In contrast, at postprandial glucose concentration (7 mM), these same two agonists induce the generation of the same signalling molecules, but under these circumstances they induce a biphasic increase in the rate of insulin secretion. The key action of glucose in determining this change in responsiveness to this agonist combination is that of regulating the plasma membrane potential. Glucose regulates membrane potential by controlling cellular ATP concentration. The concentration of ATP determines the rate of $K^+$ flux through specific ATP-sensitive $K^+$ channels in the plasma membrane of the β-cell (Cook et al 1988). At a preprandial glucose concentration the ATP level is low enough to keep $K^+_{ATP}$ channels open and the membrane potential high. Hence, even though acetylcholine causes a depolarization of the membrane, the fall in membrane potential is not sufficient to activate voltage dependent (L-type) $Ca^{2+}$ channels (Smith et al 1989). However, at the higher glucose concentration the resting membrane potential is low enough that a further fall in membrane potential, induced by acetylcholine, is sufficient to open L-type channels. Simultaneously, cAMP, acting via protein kinase A, brings about the phosphorylation of this class of calcium channel. Phosphorylation of the channels increases their mean open time, thus increasing the rate of $Ca^{2+}$ influx through them.

## The neurohumoral- and glucose-dependent regulation of insulin secretion

The mechanism by which neurohumoral agents and changes in extracellular glucose concentration interact to regulate insulin secretion exemplifies the complexities of the interactions that can occur between the $Ca^{2+}$ and cAMP messenger systems in the control of a single cellular response (Rasmussen et al 1990b, Zawalich & Rasmussen 1990). In order to discuss these interactions, it is first necessary to consider the mechanisms by which glucose, neurohumoral agonists or pharmacological agents stimulate insulin secretion.

There are three different ways by which an equivalent increase in insulin secretion can be produced in perifused rat islets of Langerhans: 10 mM extracellular glucose alone; a combination of acetylcholine and glucagon-like peptide 1 acting on islets incubated in medium containing 7.0 mM glucose; and a combination of phorbol ester (12-*O*-tetradecanoylphorbol-13-acetate, TPA), forskolin and tolbutamide in the presence of 2.75 mM glucose. In each case, a biphasic pattern of insulin secretion is seen, and four signalling events are observed: a transient increase in $[Ca^{2+}]_c$, a sustained increase in $Ca^{2+}$ influx rate, a rise in cAMP concentration and an activation of PKC.

It is our working hypothesis that the two phases of insulin secretion are regulated by the sequential activation of the two different types of protein kinase, $Ca^{2+}$/calmodulin-dependent kinases during the first phase and PKC during the second. Further work is necessary to validate the role of calmodulin-regulated kinases in the first phase. However, recent studies from our own and other laboratories have provided strong support for the role of PKC in regulating the sustained or second phase of insulin secretion. Our results show that only the $\alpha$ isoform of PKC is present in β-cells, that this PKC isoform is translocated to the plasma membrane in response to high glucose, carbachol or phorbol ester (Ganesan et al 1990), and that staurosporine, an inhibitor of PKC activity, blocks the second but not the first phase of (10 mM) glucose-induced insulin secretion (Zawalich et al 1991). Easom et al (1990) have shown that carbachol, or a combination of carbachol and 10 mM glucose, stimulates the phosphorylation of a specific PKC substrate in isolated islets, the 80 kDa MARCKS (myristoylated, alanine-rich C kinase substrate) protein. Phosphorylation of this protein is blocked by staurosporine. Taken together, these findings provide strong support for the concept that PKC plays a central role in regulation of the second or sustained phase of insulin secretion (Zawalich et al 1984).

Additional evidence for the crucial role of PKC in insulin secretion has been provided by our recent studies of the immunocytochemical localization of PKC in perifused islets (S. Ganesan, R. Calle, K. Zawalich & H. Rasmussen, unpublished work 1991). Using monoclonal antibodies specific for the $\alpha$, β and $\gamma$ isoforms of PKC we identified the distribution of these isoforms in the cell. $\alpha$-PKC was localized exclusively in the β-cells, the β isoform in the $\alpha$-cells, and no $\gamma$-PKC was found. Localization of insulin with $\alpha$-PKC was demonstrated in the β-cells. In islets perifused with 2.75 mM glucose, the $\alpha$-PKC was distributed throughout the cytosol. In islets perifused with 20 mM glucose the $\alpha$-PKC was found associated with the plasma membrane. The increase in insulin secretion and the translocation of $\alpha$-PKC in response to 20 mM glucose were both inhibited by simultaneous perifusion with 20 mM mannoheptulose, a known inhibitor of glucose metabolism in this tissue.

A model which accounts for many of the results concerned with the regulation of insulin secretion is illustrated in Fig. 1. The key regulatory importance of the rate of $Ca^{2+}$ influx across the plasma membrane is highlighted. This influx occurs largely through L-type voltage-dependent $Ca^{2+}$ channels. Hence, the membrane potential is a critical determinant of insulin secretion. Glucose, by regulating the ATP concentration, causes a decrease in $K^+$ efflux (Cook et al 1988), and hence a fall in membrane potential. Tolbutamide can mimic this key effect of glucose by acting directly on the same channel. If the glucose- or tolbutamide-induced depolarization is sufficiently large to activate L-type $Ca^{2+}$ channels, then an increase in $Ca^{2+}$ influx occurs. The influx of $Ca^{2+}$ has four effects: (a) it activates PtdIns-PLC; (b) it activates adenylate cyclase; (c) it activates membrane-associated PKC; and (d) it acts on $Ca^{2+}$-sensitive $K^+$ channels to increase $K^+$ efflux and cause the membrane to become repolarized. Hence, when high glucose acts it induces repetitive bursts of regenerative $Ca^{2+}$ currents across the plasma membrane. In addition to its effect on $K^+$ efflux (causing membrane depolarization), glucose, via its $Ca^{2+}$-dependent effects on adenylate cyclase, causes an increase in cAMP, which acts in a positive feedback sense to increase $Ca^{2+}$ influx through operative L-type channels. In addition, high glucose has a more direct effect on $Ca^{2+}$ flux through these channels (Smith et al 1989). Nitrendipine, by blocking $Ca^{2+}$ flux, inhibits the effects of high glucose on insulin secretion. Acetylcholine stimulates PtdIns-PLC, through a G protein, in a $Ca^{2+}$-independent manner. Hence, it enhances the effect of glucose on PtdIns-PLC activity. Acetylcholine has an additional effect, that of inducing a depolarization of the plasma membrane, probably as a result of increasing $Na^+$ influx. This effect enhances the glucose-induced membrane depolarization. These effects of acetylcholine are blocked by atropine, indicating that the transmitter is acting via a muscarinic receptor. Glucagon-like peptide 1 or gastric inhibitory peptide (GIP) stimulate adenylate cyclase via $G_s$, and enhance the effect of glucose on cAMP production. Forskolin has a similar ability to stimulate adenylate cyclase. In addition to enhancing $Ca^{2+}$ influx, cAMP also acts to inhibit PtdIns-PLC activity. Even though cAMP inhibits PtdIns-PLC activity, its overall effect is to enhance both phases of insulin secretion—either through an enhancement of the activity of specific protein kinases, or by inhibiting the activity of phosphoprotein phosphatase(s).

The complexity of the relationships between the $Ca^{2+}$ and cAMP messenger systems in the β-cell extends to inhibitory as well as stimulatory signals. Noradrenaline, acting via $G_i$, has two effects: it inhibits adenylate cyclase and activates a low conductance $K^+$ channel (Rorsman et al 1991). As a result, cAMP concentration falls and the membrane potential rises. These changes appear to be the means by which noradrenaline inhibits glucose-induced insulin secretion.

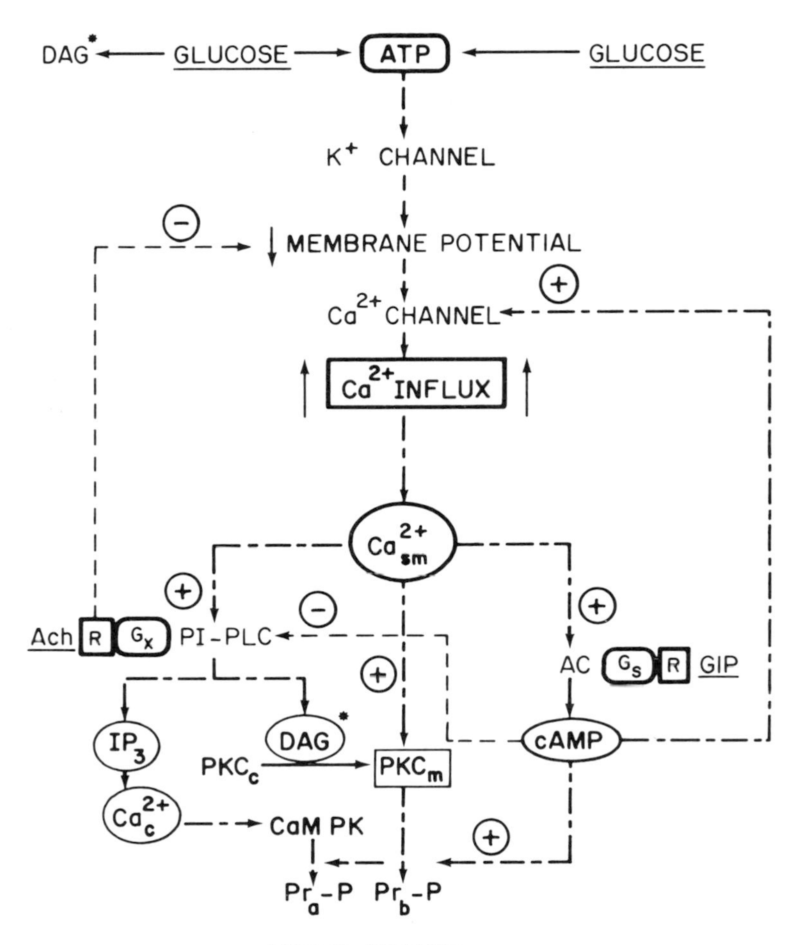


FIG. 1. The control of insulin secretion. A model of how glucose, acetylcholine (Ach) and gastric inhibitory peptide (GIP) interact to regulate insulin secretion. See text for discussion. DAG, diacylglycerol; $Ca^{2+}{}_{c}$, cytosolic free $Ca^{2+}$ concentration; $Ca^{2+}{}_{sm}$, submembrane $Ca^{2+}$ concentration; PI-PLC, phosphatidylinositol-specific phospholipase C; AC, adenylate cyclase; $IP_3$, inositol 1,4,5-trisphosphate; $PKC_c$, cytosolic protein kinase C; $PKC_m$, membrane-associated protein kinase C; CaMPK, calmodulin-dependent protein kinase; $P_ra$-P and $P_rb$-P, phosphorylated protein appearing during the first and second phases of insulin secretion, respectively; R, receptor. – – – –, negative effect; - - - - - -, positive effect; *indicates that the same DAG pool is involved.

**Time-dependent potentiation and time-dependent suppression**

Any agonist that stimulates PtdIns-PLC activity, whether or not it simultaneously stimulates insulin secretion, can induce a time-dependent potentiation of insulin secretion (Zawalich et al 1989a, Zawalich & Rasmussen 1990). For example, if isolated islets incubated in 2.75 mM glucose are perifused with 1 mM acetylcholine and challenged with 10 mM glucose 20 minutes after removal of acetylcholine, the insulin secretory response of the pretreated islets is considerably greater than that of naive ones (Zawalich & Rasmussen 1990). This enhanced responsiveness is believed to depend on a persistent association of PKC with the plasma membrane after the removal of the acetylcholine stimulus. Noteworthy is the fact that agents that stimulate adenylate cyclase do not cause this time-dependent potentiation.

What is remarkable is that any agent which induces time-dependent potentiation has a different effect if it acts on β-cells for several hours. After such treatment islets display a suppressed insulin secretory response when subsequently challenged with high glucose (Rasmussen et al 1990b, Zawalich & Rasmussen 1990). Such time-dependent suppression can also occur as a result of a prolonged exposure to agonists that activate adenylate cyclase. In all cases of time-dependent suppression there is a marked suppression of glucose-induced PtdIns hydrolysis in addition to a suppression of insulin secretion (Zawalich et al 1989b). It thus appears that these various agents bring about an inhibition of PtdIns-PLC activity through some type of negative control. The results from studies in other cell types complicate any simple interpretation of these data. The results show that activation of either PKC or protein kinase A can lead to a suppression of PtdIns-PLC activity, but each of these protein kinases seems to bring about the phosphorylation (and hence inhibition) of a different isoform of PtdIns-PLC (Kim et al 1989, Ryu et al 1990). If a similar selectively of kinase function operates in β-cells, then the results argue that both the β and γ isoforms of PtdIns-PLC are involved in glucose-induced changes in PtdIns metabolism.

**Conclusions**

The more we learn about the control of specific cellular responses, the more often we find that interactions between the $Ca^{2+}$ and cAMP messenger systems determine the responsiveness of cells to a variety of stimuli. One important lesson is that sustained elevations of second messenger concentrations often lead to desensitization or suppression of cellular responses. Thus, second messenger systems may, in many cases, operate more effectively in response to pulsatile rather than continuous exposure to appropriate extracellular signals.

*Acknowledgements*

The work from our laboratories was supported by grants (DK-19813 and DK-04123) from the National Institute of Diabetes and Digestive and Kidney Diseases. Dr Isales is supported by a Physician-Scientist Award (DK-01825).

## References

Billah MM, Anthes JC 1990 The regulation and cellular functions of phosphatidylcholine hydrolysis. Biochem J 269:281–291

Barrett PQ, Kojima I, Kojima K, Zawalich K, Isales CM, Rasmussen H 1986 Temporal patterns of protein phosphorylation after angiotensin II, A23187 and/or 12-*O*-tetradecanoylphorbol-13-acetate in adrenal glomerulosa cells. Biochem J 238:893–903

Barrett PQ, Bollag WB, Isales CM, McCarthy RT, Rasmussen H 1989 Role of calcium in angiotensin II-mediated aldosterone secretion. Endocr Rev 10:496–518

Bollag WB, Barrett PQ, Isales CM, Rasmussen H 1991 Angiotensin II-induced changes in diacylglycerol levels and their potential role in modulating the steroidogenic response. Endocrinology 128:231–241

Brami B, Vilgrain I, Chambaz EM 1987 Sensitization of adrenocortical cell adenylate cyclase activity to ACTH by angiotensin II and activators of protein kinase C. Mol Cell Endocrinol 50:131–137

Cohen CJ, McCarthy RT, Barrett PQ, Rasmussen H 1988 Ca channels in adrenal glomerulosa cells: $K^+$ and angiotensin II increase T-type Ca channel current. Proc Natl Acad Sci USA 85:2412–2416

Cook DL, Satin LS, Ashford MLJ, Hales CN 1988 ATP-sensitive $K^+$ channels in pancreatic β-cells: spare channel hypothesis. Diabetes 37:495–498

Douglas WW, Rubin RP 1961 The role of calcium in the secretory response of the adrenal medulla to acetylcholine. J Physiol (Lond) 159:40–57

Easom RA, Landt M, Colca JR, Hughes JH, Turk J, McDaniel M 1990 Effects of insulin secretagogues on protein kinase C-catalyzed phosphorylation of an endogenous substrate in isolated pancreatic islets. J Biol Chem 265:14938–14946

Exton JH 1990 Signalling through phosphatidylcholine breakdown. J Biol Chem 265:1–4

Exton JH, Taylor SJ, Blank JS, Bockino SB 1992 Regulation of phosphoinositide and phosphatidylcholine phospholipases by G proteins. In: Interactions among cell signalling systems. Wiley, Chichester (Ciba Found Symp 164) p 36–49

Ganesan S, Calle R, Zawalich K, Smallwood JI, Zawalich WS, Rasmussen H 1990 Glucose-induced translocation of protein kinase C in rat pancreatic islets. Proc Natl Acad Sci USA 87:9893–9897

Katz B, Miledi R 1969 Tetrodotoxin-resistant electric activity in presynaptic terminals. J Physiol (Lond) 203:459–487

Kim U-H, Kim JW, Rhee SG 1989 Phosphorylation of phospholipase C-γ by cAMP-dependent protein kinase. J Biol Chem 264:20167–20170

Kojima I, Kojima K, Rasmussen H 1985a Role of calcium and cAMP in the action of adrenocorticotropin on aldosterone secretion. J Biol Chem 260:4248–4256

Kojima I, Kojima K, Rasmussen H 1985b Intracellular calcium and adenosine-3′5′-cyclic monophosphate as mediators of potassium-induced aldosterone secretion. Biochem J 228:69–76

Limbird LL 1988 Receptors linked to inhibition of adenylate cyclase. Additional signalling mechanisms. FASEB (Fed Am Soc Exp Biol) J 2:2686–2695

Peter-Reisch B, Fathi M, Schlegel W, Wollheim CB 1988 Glucose and carbachol generate 1,2-diacylglycerols by different mechanisms in pancreatic islets. J Clin Invest 81:1154–1161

Pfeilschifter J 1989 Cross-talk between transmembrane signalling systems: a prerequisite for the delicate regulation of glomerulosa haemodynamics by mesangial cells. Eur J Clin Invest 19:347–361
Rasmussen H 1970 Cell communication, calcium ion, and cyclic adenosine monophosphate. Science (Wash DC) 170:404–412
Rasmussen H 1981 Calcium and cAMP as synarchic messengers. Wiley, New York
Rasmussen H, Takuwa Y, Park S 1987 Protein kinase C in the regulation of smooth muscle contraction. FASEB (Fed Am Soc Exp Biol) J 1:177–186
Rasmussen H, Kelley G, Douglas JS 1990a Interactions between $Ca^{2+}$ and cAMP messenger systems in regulation of airway smooth muscle. Am J Physiol 258 (Lung Cell Mol Physiol 2):L279–L288
Rasmussen H, Zawalich KC, Ganesan S, Calle R, Zawalich WS 1990b Physiology and pathophysiology of insulin secretion. Diabetes Care 13:655–666
Rorsman P, Bokvist K, Ammärä C et al 1991 Activation by adrenaline of a low conductance G protein-dependent $K^+$ channel in mouse pancreatic B cells. Nature (Lond) 349:77–79
Ryu SH, Kim U-H, Wahl MI et al 1990 Feedback regulation of phospholipase C-β by protein kinase C. J Biol Chem 265:17941–17945
Smith PA, Rorsman P, Ashcroft FM 1989 Modulation of dihydropyridine-sensitive $Ca^{2+}$ channels by glucose metabolism in mouse pancreatic B-cells. Nature (Lond) 342:550–553
Sutherland ER Jr 1961 The biological role of adenosine-3′5′-phosphate. Harvey Lect 57:12–33
Takuwa Y, Kelley G, Takuwa N, Rasmussen H 1988a Protein phosphorylation changes in bovine carotid artery smooth muscle during contraction and relaxation. Mol Cell Endocrinol 60:71–86
Takuwa Y, Takuwa N, Rasmussen H 1988b The effect of isoproterenol on intracellular calcium concentrations. J Biol Chem 263:762–768
Zawalich WS, Rasmussen H 1990 Control of insulin secretion: a model involving $Ca^{2+}$, cAMP and diacylglycerol. Mol Cell Endocrinol 70:119–137
Zawalich WS, Zawalich KC, Rasmussen H 1984 Insulin secretion: combined tolbutamide, forskolin and TPA mimic action of glucose. Cell Calcium 5:551–558
Zawalich WS, Zawalich KC, Rasmussen H 1989a Cholinergic agonists prime the β-cell to glucose stimulation. Endocrinology 125:2400–2406
Zawalich WS, Zawalich KC, Rasmussen H 1989b The conditions under which rat islets are labeled with [$^3$H]-inositol alter the subsequent responses of islets to a high glucose concentration. Biochem J 259:743–749
Zawalich WS, Ganesan S, Calle R, Zawalich K, Rasmussen H 1991 The effect of staurosporine on glucose-induced insulin secretion. Biochem J 279:807–813

## DISCUSSION

*Tsien:* Do you think that the target of action of the translocated protein kinase C could be a $Ca^{2+}$ channel? Is there any evidence for that?

*Rasmussen:* There are two papers which have suggested that protein kinase C can increase calcium influx into β-cells by closing the potassium channels (Yada et al 1989, Wollheim et al 1988). Both studies were in insulinoma cells and no such data are available from studies with fresh islets. Hence, it is not clear whether or not such a mechanism operates in normal β-cells. The only study which possibly links protein kinase C activation to changes in calcium influx in fresh islets is that by Thams et al (1988), in which stimulation of islets with glucose was shown to increase the accumulation of cyclic AMP. The increase in cyclic AMP was thought to be due to activation of adenylate cyclase by protein kinase C. Because cyclic AMP is known to increase calcium influx, it is possible that PKC increases calcium influx by this indirect route. However, many other studies suggest this may not be the case.

*Tsien:* There are two pieces of evidence from other systems that might not be directly relevant to your work but are interesting. The first is the work that we did with Len Kaczmarek on bag cells in *Aplysia*, in which protein kinase C recruits a novel type of $Ca^{2+}$ channel (Strong et al 1987). The second is John Connor's work in neurons which indicates that calcium can be caused to latch up (switch from the lower stable state to the higher one) after activation of protein kinase C (Connor et al 1988). Is the translocation of protein kinase sensitive to calcium?

*Rasmussen:* That's an interesting question to which I don't yet know the answer. We are trying to use tolbutamide to look at this but the results are inconclusive.

*Nishizuka:* What indication do you have of the activation of protein kinase C other than translocation?

*Rasmussen:* Dr Roberto Calle has looked at phosphorylation of the MARCKS protein, and has shown that it increases in response to 20 mM glucose.

*Fredholm:* There is evidence for oscillations of intracellular calcium in β cells (Valdeolmillos et al 1989, Li et al 1990) and it has been shown that there are true calcium waves within and between these cells (Gylfe et al 1991). Is there any evidence that activation of protein kinase C modifies the frequency of calcium spiking even though total $InsP_3$ formation or the net amount of calcium may not be modified?

*Rasmussen:* To my knowledge there is no good evidence for that.

*Fredholm:* Alkalinization enhances insulin release in these cells (Lindström & Sehlin 1986, Pace 1984). Could protein kinase C be involved in that?

*Rasmussen:* I think the effects of alkalinization are probably artifactual. I don't think that protein kinase C works by alkalinizing the cells. When you alkalinize cells there is a dramatic increase in calcium release from some pool, though I am not sure it is the $InsP_3$ pool, so it doesn't surprise me that alkalinization causes a burst of insulin secretion.

*Michell:* In your model of insulin secretion (Fig. 1) diglyceride was persistently present and protein kinase C was persistently translocated. The obvious inference from that is that perhaps they are locked together, to protect the diacylglycerol and hold the PKC at the membrane. Have you done any experiments to try to abstract the enzyme, from highly labelled cells for example, to see whether it is actually holding on to a diacylglycerol?

*Rasmussen:* We can't do that kind of experiment in pancreatic islets because there is insufficient tissue. Cultured adrenal cells show the same kind of memory phenomenon. What surprised us in these cells was that diacylglycerol concentration remained raised for a considerable period of time, up to 40 minutes after we had removed the hormone (Bollag et al 1991). If at any point during that time we challenged the cells with a calcium channel agonist, BK-8644, they showed an enhanced aldosterone secretory response. The decay of responsiveness was more or less parallel to the decay in diacylglycerol content. Unfortunately, for technical reasons, we haven't yet been able to measure protein kinase C translocation under those circumstances. There are at least two isoforms of the enzyme in cultured adrenal cells, which complicates such measurements.

*Nishizuka:* Is it possible to distinguish between the signal-induced diacylglycerol formation and that resulting from lipolysis in your system?

*Rasmussen:* As far as I know there's no evidence for lipolysis in the islet cell. What does complicate matters is the *de novo* synthesis of diacylglycerol from glucose (Peter-Reisch et al 1988). When you give high glucose, there is an increase in diacylglycerol synthesis, but we don't know where the diacylglycerol is made, or how rapidly it gets incorporated into plasma membrane lipids.

*Krebs:* Why is the metabolism of glucose essential for insulin secretion? Does it serve simply as an oxidizable substrate? Can some other energy source be used?

*Rasmussen:* Glucose metabolism is absolutely crucial. At low extracellular glucose concentrations the cell is essentially non-responsive to any of these signals. Very high levels of ketoisocaproate enhance insulin secretion, but are much more effective when there is a 2.75 mM glucose concentration than when glucose is absent. α-Ketoisocaproate can maintain the ATP concentration but it cannot substitute for glucose.

*Parker:* Does ketoisocaproate increase the *de novo* synthesis of diacylglycerol?

*Rasmussen:* No.

*Parker:* So is that the distinction between glucose and ketoisocaproate, as they presumably can both maintain the ATP concentration?

*Rasmussen:* The striking thing is that high concentrations of α-ketoisocaproate will induce a biphasic increase in secretion of insulin, without *de novo* synthesis of diacylglycerol—that's the paradox. I don't know how that happens. All I know is that 15 mM glucose and 15 mM ketoisocaproate cause approximately the same increase in insulin secretion. At a glucose concentration of 7 mM a combination of gastric inhibitory peptide (GIP) and acetylcholine causes a biphasic increase in insulin secretion, but with 7 mM ketoisocaproate these two agents don't have this effect.

*Fischer:* Is there any evidence that protein kinase C binds to an anchoring molecule in the membrane?

*Rasmussen:* We have no evidence of that. We are now doing some immunocytochemistry at the electron microscope level to get further information.

*Tsien:* Staurosporine abolished the second phase of glucose-induced insulin secretion, but not the first. What happens if you apply staurosporine during the first phase or after the second phase?

*Rasmussen:* When it is applied after the second phase the insulin secretion rate falls progressively.

*Thomas:* Does staurosporine totally inhibit the second phase of glucose-induced secretion?

*Rasmussen:* No; there is a component of the second phase, about 15%, that is not inhibited by staurosporine.

*Thomas:* Is phosphorylation of the MARCKS protein completely blocked by staurosporine treatment?

*Rasmussen:* According to the data of Easom et al (1990) it is almost completely blocked by staurosporine.

*Thomas:* Are both phases of secretion dependent on protein synthesis?

*Rasmussen:* I don't think either is. Insulin is very abundant in β-cells, so immediate synthesis is not necessary for sustained secretion.

*Thomas:* Presumably insulin secretion is a regulated process, so could this step itself be protein synthesis dependent?

*Rasmussen:* The cells are full of insulin secretory granules and protein synthesis is not needed for exocytosis.

*Fredholm:* One possible explanation for the two phases of secretion is that the first is from vesicles very close to the membrane and the second involves mobilization of granules deeper in the cell.

*Rasmussen:* The argument against that hypothesis is that by manipulating the system we can get an enormous first phase, so there doesn't seem to be a pool of restricted size.

*Collingridge:* Do you have any evidence about which subtypes of muscarinic acetylcholine receptors are involved?

*Rasmussen:* No. Very little has been done on that question.

*Hunter:* Does phosphorylation of MARCKS continue after removal of the stimulus?

*Rasmussen:* We haven't looked at that.

*Hunter:* So you don't know whether protein kinase C remains in an active state or not.

*Rasmussen:* There's a real paradox. We are arguing that protein kinase is both the signal for the second phase of secretion and an inhibitory signal which leads to time-dependent suppression of secretion under conditions where there is virtually no increase in insulin secretion. When we treat the cells in low glucose with cholecystokinin or acetylcholine we are surely turning on PtdIns metabolism and protein kinase C translocates to the membrane—but no insulin secretion is seen; yet, the activation of PtdIns turnover is doing something. If the mechanism of suppression really involves protein kinase C activity, some key protein or proteins must be phosphorylated. What we are really saying is that if that's correct, there must be different pools of proteins, possibly a pool in the membrane that is phosphorylated in a calcium-independent manner, and another set involved in the secretory process that is phosphorylated in a calcium-dependent fashion.

## References

Bollag WB, Barrett PQ, Isales CM, Rasmussen H 1991 Angiotensin II-induced changes in diacylglycerol levels and their potential role in modulating the steroidogenic response. Endocrinology 128:231–241

Connor JA, Wadman WJ, Hockberger PE, Wong RK 1988 Sustained dendritic gradients of $Ca^{2+}$ induced by excitatory amino acids in CA1 hippocampal neurons. Science (Wash DC) 240:649–653

Easom RA, Landt M, Colca JR, Hughes JH, Turk J, McDaniel M 1990 Effects of insulin secretagogues on protein kinase C-catalyzed phosphorylation of an endogenous substrate in isolated pancreatic islets. J Biol Chem 265:14938–14946

Gylfe E, Grapengiesser E, Hellman B 1991 Propagation of cytoplasmic $Ca^{2+}$ oscillations in clusters of pancreatic β-cells exposed to glucose. Cell Calcium 12:91–102

Li G, Regazzi R, Ullrich S, Pralong W-F, Wollheim CB 1990 Potentiation of stimulus-induced insulin secretion in protein kinase C-deficient RINm5F cells. Biochem J 272:637–645

Lindström P, Sehlin J 1986 Effect of intracellular alkalinization on pancreatic islet calcium uptake and insulin secretion. Biochem J 239:199–204

Pace CS 1984 Role of pH as a transduction device in triggering electrical and secretory responses in islet B cells. Fed Proc 43:2379– 2384

Peter-Reisch B, Fathi M, Schlegel W, Wollheim CB 1988 Glucose and carbachol generate 1,2-diacylglycerols by different mechanisms in pancreatic islets. J Clin Invest 81:1154–1161

Strong JA, Fox AP, Tsien RW, Kaczmarek LK 1987 Stimulation of protein kinase C recruits covert calcium channels in Aplysia bag cell neurons. Nature (Lond) 325:714–717

Thams P, Capito K, Hedeskov CJ 1988 Stimulation by glucose of cyclic AMP accumulation in mouse pancreatic islets is mediated by protein kinase C. Biochem J 253:229–234

Valdeolmillos M, Santos RM, Contreras D, Soria B, Rosario LM 1989 Glucose-induced oscillations of intracellular $Ca^{2+}$ concentration resembling bursting electrical activity in single mouse islets of Langerhans. FEBS (Fed Eur Biochem Soc) Lett 259:19–23

Wollheim CB, Dunne MJ, Peter-Riesch B, Bruzzone R, Pozzan T, Petersen OH 1988 Activators of protein kinase C depolarize insulin-secreting cells by closing $K^+$ channels. EMBO (Eur Mol Biol Organ) J 7:2443–2449

Yada T, Russo LL, Sharp GW 1989 Phorbol ester-stimulated insulin secretion by RINm5F insulinoma cells is linked with membrane depolarization and an increase in cytosolic free $Ca^{2+}$ concentration. J Biol Chem 264:2455–2462

# Growth factor-stimulated phosphorylation cascades: activation of growth factor-stimulated MAP kinase

Natalie G. Ahn*, Rony Seger†‡, Rebecca L. Bratlien† and Edwin G. Krebs*†‡

*Departments of Biochemistry* and Pharmacology† and Howard Hughes Medical Institute‡, University of Washington, Seattle, WA 98185, USA*

*Abstract.* Protein phosphorylation is an important mechanism in the response of cells to growth factors by which signals can be conveyed from cell surface receptors to intracellular targets. In addition to stimulation of protein tyrosine phosphorylation, activation of growth factor receptors having protein tyrosine kinase activity leads to dramatic alterations in the levels of protein serine/threonine phosphorylation. Several growth factor-stimulated serine/threonine-specific kinases have been identified as potential mediators of such signalling. MAP (microtubule-associated protein) kinase has emerged as a very interesting member of this group, because it activates a separate kinase, $pp90^{rsk}$, which is also growth factor-stimulated. MAP kinase itself appears to be regulated by protein phosphorylation, because it can be inactivated by protein phosphatases. We have identified two 60 kDa proteins that promote the phosphorylation and full activation of MAP kinase in a manner paralleling its activation by growth factors in intact cells. These 'MAP kinase activators' are themselves stimulated by growth factors, suggesting that they function as intermediates between the MAP kinase and cell surface receptors in a growth factor-stimulated kinase cascade. Identification of the components of this protein kinase cascade reveals a mechanism by which at least some of the effects of receptor tyrosine kinases can be mediated through serine/threonine phosphorylation.

*1992 Interactions among cell signalling systems. Wiley, Chichester (Ciba Foundation Symposium 164) p 113–131*

One of the key events in the response of cells to external signals, such as growth factors and mitogenic stimuli, is a change in the state of phosphorylation of a large number of intracellular protein targets. Protein phosphorylation represents a major mechanism by which the molecular properties of these targets are altered in response to external signals, ultimately leading to changes in cellular processes governing metabolism, proliferation and differentiation.

Many cell surface receptors for growth and differentiation factors have intrinsic tyrosine-specific protein kinase activity that increases after ligand binding, suggesting that tyrosine phosphorylation is the first step in many signalling pathways. However, the majority of proteins that are targets in these pathways are regulated not by tyrosine phosphorylation, but by serine or threonine phosphorylation. Thus, the mechanisms by which protein kinases communicate with protein serine/threonine kinases is of primary importance in understanding how cells respond to external stimuli.

An increasing number of serine/threonine kinases that are stimulated in the response of cells to growth factors and other mitogenic stimuli have been reported. Some of these enzymes are regulated by the levels of second messengers, such as diacylglycerol, calcium or cyclic nucleotides. Others seem to be independent of second messengers. Many of these 'messenger-independent' protein kinases appear to be regulated by phosphorylation; in some instances, inactivation by protein phosphatases has provided definite evidence for this. Table 1 shows a partial list of messenger-independent protein serine/threonine kinases, activated by growth factors, for which there is evidence of regulation by phosphorylation.

An attractive hypothesis is that these phosphatase-sensitive enzymes are activated by other protein kinases, which may themselves be growth factor-stimulated enzymes. One might envision a cascade of phosphorylation events

**TABLE 1 Second messenger-independent growth factor-stimulated protein kinases that may be regulated by protein phosphorylation**

| *Kinase* | *Reference* |
|---|---|
| Ribosomal protein S6 kinase (70 kDa) | Jenö et al 1988<br>Ballou et al 1988a,b<br>Price et al 1989, 1990 |
| S6 kinase II (pp90$^{rsk}$) | Maller 1987<br>Erikson & Maller 1989<br>Gregory et al 1989<br>Sweet et al 1990b<br>Chen & Blenis 1990<br>Ahn et al 1990 |
| MAP kinase (ERK1 and ERK2) | Ray & Sturgill 1988b<br>Sturgill et al 1988<br>Hoshi et al 1989<br>Anderson et al 1990<br>Ahn et al 1990<br>Gomez et al 1990 |
| pp54 MAP2 kinase | Kyriakis & Avruch 1990 |
| Casein kinase II | Ackerman et al 1990 |
| Raf-1 | Morrison et al 1989 |

following ligand binding, beginning with the activation of a tyrosine kinase and continuing with the sequential activation of a number of protein kinases, by either direct or indirect means. Ultimately, such a cascade would lead to the phosphorylation of target proteins on serine or threonine residues, and might account for the observed amplification in serine/threonine phosphorylation that occurs in the response of cells to growth factors.

Evidence for a growth factor-linked kinase cascade was first provided by studies of the enzyme MAP kinase, which was first reported as an activity derived from soluble extracts of 3T3-L1 adipocytes treated with insulin that phosphorylated microtubule-associated protein-2 (Ray & Sturgill 1987). After partial purification, it was found that the insulin-stimulated MAP kinase could phosphorylate and activate another insulin-stimulated kinase, the *Xenopus laevis* S6 kinase II, which had been previously inactivated by protein phosphatase 2A (Sturgill et al 1988). This was the first example of two serine/threonine-specific kinases, both growth factor-stimulated, being reconstituted to form a kinase cascade.

Other laboratories reported a growth factor-stimulated activity that phosphorylated myelin basic protein (MBP) (Cicirelli et al 1988, Pelech et al 1988, Ahn et al 1990); this activity was found to be identical to MAP kinase (Ahn et al 1990, Anderson et al 1990, Boulton et al 1991a, Boulton & Cobb 1991). MAP kinase activity has been shown to be activated in response to a wide range of stimuli, including insulin, epidermal growth factor, platelet-derived growth factor, nerve growth factor, serum, phorbol esters, nicotine, okadaic acid and hormones that induce oocyte maturation (Ray & Sturgill 1987, 1988a, Hoshi et al 1988, 1989, Sanghera et al 1990, Ely et al 1990, Miyasaka et al 1990, Ahn et al 1990, Haystead et al 1990, Gomez et al 1990, Boulton et al 1991b) and in T cell activation (Hanekom et al 1989). In many systems, two forms of MAP kinase can be resolved; these differ slightly in molecular mass but otherwise show similar properties (Ahn et al 1990, Haystead et al 1990, Gomez et al 1990, Boulton & Cobb 1991). Cloning of MAP kinase led to the discovery of several homologous forms of the enzyme, named ERK, for 'extracellular signal regulated kinase' (Boulton et al 1990). Two of these enzymes (ERK1 and ERK2) are 90% identical in sequence (Boulton et al 1991a). Specific antipeptide antibodies have been used to demonstrate clearly that ERK1 and ERK2 are identical to the two forms of MAP kinase found in extracts from growth factor-stimulated cells (Boulton & Cobb 1991).

The activation of S6 kinases by MAP kinases has also been demonstrated using rabbit liver S6 kinase stimulated by MAP kinase derived from rat fibroblasts (Gregory et al 1989), and using both components derived from a single source, Swiss 3T3 cells (Ahn & Krebs 1990). Thus, the pathway seems to be relevant in mammalian systems. Importantly, the MAP kinase-activated S6 kinases in each case were clearly distinguishable from another growth factor-stimulated S6 kinase, an enzyme that migrates as a 70 kDa polypeptide on

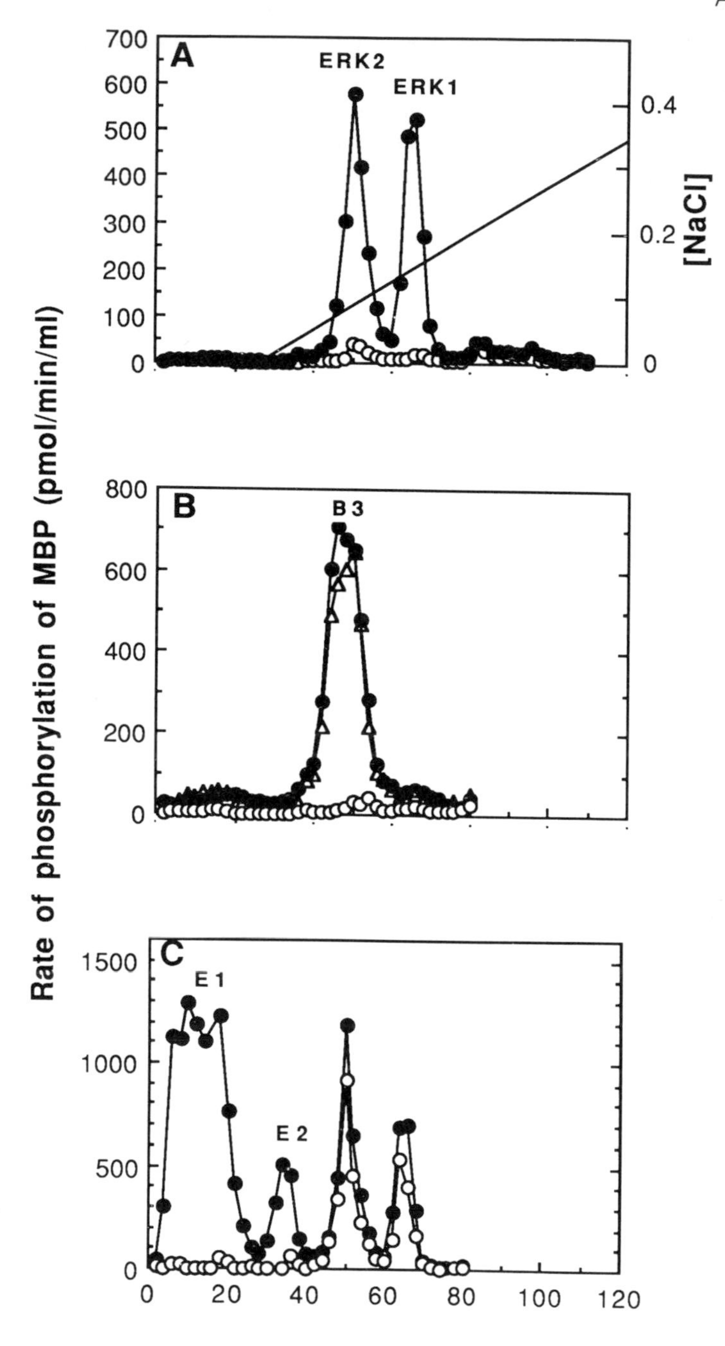
A
ERK2
ERK1
700
600
500
400
300
200
100
0
0.4
0.2
0
[NaCl]
Rate of phosphorylation of MBP (pmol/min/ml)
B
B3
800
600
400
200
0
C
E1
E2
1500
1000
500
0
0
20
40
60
80
100
120
Fraction No.

SDS–PAGE (Jenö et al 1989, Price et al 1989). Molecular cloning of S6 kinases from mammalian and avian sources has revealed at least two distinct classes of enzymes, one homologous to *Xenopus laevis* S6 kinase II, referred to as $pp90^{rsk}$ (Jones et al 1988, Alcorta et al 1989), and a separate form corresponding to the 70 kDa enzyme (Banerjee et al 1990, Kozma et al 1990). Growth factor stimulation and decay of $pp90^{rsk}$ activity are more rapid than they are for the 70 kDa S6 kinase (Sweet et al 1990a, Ahn et al 1990). Furthermore, activation of the 70 kDa S6 kinase by MAP kinase could not be demonstrated (Ahn & Krebs 1990, Price et al 1990, Ballou et al 1991). Thus, the 90 kDa and 70 kDa forms of S6 kinase appear to be regulated by divergent pathways after growth factor binding.

That MAP kinase itself might be a target in a kinase cascade and regulated by protein phosphorylation was suggested initially by its phosphorylation on threonine and tyrosine residues in intact cells in response to growth factors (Ray & Sturgill 1988b) and its inactivation by protein phosphatases (Sturgill et al 1988, Hoshi et al 1989, Hanekom et al 1989, Ahn et al 1990). Subsequently, Anderson et al (1990) found that MAP kinase undergoes dephosphorylation on phosphothreonine when treated with the serine/threonine-specific protein phosphatase 2A, whereas treatment with the tyrosine-specific phosphatase CD45 led to specific dephosphorylation of phosphotyrosine residues. In both cases, complete inactivation of MAP kinase activity was observed, leading to the conclusion that both threonine and tyrosine residues must be phosphorylated for the enzyme to be active. Similar results were obtained with MAP kinase from PC12 cells stimulated by nerve growth factor (Gomez et al 1990).

We began our studies on the mechanism of activation of the MAP kinase by reasoning that the inactive form of MAP kinase, derived from untreated-cell

---

FIG. 1. Identification of synergism in MAP kinase activity. (A) Cytosolic extracts of cells treated for 5 min with 100 ng/ml EGF (●) or bovine serum albumin carrier (○) were fractionated by Mono-Q FPLC using an increasing gradient of NaCl. Two peaks of MAP kinase activity, measured using MBP as substrate, were found to be stimulated by EGF. (B) Alternate fractions from untreated cells were incubated with pooled fractions from EGF-treated cells (capable of showing synergy of MBP phosphorylation with untreated extracts; see text) for 15 min in the presence of MgATP before addition of MBP: (Δ) MAP kinase activity in fractions from untreated cells incubated with fractions 6–18 (termed peak E1) from EGF-treated cells; (●) MAP kinase activity in fractions from untreated cells incubated with fractions 32–37 (peak E2) from treated cells; (○) MAP kinase activity in fractions from untreated cells incubated with buffer in place of treated-cell extracts. After incubation with either E1 or E2, untreated cell fractions 42–54 (termed pool B3) showed a high rate of MBP phosphorylation. (C) (●) Alternate fractions from EGF-treated cells were incubated with pool B3 (fractions 42–45 from untreated cells) in the presence of MgATP for 15 min before addition of MBP. Synergism in MBP phosphorylation is seen in two peaks, E1 and E2; the remaining peaks are also seen when buffer was used in place of B3 (○), as expected, because these peaks correspond to active ERK2 and ERK1, respectively, from EGF-treated cells, as in (A).

extracts, might be stimulated when incubated in the presence of ATP with a hypothetical activating factor, derived from growth factor-treated cell extracts. We fractionated extracts of Swiss 3T3 cells treated for five minutes with epidermal growth factor (EGF) by anion exchange (Mono-Q) FPLC and compared the fractionated extracts with those from untreated cells. This revealed two peaks (fractions 48–54 and 62–68) of kinase activity towards the substrate myelin basic protein (MBP) that were stimulated by treatment of the cells with EGF (Fig. 1A). Antibodies raised to synthetic peptides based on sequences from ERK1, which recognize both ERK1 and ERK2, show that the first and second peaks eluting from the Mono-Q column correspond to the two forms of MAP kinase, polypeptides of 42 kDa (ERK2) and 44 kDa (ERK1), respectively (Boulton et al 1991a, Boulton & Cobb 1991, N. Ahn, T. Boulton, R. Seger, M. Cobb & E. Krebs, unpublished work).

To identify a hypothetical activation factor we used a coupled assay, looking for fractions containing this factor which, when mixed with the inactive form of MAP kinase, would yield a rate of MBP phosphorylation that was greater than that seen in each pool measured separately. Fractions 6–18 (termed pool E1) or fractions 32–37 (pool E2) from the Mono-Q profile of EGF-treated cells were preincubated with individual fractions from non-treated cells in the presence of MgATP for 15 minutes before MBP was added and the rate of MBP phosphorylation measured. Synergy in MBP phosphorylation was observed with fractions 42–54 (pool B3), compared with controls in which E1 or E2 were omitted (Fig. 1B). When pool B3 fractions (42–54) from untreated cells were preincubated with individual fractions from the EGF-treated cells, two peaks showing synergy in MBP phosphorylation were observed, defining the elution positions of E1 and E2 (Fig. 1C). Thus activation of MAP kinase activity was found in two combinations of pooled fractions—B3 and E1, and B3 and E2—mixed and incubated in the presence of MgATP.

The components within the E1 and E2 pools that act in synergy with the components of B3 in each case had apparent molecular masses of 60 kDa by sizing gel filtration, whereas the components within B3 eluted as a heterogeneous peak with molecular masses in the range of 30–50 kDa (Ahn et al 1991). The mobility of B3 was identical to that of activated MAP kinase, indicating that B3 contains the inactive forms of MAP kinase; thus, E1 and E2 contain the MAP kinase-activating factors. Importantly, the MAP kinase activators could only be found in extracts of the EGF-treated cells, suggesting that they are growth factor-stimulated as well.

When the E1 and B3 pools were preincubated with MgATP before fractionation by anion exchange chromatography, MAP kinase activity resolved into two peaks (Fig. 2A), which eluted at salt concentrations similar to those at which ERK2 and ERK1 stimulated in intact cells with EGF eluted. Fractionation by gel sizing chromatography indicated molecular masses of the first and second peaks that were consistent with those seen with ERK2 and

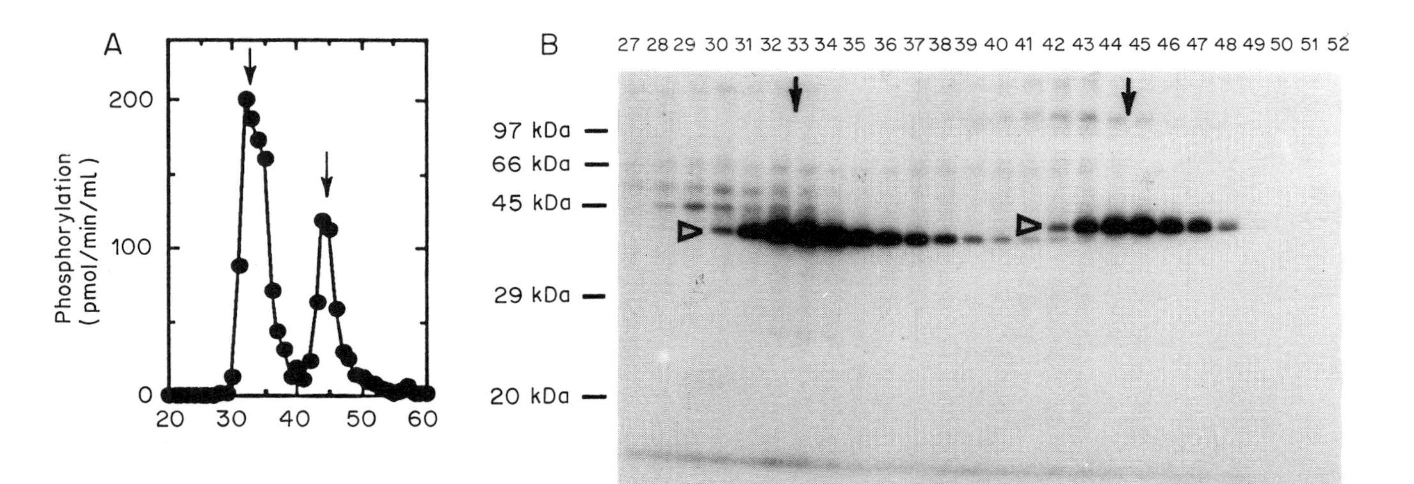


FIG. 2. Fractionation of MAP kinase activity by anion exchange chromatography after *in vitro* activation. (A) Pools B3 and E1 were incubated in the presence of 10 mM $MgCl_2$ and 100 µM [$\gamma$-$^{32}$P]ATP for 15 min before fractionation by Mono-Q FPLC. Two peaks of MAP kinase activity were resolved, shown using MBP as substrate (●). (B) Fractions were analysed by SDS–PAGE autoradiography. Phosphorylated polypeptides of 42 and 44 kDa (marked with triangles) correlated respectively with the first and second peaks of MAP kinase activity (marked with arrows). Modified from Ahn et al 1991.

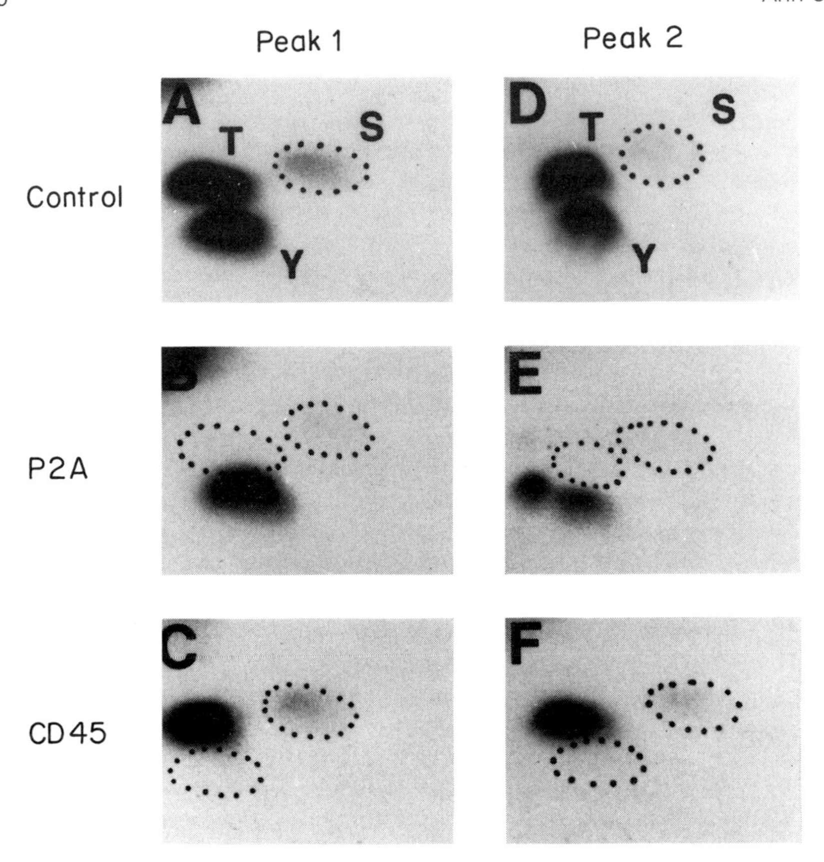


FIG. 3. Phosphoamino acid analysis of *in vitro*-activated MAP kinases. (A) 42 kDa and (D) 44 kDa polypeptides corresponding to the first and second peaks of MAP kinase activity, respectively, (as in Fig. 2) were phosphorylated primarily on threonine and tyrosine residues, with trace phosphorylation on serine. Preferential loss of phosphothreonine was seen after treatment with (B,E) phosphatase 2A (100 U/ml), whereas preferential loss of phosphotyrosine was seen after treatment with (C,F) CD45 (250 U/ml). In each case, treatment with phosphatase led to complete inactivation of MAP kinase activity. Modified from Ahn et al 1991.

ERK1, respectively. The *in vitro* stimulated kinases were capable of activating S6 kinase activity, indicating that they are also functionally similar to those stimulated in intact cells (Ahn et al 1991).

When [$\gamma$-$^{32}$P]ATP was included in the *in vitro* activation, phosphorylated polypeptides of 42 and 44 kDa were found to coelute with the first and second peaks of MAP kinase activity (Fig. 2); these phosphorylated peptides were recognized by Western blotting using antibodies specific to MAP kinase (N. Ahn, T. Boulton, R. Seger, M. Cobb & E. Krebs, unpublished results). Phosphoamino acid analysis showed that the 42 and 44 kDa bands both

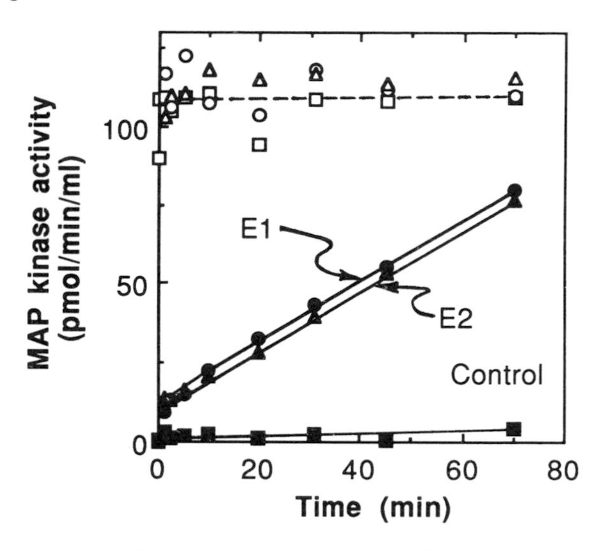


FIG. 4. Reactivation of inactivated MAP kinase by E1 and E2. The first MAP kinase peak (42 kDa) was inactivated by preincubation with a combination of phosphatase 2A (100 U/ml) and CD45 (250 U/ml) (closed symbols); controls in which buffer was added in place of phosphatase (open symbols) were not inactivated. After 20 min the phosphatases were inactivated with okadaic acid (1 μM) and sodium vanadate (2 mM). Reactivation of MAP kinase with time was then observed after the MAP kinase activators in E1 (●), or in E2 ( ▲ ) were added. No reactivation was seen in controls in which buffer ( ■ ) was added in place of the activators. No effect of E1 (○), E2 ( △ ) or buffer ( □ ) was seen in controls that were not phosphatase-inactivated. Similar results were obtained using the second MAP kinase peak (44 kDa) (data not shown).

contained phosphothreonine and phosphotyrosine (Fig. 3A,D). Complete inactivation of MAP kinase activity resulted from treatment of the 42 or 44 kDa enzymes with either phosphatase 2A or the protein tyrosine phosphatase CD45 (Ahn et al 1991). Under these conditions, treatment with phosphatase 2A led to preferential loss of phosphothreonine (Fig. 3B,E), whereas treatment with CD45 led to selective loss of phosphotyrosine (Fig. 3C,F). Thus, the MAP kinase activator promotes the incorporation of phosphate on both tyrosine and threonine residues and both phosphotyrosine and phosphothreonine are necessary for MAP kinase activity. These results *in vitro* parallel those seen with the MAP kinases that are stimulated by growth factors in intact cells (Anderson et al 1990), suggesting that the MAP kinase activators catalyse physiologically important reactions.

After inactivation by either phosphatase 2A or CD45, complete reactivation of MAP kinase could be obtained by addition of either E1 or E2 (Ahn et al 1991). Reactivation of MAP kinase could also be obtained after inactivating it with a combination of phosphatase 2A and CD45 (Fig. 4), which, when used together, remove phosphate from both phosphothreonine and phosphotyrosine.

Thus, the MAP kinase activators within E1 and E2 were each able to reverse the phosphatase-catalysed inactivation of MAP kinase, indicating that phosphorylation plays an important role in the mechanism of activation *in vitro*. Both ATP and magnesium ions are required as well, consistent with this model. Interestingly, reactivation could be achieved with either E1 or E2, regardless of which pool was initially used to activate MAP kinase, indicating that these two pools of MAP kinase activator behave interchangeably and most probably activate MAP kinase through the same mechanism.

Recent work in our laboratory has led to the purification of a MAP kinase-activating factor from A431 epidermal carcinoma cells that is stimulated by EGF and has properties similar to those of the activators originally found in Swiss 3T3 cells. The purified MAP kinase activator is capable of activating MAP kinase (ERK1) purified from growth factor-stimulated cells (Boulton et al 1991b) and inactivated by protein phosphatase treatment, and MAP kinase (ERK2) expressed recombinantly in *E. coli* (Boulton et al 1991a, R. Seger, N. G. Ahn, M. H. Cobb & E. G. Krebs, unpublished work 1991). Further characterization of the MAP kinase activator is underway.

At this stage, we can envision several simple models that may account for the mechanism of *in vitro* activation of the MAP kinase activator. First, the activator may itself be a kinase that phosphorylates MAP kinase on tyrosine residues, with this being followed by autophosphorylation of MAP kinase on threonine residues, and activation of the enzyme. Alternatively, the MAP kinase activator may be composed of two co-purifying kinases, one capable of phosphorylating tyrosine residues and the other having specificity towards threonine residues on MAP kinase. As a third possibility, the MAP kinase activator may phosphorylate threonine residues on MAP kinase, which then undergoes autophosphorylation on tyrosine residues. Or, the MAP kinase activator may be a kinase with the ability to phosphorylate both tyrosine and threonine residues on MAP kinase. Finally, the activator might not be a kinase at all, but may interact with MAP kinase in such a way as to induce autophosphorylation of MAP kinase on *both* tyrosine and threonine residues. We have not been able to identify an alternative substrate for the MAP kinase activator, thus, because of the possibility that phosphate incorporation into MAP kinase might be due to autophosphorylation, no direct evidence exists that the activator has kinase activity. We have recently found that MAP kinase is capable of phosphorylating itself both on tyrosine and on threonine residues (Seger et al 1991). The autophosphorylation reaction is very slow and the highest phosphate incorporation we have seen thus far has not exceeded 2% mol/mol stoichiometry, but this is nonetheless correlated with a proportionate increase in MAP kinase activity, suggesting that autophosphorylation may be relevant to the mechanism of activation. However, because the rate and extent of phosphorylation and activation of MAP kinase are much greater in the presence of the MAP kinase activator, it remains to

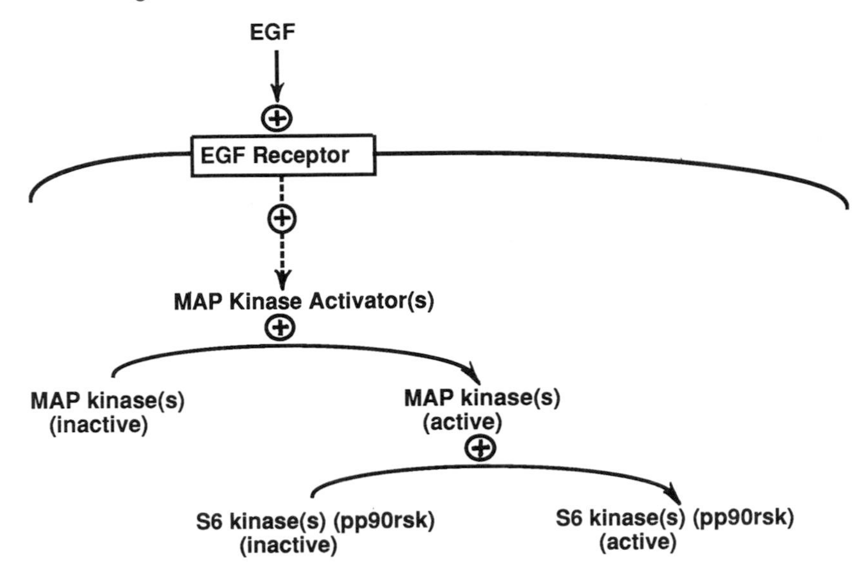


FIG. 5. Model for activation of MAP kinase and S6 kinase in response to EGF. Shown is a hypothetical scheme of an EGF-stimulated kinase cascade, involving activation of MAP kinases (ERK1 and ERK2) by the MAP kinase activators in peaks E1 and E2, and the subsequent activation of S6 kinases by the MAP kinases. Modified from Ahn et al 1991.

be seen whether autophosphorylation alone accounts for the catalytic component of the activation process.

The last three of the possible mechanisms described above challenge the dogma that protein kinases are specific for either tyrosine or serine/threonine residues, but not both. Recently, three novel kinases have been reported that can each be classified by sequence comparison as serine/threonine protein kinases, but which have the capability to either autophosphorylate or to phosphorylate external substrates on tyrosine residues (Dailey et al 1990, Howell et al 1991, Stern et al 1991). Interestingly, we have found striking sequence similarities between MAP kinase and each of these enzymes (Seger et al 1991), occurring within subdomain XI of the conserved protein kinase catalytic domain, as defined by Hanks et al (1988). This subdomain contains two residues (out of 30) that are conserved among all kinases. There is >50% sequence similarity in this domain between rat MAP kinase (Boulton et al 1990, 1991a), the yeast enzymes YPK1 (Dailey et al 1990) and SPK1 (Stern et al 1991), and the mammalian STY (Howell et al 1991). This similarity may be related to the common ability of these enzymes to recognize tyrosine as well as serine/threonine residues as substrates.

In summary, we have identified two components of EGF-stimulated cells that can effect the phosphorylation and activation of two forms of the MAP kinase in a manner that is consistent with growth factor stimulation of MAP kinase in intact cells. We envision that these activating factors are intermediates in a growth factor-stimulated kinase cascade, functioning in the pathway between growth factor receptor tyrosine kinases and the MAP kinases (Fig. 5). The MAP kinase activators appear to be unique as catalytic factors—if they are kinases, they show a narrow substrate specificity that includes only MAP kinase, as far as we have been able to determine; if they are not kinases, they facilitate an unusual process allowing autophosphorylation of a serine/threonine kinase on tyrosine as well as serine/threonine residues. Understanding the mechanistic details of the activation of MAP kinase by the MAP kinase activators will offer a novel perspective on the regulation of phosphorylation during cell signalling.

## References

Ackerman P, Glover CVC, Osheroff N 1990 Stimulation of casein kinase II by epidermal growth factor: relationship between the physiological activity of the kinase and the phosphorylation state of its β subunit. Proc Natl Acad Sci USA 87:821–825

Ahn NG, Krebs EG 1990 Evidence for an epidermal growth factor-stimulated protein kinase cascade in Swiss 3T3 cells. Activation of serine peptide kinase activity by myelin basic protein kinases *in vitro*. J Biol Chem 265:11495–11501

Ahn NG, Weiel JE, Chan CP, Krebs EG 1990 Identification of multiple epidermal growth factor-stimulated protein serine/threonine kinases from Swiss 3T3 cells. J Biol Chem 265:11487–11494

Ahn NG, Seger R, Bratlien RL, Diltz CD, Tonks NK, Krebs EG 1991 Multiple components in an epidermal growth factor-stimulated protein kinase cascade. J Biol Chem 266:4220–4227

Alcorta DA, Crews CM, Sweet LJ, Bankston L, Jones SW, Erikson RL 1989 Sequence and expression of chicken and mouse *rsk*; homologs of *Xenopus laevis* ribosomal S6 kinase. Mol Cell Biol 9:3850–3859

Anderson NG, Maller JL, Tonks NK, Sturgill TW 1990 Requirement for integration of signals from two distinct phosphorylation pathways for activation of MAP kinase. Nature (Lond) 343:651–653

Ballou LM, Jenö P, Thomas G 1988a Protein phosphatase 2A inactivates the mitogen-stimulated S6 kinase from Swiss mouse 3T3 cells. J Biol Chem 263:1188–1194

Ballou LM, Siegmann M, Thomas G 1988b S6 kinase in quiescent Swiss mouse 3T3 cells is activated by phosphorylation in response to serum treatment. Proc Natl Acad Sci USA 85:7154–7158

Ballou LM, Luther H, Thomas G 1991 MAP2 kinase and 70K S6 kinase lie on distinct signalling pathways. Nature (Lond) 349:348–350

Banerjee P, Ahmad MF, Grove JR, Kozlosky C, Price DJ, Avruch J 1990 Molecular structure of a major insulin/mitogen-activated 70-kDa S6 protein kinase. Proc Natl Acad Sci USA 87:8550–8554

Boulton TG, Cobb MH 1991 Identification of multiple extracellular signal-regulated kinases (ERKs) with antipeptide antibodies. Cell Regul 2:357–371

Boulton TG, Yanacopoulos GD, Gregory JS et al 1990 An insulin-stimulated protein kinase homologous to yeast kinases involved in pheromone-regulated cell cycle control. Science (Wash DC) 249:64–67

Boulton TG, Nye SH, Robbins DJ et al 1991a ERKs: a family of protein-serine/threonine kinases that are activated and tyrosine phosphorylated in response to insulin and NGF. Cell 65:663–666

Boulton TG, Gregory JS, Cobb MH 1991b Purification and properties of extracellular signal-regulated kinase 1, an insulin-stimulated microtubule-associated protein 2 kinase. Biochemistry 30:278–286

Chen R-H, Blenis J 1990 Identification of *Xenopus* S6 protein kinase homologs ($pp90^{rsk}$) in somatic cells: phosphorylation and activation during initiation of cell proliferation. Mol Cell Biol 10:3204–3215

Cicirelli MF, Pelech SL, Krebs EG 1988 Activation of multiple protein kinases during the burst of protein phosphorylation that precedes the final meiotic cell division in *Xenopus* oocytes. J Biol Chem 263:2009–2019

Dailey D, Schieven GL, Lim ML et al 1990 Novel yeast protein kinase (YPK1 gene product) is a 40-kilodalton phosphotyrosyl protein associated with protein-tyrosine kinase activity. Mol Cel Biol 10:6244–6256

Ely CM, Oddie KM, Litz JS et al 1990 A 42-kD tyrosine kinase substrate linked to chromaffin cell secretion exhibits an associated MAP kinase activity and is highly related to a 42-kD mitogen-stimulated protein in fibroblasts. J Cell Biol 110:731–742

Erikson E, Maller JL 1989 *In vivo* phosphorylation and activation of ribosomal protein S6 kinases during *Xenopus* oocyte maturation. J Biol Chem 264:13711–13717

Gomez N, Tonks NK, Morrison C, Harmar T, Cohen P 1990 Evidence for communication between nerve growth factor and protein tyrosine phosphorylation. FEBS (Fed Eur Biochem Soc) Lett 271:119–122

Gregory JS, Boulton TG, Sang BC, Cobb MH 1989 An insulin-stimulated ribosomal protein S6 kinase from rabbit liver. J Biol Chem 264:18397–18401

Hanekom C, Nel A, Gittinger C, Rhecker A, Landreth G 1989 Complexing of the CD-3 subunit by a monoclonal antibody activates a microtubule-associated (MAP-2) serine kinase in Jurkat cells. Biochem J 262:449–456

Hanks SK, Quinn AM, Hunter T 1988 The protein kinase family: conserved features and deduced phylogeny of the catalytic domains. Science (Wash DC) 241:42–52

Haystead TAJ, Weiel JE, Litchfield DW, Tsukitani Y, Fischer EH, Krebs EG 1990 Okadaic acid mimics the action of insulin in stimulating protein kinase activity in isolated adipocytes: the role of protein phosphatase 2A in attenuation of the signal. J Biol Chem 265:16571–16580

Hoshi M, Nishida E, Sakai H 1988 Activation of a $Ca^{2+}$-inhibitable protein kinase that phosphorylates microtubule-associated protein 2 *in vitro* by growth factors, phorbol esters, and serum in quiescent cultured human fibroblasts. J Biol Chem 263:5396–5401

Hoshi M, Nishida E, Sakai H 1989 Characterization of a mitogen-activated, $Ca^{2+}$-sensitive microtubule-associated protein-2-kinase. Eur J Biochem 184:477–486

Howell BW, Afar DEH, Lew J et al 1991 STY, a tyrosine-phosphorylating enzyme with sequence homology to serine/threonine kinases. Mol Cell Biol 11:568–572

Jenö P, Ballou LM, Novak-Hofer I, Thomas G 1988 Identification and characterization of a mitogen-activated S6 kinase. Proc Natl Acad Sci USA 85:406–410

Jenö P, Jaggi N, Luther H, Siegmann M, Thomas G 1989 Purification and characterization of a 40S ribosomal protein S6 kinase from vanadate-stimulated Swiss 3T3 cells. J Biol Chem 264:1293–1297

Jones SW, Erikson E, Blenis J, Maller JL, Erikson RL 1988 A Xenopus ribosomal protein S6 kinase has two apparent kinase domains that are each similar to distinct protein kinases. Proc Natl Acad Sci USA 85:3377–3381

Kozma SC, Ferrari S, Bassand P, Siegmann M, Totty N, Thomas G 1990 Cloning of the mitogen-activated S6 kinase from rat liver reveals an enzyme of the second messenger subfamily. Proc Natl Acad Sci USA 87:7365–7369

Kyriakis JM, Avruch J 1990 pp54 Microtubule-associated protein 2 kinase. J Biol Chem 265:17355–17363

Maller JL 1987 Mitogenic signalling and protein phosphorylation in *Xenopus* oocytes. J Cyclic Nucleotide Protein Phosphorylation Res 11:543–555

Miyasaka T, Chao MV, Sherline P, Saltiel AR 1990 Nerve growth factor stimulates a protein kinase in PC-12 cells that phosphorylates microtubule-associated protein-2. J Biol Chem 265:4730–4735

Morrison DK, Kaplan DR, Escobedo JA, Rapp UR, Roberts TM, Williams LT 1989 Direct activation of the serine/threonine kinase activity of Raf-1 through tyrosine phosphorylation by the PDGF β-receptor. Cell 58:649–657

Pelech SL, Tombes RM, Meijer L, Krebs EG 1988 Activation of myelin basic protein kinases during echinoderm oocyte maturation and egg fertilization. Dev Biol 130:28–36

Price DJ, Nemenoff RA, Avruch J 1989 Purification of a hepatic S6 kinase from cycloheximide-treated rats. J Biol Chem 264:13825–13833

Price DJ, Gunsalus JR, Avruch J 1990 Insulin activates a 70-kDa S6 kinase through serine/threonine specific phosphorylation of the enzyme polypeptide. Proc Natl Acad Sci USA 87:7944–7948

Ray LB, Sturgill TW 1987 Rapid stimulation by insulin of a serine/threonine kinase in 3T3-L1 adipocytes that phosphorylates microtubule-associated protein 2 *in vitro*. Proc Natl Acad Sci USA 84:1502–1506

Ray LB, Sturgill TW 1988a Characterization of insulin-stimulated microtubule-associated protein kinase. J Biol Chem 263:12721–12727

Ray LB, Sturgill TW 1988b Insulin-stimulated microtubule-associated protein kinase is phosphorylated on tyrosine and threonine *in vivo*. Proc Natl Acad Sci USA 85:3753–3757

Sanghera JS, Paddon HB, Bader SA, Pelech SL 1990 Purification and characterization of a maturation-activated myelin basic protein kinase from sea star oocytes. J Biol Chem 265:52–57

Seger R, Ahn NG, Boulton TG et al 1991 Microtubule-associated protein 2 kinases ERK1 and ERK2, undergo autophosphorylation on both tyrosine and threonine residues: implications for their mechanisms of activation. Proc Natl Acad Sci USA 88:6142–6147

Stern DF, Zheng P, Beidler DR, Zerillo C 1991 Spk1, a new kinase from *Saccharomyces cerevisiae*, phosphorylates proteins on serine, threonine and tyrosine. Mol Cell Biol 11:987–1001

Sturgill TW, Ray LB, Erikson E, Maller JL 1988 Insulin-stimulated MAP-2 kinase phosphorylates and activates ribosomal protein S6 kinase II. Nature (Lond) 334:715–718

Sweet LJ, Alcorta DA, Erikson RL 1990a Two distinct enzymes contribute to biphasic S6 phosphorylation in serum-stimulated chicken embryo fibroblasts. Mol Cell Biol 10:2787–2792

Sweet LJ, Alcorta DA, Jones SW, Erikson E, Erikson RL 1990b Identification of mitogen-responsive ribosomal protein S6 kinase pp90 *rsk*, a homolog of *Xenopus* S6 kinase II, in chicken embryo fibroblasts. Mol Cell Biol 10:2413–2417

## DISCUSSION

*Carpenter:* The autophosphorylation of MAP kinase on tyrosine is slow, so do you think that the *in vivo* formation of phosphotyrosine is, in fact, due to autophosphorylation, or is it still possible that there is a tyrosine kinase responsible for the phosphorylation?

*Krebs:* I think it is possible that the MAP kinase activator is a kinase. However, most protein kinases can phosphorylate more than one substrate, and thus far our activator does not appear to catalyse the phosphorylation of any proteins other than MAP kinase. That of course is negative evidence.

*Parker:* You could inactivate any potential kinase activity that may be associated with the activator by incubation with an ATP analogue such as fluorylsulphonylbenzoyl adenosine (FSBA) that will cross-link and inactivate. Does that block the ability of the activator to interact with and activate MAP kinase?

*Krebs:* We have evidence that there is an ATP-binding site in the activator, but we have not successfully carried out the blocking experiment.

*Tsien:* How surprised should we be to find autophosphorylation of both tyrosine and serine residues? Does the kinase have one catalytic site which is relatively undiscriminating, or are there two catalytic sites?

*Krebs:* I don't know the answer to your question, but during the past 12 months or so a number of enzymes that have both tyrosine kinase activity and serine/threonine kinase activity have been described. These enzymes display a fairly high degree of sequence identity in one particular region of what is referred to as subdomain 11 of the kinase domain. Examples of such enzymes include STY protein, MCK1, SPK1 and CDC2. CDC2 is not known to be an autophosphorylating enzyme, but I don't think anyone has really identified the tyrosine kinase that phosphorylates it on tyrosine, so it might belong to this group. We think this homology might be associated with the ability to catalyse the phosphorylation of both serine and tyrosine residues. I don't know, however, if this particular region of the kinase domain is involved in the catalytic centre of any of these enzymes.

*Hunter:* Now that the crystal structure of the cyclic AMP-dependent protein kinase catalytic subunit has been solved by Knighton et al (1991) it should be possible to see whether this region lies in the catalytic domain. There are other examples of this bipotential activity; p107$^{wee1}$, the putative CDC2 tyrosine kinase, although it looks like a serine kinase, has been shown to phosphorylate angiotensin on tyrosine residues and to have autophosphorylating activity on both tyrosine and threonine (Featherstone & Russell 1991).

*Krebs:* Tony Hunter and I are on a kinase nomenclature committee and we once thought that we could at least use specificity to separate serine/threonine kinases from tyrosine kinases. It now appears that may not be possible.

*Nairn:* Are these threonine and tyrosine phosphorylations or threonine/serine and tyrosine phosphorylations?

*Krebs:* They are threonine/serine and tyrosine phosphorylations.

*Nishizuka:* What is the time course of phosphorylation of MAP kinase in response to TPA (12-*O*-tetradecanoyl phorbol 13-acetate)?

*Hunter:* Tyrosine phosphorylation of p42 (MAP kinase) occurs early in the response to TPA, within about 5 min (Cooper et al 1984).

*Krebs:* MAP kinase activity peaked within 5–10 min after exposure of the cells to TPA and fell off gradually over the next hour.

I omitted to mention that as early as 1984 it was observed that TPA caused tyrosine phosphorylation of a protein which for years was known as p42 (Cooper et al 1984); in the last year or so it has been firmly established that p42 is MAP kinase.

*Carpenter:* Will MAP kinase phosphorylate exogenous substrates on tyrosine residues or will it only catalyse autophosphorylation of tyrosine?

*Krebs:* We have tried all kinds of peptides and other classical tyrosine kinase substrates, such as poly(Glu,Tyr) polymers, and have not obtained any evidence for phosphorylation of external substrates.

*Thomas:* We have been trying to identify the domain in S6 that is recognized by the 70 kDa kinase and have found that the kinase requires a basic block of four amino acids (Lys230-Arg-Arg-Arg233). If you alter either the second or fourth residues, Arg-231 or Arg-233, the $K_m$ of the kinase for the substrate is increased more than 1000-fold. This probably explains why the octapeptide (Arg232-Arg-Leu-Ser-Ser-Leu-Arg-Ala239) is not very sensitive in detecting the 70 kDa kinase in comparison with 40S ribosomal subunits.

In your reactivation studies you restored almost complete S6 kinase activity with MAP-2 kinase. In contrast, Sturgill et al (1988) saw only a 15–30% reactivation. Have you actually determined the sites of phosphorylation *in vivo*?

*Krebs:* I don't know why our results differ. We have always had remarkably good luck with the inactivation–reactivation experiments. We have not determined whether or not the *in vivo* phosphorylation pattern of the activated enzyme is the same as that which we get *in vitro*.

*Thomas:* Is it clear which of the two 90 kDa S6 kinases you are studying?

*Krebs:* A polyclonal antibody to the 90 kDa S6 kinase, that is, the *rsk* gene product (Sweet et al 1990), recognized all of the S6 peptide kinase bands.

*Thomas:* Does that antibody recognize both S6 kinases, I and II?

*Krebs:* I know that it recognizes *Xenopus* S6 kinase II, but I'm not sure about S6 kinase I.

*Carpenter:* Okadaic acid activates MAP kinase, so is it possible that your activator is a phosphatase inhibitor?

*Krebs:* I don't think so. We have purified the activator extensively and have found no evidence that the preparation contained any phosphatase activity. It's difficult to be completely sure that there isn't a very specific phosphatase involved.

*Tsien:* The picture that is emerging is of a whole cascade of very specific kinases, but there is no such emphasis on specific phosphatases. Could you reverse your strategy to hunt for phosphatases that might work physiologically at specific steps? Or, is the phosphatase 2A used at a low, physiologically reasonable concentration so that this enzyme might be the biologically relevant one?

*Krebs:* What's known about the serine/threonine phosphatases from many different cell types suggests that there are relatively few phosphatases compared to the number of kinases; thus, it stands to reason that the phosphatases have a much broader specificity than the kinases. The concentrations of phosphatase 2A that we used were moderately high, 100 units/ml, for example, but it's difficult to say whether or not this is 'physiologically reasonable'.

*Nishizuka:* I am curious about the properties of the activator. Does sequencing suggest that it is a kinase?

*Krebs:* We aren't 100% sure that what we are examining is the activator itself, but in the sequences that we have obtained from our most pure preparation we don't find any recognizable kinase motifs.

*Nishizuka:* What is the molecular weight of the activator?

*Krebs:* The major component in our most pure preparation is around 65–70 kDa and the second most abundant protein is around 48 kDa.

*Parker:* From the stoichiometry of the interaction, do you know whether the activator acts catalytically?

*Krebs:* We don't have enough substrate, purified ERK2, to do good kinetic experiments of that nature. It's our impression that the activator acts catalytically, and not in a stoichiometric manner.

*Thomas:* To see if the activator is dependent on protein kinase C you could try to down-regulate PKC with TPA and then challenge the cells with EGF to see whether MAP-2 kinase can still be activated. You can down-regulate PKC very well in Swiss 3T3 cells.

*Krebs:* We have not done such experiments recently. A number of years ago we down-regulated PKC in 3T3 cells using TPA. We lost the TPA effect but retained the growth factor effect on MAP kinase and S6 peptide kinase activation. I don't know, however, whether down-regulation is a foolproof way of wiping out PKC activity.

*Hunter:* That experiment has been done (Kazlauskas & Cooper 1988, Vilat & Weber 1988). A considerable amount of the EGF-induced MAP kinase activation, or tyrosine phosphorylation, is lost in cells that are down-regulated for PKC (L'Allemain et al 1991).

*Thomas:* They were examining phosphorylation of a 42 kDa protein, not MAP-2 kinase activity. In Jonathan Cooper's experiment, it's not clear to me that the protein he was looking at was MAP-2 kinase, because he detected a serine phosphorylation rather than threonine phosphorylation.

*Hunter:* He was looking at tyrosine phosphorylation of p42.

*Thomas:* But the protein was phosphorylated on serine as well as tyrosine, and not on threonine.

*Hunter:* There are several isoforms of p42 seen by two-dimensional gel analysis, one of which is serine/threonine/tyrosine phosphorylated and one of which is serine/tyrosine phosphorylated (Cooper et al 1984).

*Thomas:* Is it clear that the MAP-2 kinase is phosphorylated on serine? Serine phosphorylation was the major phosphorylation in the 42 kDa protein, so it's not clear to me whether they were really looking at MAP-2 kinase or whether they were looking at another protein.

*Parker:* It's clear in the U937 cell, a related system, that the 42 kDa protein that is threonine and tyrosine phosphorylated in response to TPA co-purifies with the MAP kinase (P. Adams and P.J. Parker, unpublished work).

*Hunter:* Mike Weber has actually shown that down-regulation of PKC abolishes some, but not all, of the activation induced by PDGF treatment (L'Allemain et al 1991). He looked at activity as well as the phosphorylation of p42, and has demonstrated that TPA regulates authentic MAP kinase.

*Nishizuka:* We know that there is some down-regulation of the $\alpha$, $\beta$ and $\gamma$ subspecies of PKC but have no evidence about what happens with the other subclass, the $\delta$, $\varepsilon$ and $\zeta$ forms.

*Thomas:* I asked the question because we have down-regulated PKC using TPA in Swiss 3T3 cells and saw no activation of S6 kinase when we re-challenged with TPA.

*Krebs:* In one sense it would be quite a surprise if there turned out not to be a direct serine kinase target for a tyrosine kinase receptor and that the main targets for tyrosine phosphorylation are in the phospholipase C category. At the moment, I think that is a possibility.

*Parker:* What is your view on the activation of Raf through receptor tyrosine kinase activation?

*Krebs:* Is solid evidence available that Raf is directly activated by tyrosine phosphorylation?

*Hunter:* I think it depends on the system. The evidence that it's tyrosine phosphorylated in fibroblasts is not as strong as originally thought, but in some haemopoietic cells, for example, there's fairly good evidence that Raf is substantially phosphorylated on tyrosine in response to stimulation by IL-2 (Turner et al 1991) or IL-3 (Carroll et al 1990).

*Krebs:* But is it also activated?

*Hunter:* It is activated, but it hasn't been shown that the tyrosine phosphorylation is responsible for the activation. Rusty Williams has shown for Raf and the PDGF receptor co-expressed in a baculovirus system in Sf9 cells that the increase in activity (Morrison et al 1989) is in part dependent on tyrosine phosphorylation, because some of the activation is lost on CD45 treatment (unpublished work). Raf *can* be activated by tyrosine phosphorylation, though in fibroblasts the major mechanism is probably serine phosphorylation. Unfortunately, no one has identified the sites of tyrosine phosphorylation and mutated them, so it's not clear exactly what's going on.

## References

Carroll MP, Clark-Lewis I, Rapp UR, May WS 1990 Interleukin-3 and granulocyte-macrophage colony-stimulating factor-mediated rapid phosphorylation and activation of cytosolic-c-Raf. J Biol Chem 265:19812–19817

Cooper JA, Sefton BM, Hunter T 1984 Diverse mitogenic agents induce the phosphorylation of two related 42,000-dalton proteins on tyrosine in quiescent chick cells. Mol Cell Biol 4:30–37

Featherstone C, Russell P 1991 Fission yeast p107$^{wee1}$ mitotic inhibitor is a tyrosine/serine kinase. Nature (Lond) 349:808–811

Kazlauskas A, Cooper JA 1988 Protein kinase C mediates platelet-derived growth factor-induced tyrosine phosphorylation of p42. J Cell Biol 106:1395–1402

Knighton DR, Zheng JH, Teneyck LF et al 1991 Crystal structure of the catalytic subunit of cyclic adenosine monophosphate dependent protein kinase. Science (Wash DC) 253:407–414

L'Allemain G, Sturgill TW, Weber MJ 1991 Defective regulation of mitogen-activated protein kinase activity in a 3T3 cell variant mitogenically nonresponsive to tetradecanoyl phorbol acetate. Mol Cell Biol 11:1002–1008

Morrison DK, Kaplan DR, Escobedo JA, Rapp UR, Roberto TM, Williams LT 1989 Direct activation of the serine/threonine kinase activity of Raf-1 through tyrosine phosphorylation by the PDGF beta-receptor. Cell 58:649–657

Sturgill TW, Ray LB, Erikson E, Maller JL 1988 Insulin-stimulated MAP-2 kinase phosphorylates and activates ribosomal protein S6 kinase II. Nature (Lond) 334:715–718

Sweet LJ, Alcorta DA, Jones SW, Erikson E, Erikson RL 1990 Identification of mitogen-reponsive ribosomal protein S6 kinase pp90$^{rsk}$, a homolog of *Xenopus* S6 kinase II, in chicken embryo fibroblasts. Mol Cell Biol 10:2413–2417

Turner B, Rapp UR, App H, Greene M, Dobashi K, Reed J 1991 Interleukin 2 induces tyrosine phosphorylation and activation of p72–74 raf-1 kinase in a T cell line. Proc Natl Acad Sci USA 88:1227–1231

Vilat J, Weber MJ 1988 Mitogen-stimulated tyrosine phosphorylation of a 42-kD cellular protein: evidence for a protein kinase-C requirement. J Cell Physiol 135:285–292

# Tyrosine phosphatases and their possible interplay with tyrosine kinases

E. H. Fischer*, H. Charbonneau*, D. E. Cool* and N. K. Tonks†

*Department of Biochemistry, SJ-70, University of Washington, Seattle, WA 98195 and †Cold Spring Laboratory, Cold Spring, NY 11724, USA

*Abstract*. Protein tyrosine phosphatases represent a new family of intracellular and receptor-linked enzymes. They are totally specific toward tyrosyl residues in proteins, and, with specific activities 10–1000-fold greater than those of the protein tyrosine kinases, they can be expected to tightly control the level of phosphotyrosine within the cell. Most transmembrane forms contain two conserved intracellular catalytic domains, as displayed by the leukocyte common antigen CD45, but highly variable external segments. Some are related to the neuronal cell adhesion molecules (NCAMs) or fasciclin II and others contain fibronectin III repeats; this suggests that these enzymes might be involved in cell–cell interaction. The intracellular enzymes appear to contain a highly conserved catalytic core linked to a regulatory segment. Deletion of the regulatory domain alters both substrate specificity and cellular localization. Likewise, overexpression of the full-length and truncated enzymes affects cell cycle progression and actin filament stability, respectively. The interplay between tyrosine kinases and phosphatases is considered. A hypothesis is presented suggesting that in some systems phosphatases might act synergistically with the kinases and elicit a physiological response, irrespective of the state of phosphorylation of the target protein.

*1992 Interactions among cell signalling systems. Wiley, Chichester (Ciba Foundation Symposium 164) p 132–144*

Over the last ten years it has been well established that tyrosine phosphorylation plays an essential role in cell growth, proliferation, differentiation and transformation. These processes are initiated by a variety of tyrosine kinases of cellular or viral origin, some of which are part of receptors for hormones and growth factors. Here, we focus on the structure and properties of protein tyrosine phosphatases (PTPs), enzymes that catalyse the reverse reaction, and the interplay that must exist between tyrosine and Ser/Thr phosphorylation in the modulation of physiological responses.

Tyrosine phosphatases are now known to represent a broad family of intracellular and transmembrane molecules involved in signal transduction, cytoskeletal reorganization and cell cycle regulation. They have specific activities up to three orders of magnitude higher than the viral or receptor-linked tyrosine kinases (PTKs) and $K_m$ values in the submicromolar range. Therefore, they would be expected to regulate the level of tyrosine phosphorylation within the cell effectively, but also, these enzymes must be tightly controlled to allow for the expression of those signals that are needed for normal cell development. They are totally specific toward tyrosyl residues in proteins, are inhibited by vanadate, molybdate or acidic polymers such as heparin, and they display all the characteristics of an enzyme with an essential SH group. When injected into *Xenopus* oocytes, the isolated placental tyrosine phosphatase PTP 1B antagonizes the short and long term responses to insulin by blocking the activation of ribosomal S6 peptide kinase and delaying oocyte maturation by up to five hours (Tonks et al 1990, Cicirelli et al 1990). Surprisingly, amino acid sequence analysis showed no similarity with any of the various types of Ser/Thr protein phosphatases, but considerable structural resemblance to the internal domain of a well-known and abundant cell surface antigen, namely the leukocyte common antigen CD45 (also known as L-CA, T200, B220 and Ly-5) (Charbonneau et al 1989). CD45 comprises a broad family of transmembrane molecules of 180–220 kDa that are found in all haemopoietic cells except mature erythrocytes. The external segments of these proteins are heavily *O*- and *N*-glycosylated, and the intracellular moiety contains two internally homologous domains of about 30 kDa each. It is these two domains that are structurally related to the placental enzyme PTP 1B. CD45 participates in lymphocyte activation and proliferation, but, depending on the ligand or conditions used, it can either enhance or inhibit cell proliferation. It has also been said to inhibit cytotoxic T lymphocyte and natural killer cell cytolysis and to modulate interleukin 2 receptor expression (Thomas 1989).

CD45 was shown to cause the dephosphorylation of the lymphocyte-specific tyrosine kinase $p56^{lck}$ both *in vitro* and *in vivo*. $p56^{lck}$ is associated with the cytoplasmic domains of CD4 and CD8 (Veillette et al 1989, Barber et al 1989); dephosphorylation of Tyr-505 derepresses its enzymic activity and may be involved in its regulation *in vivo*. Substantial evidence suggests that T cell receptor-mediated induction of PtdIns turnover involves the activation of protein tyrosine kinases, one of which may be Lck. Direct evidence in support of the involvement of CD45 in T cell activation comes from recent work on $CD45^-$ mutant T cell lines. In these cells, no PtdIns turnover occurred in response to T cell receptor stimulation and the overall level of tyrosine phosphorylation was greatly reduced (Koretzky et al 1990, 1991). These data suggest that in T cells, at least, CD45 operates mainly by activating some of the *src* family tyrosine kinases present (Lck, Fyn, Lyn, Hck, Blk etc). Whatever its mode of action, it clearly displays all the characteristics of a receptor having tyrosine phosphatase activity and involved directly in transmembrane signalling.

Amino acid sequences are now available for almost 50 catalytic domains from low $M_r$ and transmembrane PTPs; from these, consensus motifs have been identified. All but one of the receptor-linked PTPs possess two catalytic domains in their cytoplasmic portion. In contrast, they show a considerable diversity in their external segments (see Fig. 1), some of which have the structural characteristics of cell adhesion molecules (reviewed in Fischer et al 1991).

Currently, PTPs can be roughly subdivided into four categories. Type I, such as CD45, appears to be restricted to cells of the myeloid or lymphoid lineage. Type II, the 'leukocyte common antigen-related' PTPs (LARs) have immunoglobulin-like and fibronectin type III repeats (Thomas 1989), reminiscent of the neural cell adhesion molecules (NCAMs), or the related proteins L1 and fasciclin II (Krueger et al 1990); this suggests that they might be involved in homophilic cell–cell interactions and perhaps also in modulation of cell morphogenesis and tissue development. Type III enzymes also contain fibronectin type III repeats and might mediate cell–cell or cell–matrix signalling. Finally, type IV receptors have a variety of external domains, some so short (only 27 residues) that the enzyme is unlikely to function as a receptor for external factors. They could serve to regulate metabolic responses to hormone and growth factor receptors.

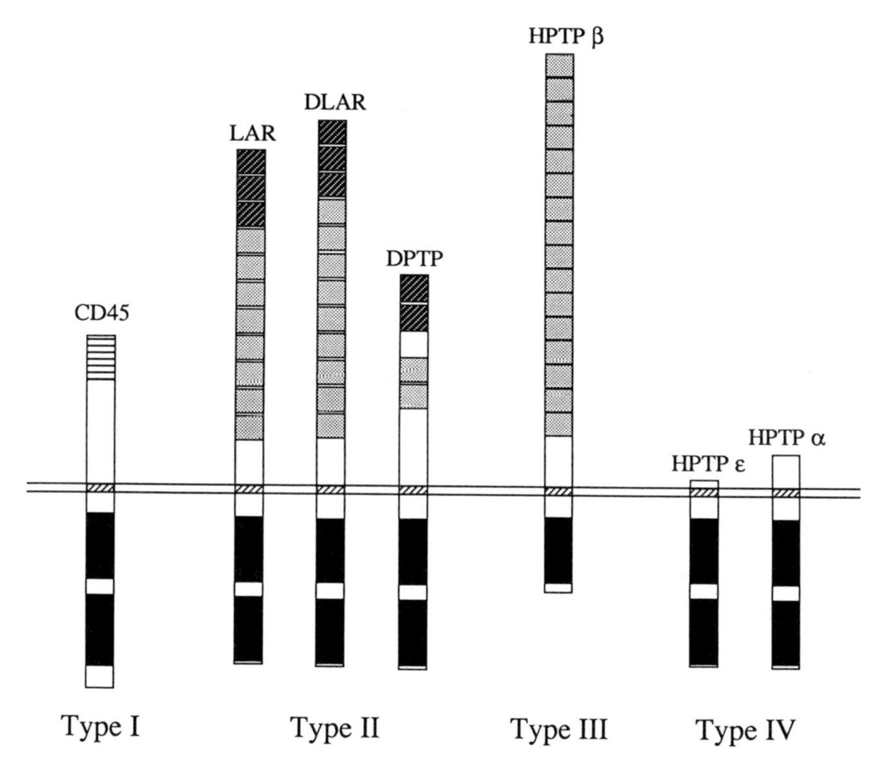


FIG. 1. Structural organization of transmembrane protein tyrosine phosphatases. Solid boxes represent conserved catalytic domains within the cytoplasm. LAR, leukocyte common antigen-related; PTP, protein tyrosine phosphatase; D, *Drosophila*; H, human. From Fischer et al 1991, copyright 1990 by the AAAS.

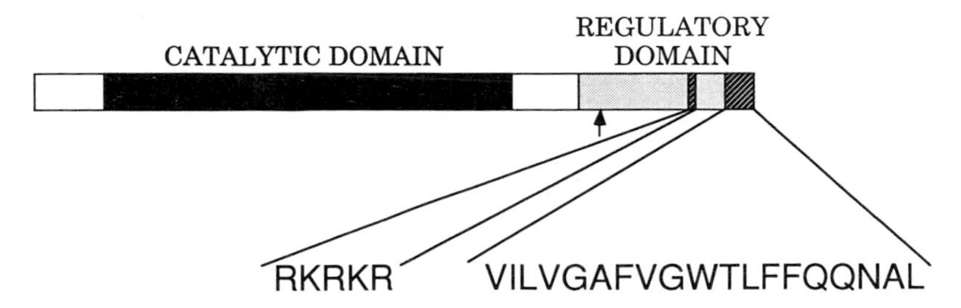


FIG. 2. Schematic representation of the non-receptor, T cell protein tyrosine phosphatase. The catalytic and putative regulatory domains are depicted. The arrow indicates the site of truncation by introduction of a premature stop codon into the cDNA (Cool et al 1990). From Fischer et al 1991, copyright 1991 by the AAAS.

Low $M_r$ PTPs from human T cell or placenta have the structure depicted in schematic form in Fig. 2. They have a highly conserved catalytic domain, linked to a regulatory segment at the C-terminus in the case of the human T cell and placental enzymes (Cool et al 1989). That this segment plays a regulatory function is indicated by the fact that when it is eliminated by the introduction of a premature stop codon in the cDNA, the resulting truncated form (a) shows a 10–20-fold increase in activity when a phosphorylated derivative of lysozyme (RCM-lysozyme; Cool et al 1990) is used as substrate, and (b) unlike the parent protein, no longer becomes localized in the particulate fraction of expressing cells. There are further suggestions that these non-catalytic domains are responsible for enzyme localization. The virulence plasmid YopH of bacteria of the genus *Yersinia* (e.g. *Y. pseudotuberculosis*, *enterocolitica* or, more importantly, *pestis*, which is responsible for the bubonic plague) encodes a protein tyrosine phosphatase (Yop51) (Guan & Dixon 1990, Bliska et al 1991). It also has such a putative regulatory domain, upstream rather than downstream from the catalytic core. It can be assumed that this domain, by determining the specific localization of this enzyme, is responsible for the virulence, because overexpression of the catalytic domain alone would not be expected to have such pathological consequences.

There is a final reason for assigning a physiological function to the non-catalytic regulatory domain. Overexpression of the truncated form of the T cell enzyme in baby hamster kidney (BHK) cells disrupts the later phases of the cell cycle: although BHK cells can still proceed through nuclear division, they will no longer undergo cytokinesis. As a consequence, multinucleated cells appear. More unexpectedly, however, these cells often contain an odd number of nuclei because of an asynchrony in mitosis. In fact, multinucleated cells containing nuclei at all stages of nuclear division can be observed. Such aberrations are not seen in cells overexpressing the wild-type enzyme. These results would suggest that a step or structural element regulated by tyrosine phosphorylation and involved in cytokinesis has been directly or indirectly affected by the phosphatase. Since cell division is an actomyosin-dependent process, a likely explanation is that a cytoskeletal protein that perhaps links actomyosin filaments to the plasma membrane has been modified.

In contrast to the effects of the truncated form, overexpression of the full-length enzyme in BHK cells had no effect on cell division, but rendered actin filaments stable to cytochalasin-induced disassembly. After addition of nocodazole, cytochalasin-treated cells displayed a massive array of actin fibres that carpeted the base of the cell. These data suggest that the PTPs might act on the complex membrane-associated structures where actin bundles terminate. These structures would include proteins such as talin, vinculin, paxilin and α-actinin, that are concentrated at regions of cell–substratum or cell–cell contact (Pasquale et al 1986, Burridge et al 1988). Talin and vinculin can be phosphorylated on tyrosyl residues and are thought to be associated with the cytoplasmic domain of the integrin receptors, which is also phosphorylated on a tyrosyl residue (Buck & Horwitz 1987). Integrins transduce information across the plasma membrane by interacting with extracellular matrix proteins such as fibronectin or vitronectin, laminin and certain types of collagen (Hynes 1987, Ruoslahti & Pierschbacher 1987). Cadherins are mostly concentrated at cell–cell junctions that contain, among other components, cortical actin bundles, vinculin and α-actinin (Takeichi 1990). Actin fibres are also anchored to the band 4.1 spectrin/ankyrin system underlying the erythrocyte plasma membrane (Bennett 1985). Fodrin, a brain homologue of spectrin, was reported to interact with the transmembrane PTP CD45 (Suchard & Bourguignon 1987). Other cytoskeletal proteins that are phosphorylated on Ser/Thr residues, such as vimentin by $p34^{cdc2}$ (Chou et al 1990) or caldesmon by the calmodulin-dependent protein kinase II, could also be indirectly affected (Ikebe & Reardon 1990). Such sites of action of the PTPs would be consistent with their localization in the particulate fraction.

## Interplay between tyrosine and serine/threonine phosphorylation

Proteins phosphorylated on tyrosyl residues are often phosphorylated on Ser and Thr as well, suggesting that these two types of modification could act in concert. Both types of phosphorylation are implicated in the regulation of $p34^{cdc2}$ (Gould & Nurse 1989), myelin basic protein kinase, (Ahn et al 1990, Ahn & Krebs, 1990), microtubule-associated protein (MAP-2) kinase (Anderson et al 1990) and growth factor receptor kinases (Yarden & Ullrich 1988). Both TPA (12-*O*-tetradecanoyl-phorbol 13-acetate), presumably through the activation of PKC (Kazlauskas & Cooper 1988), and okadaic acid (Haystead et al 1990) cause an increase in Tyr as well as Ser/Thr phosphorylation. Ser/Thr phosphorylation could activate a tyrosine kinase or target proteins for tyrosine phosphorylation. Conversely, it could inactivate PTPs or render substrates more resistant to their action, for example, through 'second-site' phosphorylation effects (Roach 1990).

However, there are obvious differences between the systems controlled by tyrosine rather than Ser/Thr phosphorylation. Because most protein tyrosine kinases are associated with the membrane, tyrosine phosphorylation has been regarded as a primary signal capable of affecting the state of activity of secondary enzymes downstream (Ahn et al 1990, Ahn & Krebs 1990, Haystead et al 1990, Boulton et al 1990, Howell et al 1991). In such a system, tyrosine kinases would serve a triggering role, relying on the more long-lasting Ser/Thr phosphorylation of secondary enzymes to maintain the memory of the signal. Indeed, more often than not, tyrosine phosphorylation appears to be transient and sub-stoichiometric, even though it can elicit a normal physiological response. Tyrosine phosphatases would assure that the activity of the tyrosine kinases is transient, thus contributing to the desensitizing process (Fischer et al 1991).

## Possible synergism between kinases and phosphatases

One could suggest another mechanism that might contribute to the low level of tyrosine phosphorylation observed within the cell. Although phosphatases have been viewed historically as counteracting the action of kinases, one may question whether this concept is necessarily correct and whether it applies to all kinase/phosphatase systems. Let's assume that the substrate of a kinase reaction (for instance, an enzyme activated by phosphorylation) remains in its active conformation for a finite period of time after undergoing dephosphorylation according to the scheme shown in Fig. 3.

This scheme describes a simple hysteretic reaction. If $k_2 \gg k_3$, the activity state of the system would depend essentially on the ratio of $E_a$ (active enzyme) to the total amount of enzyme, and would be independent of the phosphorylation state of E. An attractive feature of this system is that the rate-limiting reaction, the $E_a \rightarrow E_i$ step (rate constant $k_3$) could be easily controlled by metabolites, protein–protein interaction or phosphorylation at a secondary site, such as on

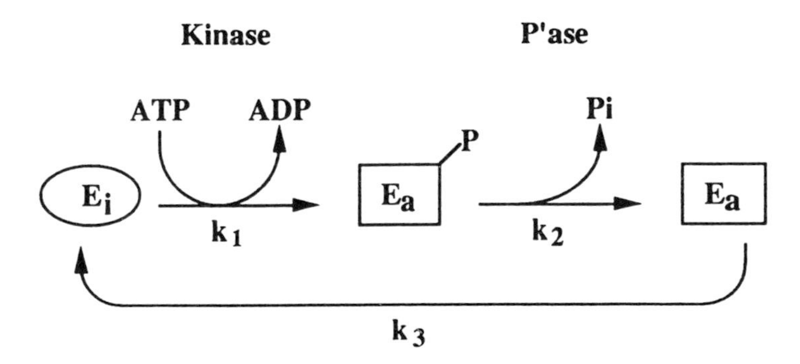


FIG. 3. Hysteretic model for kinase-phosphatase coupling. E, enzyme; a, active, i, inactive; P'ase, phosphatase.

Ser/Thr residues. A similar hysteretic reaction occurs in the activation of the cytosolic form of type 1 (Ser/Thr) phosphatase, where no correlation exists between the phosphorylation state of the complex and its activity (Villa Moruzzi et al 1984, Ballou et al 1985, Ballou & Fischer 1986, DePaoli-Roach 1989). However, one can envisage a further situation in which the phosphatase could play an even more direct role in amplifying the kinase reaction by assuming that the product of the phosphatase reaction inhibits the kinase. In this case, inhibition would be relieved by the phosphatase and the cellular response would actually increase while the level of substrate phosphorylation decreases. Kinases and phosphatases would act in concert, serving as the 'hand crank' that at every turn winds up the enzyme into its active state. Leaving aside the direct activation of *src* family kinases and $p34^{cdc2}$ by tyrosine dephosphorylation, this hypothesis explains how physiological responses could occur even though the specific activity of the PTPs is much higher than that of the PTKs.

In conclusion, while there are obvious structural similarities between the receptors for hormones and growth factors and those linked to tyrosine phosphatase, many questions remain. We don't know: (a) the identity of the ligands for the tyrosine phosphatase receptors, nor how the signal is transduced; (b) why they have two catalytic domains and whether one or both are active under physiological conditions; (c) what their physiological substrates are—one could suggest intracellular or receptor-linked protein tyrosine kinases, the ζ chain of CD3, cytoskeletal elements such as vinculin, talin, ezrin, paxilin or the β chain of the integrin receptor; and (d) why there are so many isoforms. The various isoforms could be directed towards specific target proteins by the structural characteristics of their non-catalytic domains.

Deregulation of protein tyrosine kinases can lead to oncogenic transformation. If the role of some of the tyrosine phosphatases is to control the activity of kinases, transformation could also result from underexpression of the phosphatases. Conversely, overexpression of the tyrosine phosphatases or some of their mutated forms might suppress, or even reverse, cell transformation—an exciting clinical prospect.

*Acknowledgements*

Supported by NIH grant DK07902 and the Muscular Dystrophy Association of America.

## References

Ahn NG, Krebs EG 1990 Evidence for an epidermal growth factor-stimulated protein kinase cascade in Swiss 3T3 cells. Activation of serine peptide kinase activity by myelin basic protein kinases *in vitro*. J Biol Chem 265:11495–11501

Ahn NG, Weiel JE, Chan CP, Krebs EG 1990 Identification of multiple epidermal growth factor-stimulated protein serine/threonine kinases from Swiss 3T3 cells. J Biol Chem 265:11487–11494

Anderson NG, Maller JL, Tonks NK, Sturgill TW 1990 Requirement for integration of signals from two distinct phosphorylation pathways for activation of MAP kinase. Nature (Lond) 343:651-653

Ballou LM, Fischer EH 1986 Phosphoprotein Phosphatases. The Enzymes 17:311-361

Ballou LM, Villa Moruzzi E, Fischer EH 1985 Subunit structure and regulation of phosphorylase phosphatase. Curr Top Cell Regul 27:183-192

Barber EK, Dasgupta JD, Schlossman SF, Trevillyan JM, Rudd CE 1989 The CD4 and CD8 antigens are coupled to a protein-tyrosine kinase ($p56^{lck}$) that phosphorylates the CD3 complex. Proc Natl Acad Sci USA 86:3277-3281

Bennett V 1985 The membrane skeleton of human erythrocytes and its implications for more complex cells. Annu Rev Biochem 54:273-304

Bliska JB, Guan K, Dixon JE, Falkow S 1991 Tyrosine phosphate hydrolysis of host proteins by an essential *Yersinia* virulence determinant. Proc Natl Acad Sci USA 88:1187-1191

Boulton TG, Gregory JS, Jong SM, Wang LH, Ellis L, Cobb MH 1990 Evidence for insulin-dependent activation of S6 and microtubule-associated protein-2 kinases via a human insulin receptor/v-ros hybrid. J Biol Chem 265:2713-2719

Buck CA, Horwitz AF 1987 Cell surface receptors for extracellular matrix molecules. Annu Rev Cell Biol 3:179-205

Burridge K, Fath K, Kelly T, Nuckolls G, Turner C 1988 Focal adhesions: transmembrane junctions between the extracellular matrix and the cytoskeleton. Annu Rev Cell Biol 4:487-525

Charbonneau H, Tonks NK, Kumav S et al 1989 Human placenta protein-tyrosine-phosphatase: amino acid sequence and relationship to a family of receptor-like proteins. Proc Natl Acad Sci USA 86:5252-5256

Chou YH, Bischoff JR, Beach D, Goldman RD 1990 Intermediate filament reorganization during mitosis is mediated by p34cdc2 phosphorylation of vimentin. Cell 62:1063-1071

Cicirelli MF, Tonks NK, Diltz CD, Weiel JE, Fischer EH, Krebs EG 1990 Microinjection of a protein-tyrosine-phosphatase inhibits insulin action in *Xenopus* oocytes. Proc Natl Acad Sci USA 87:5514-5518

Cool DE, Tonks NK, Charbonneau H, Walsh KA, Fischer EH, Krebs EG 1989 cDNA isolated from a human T-cell library encodes a member of the protein-tyrosine-phosphatase family. Proc Natl Acad USA 86:5257-5261

Cool DE, Tonks NK, Charbonneau H, Fischer EH, Krebs EG 1990 Expression of a human T-cell protein tyrosine phosphatase in baby hamster kidney cells. Proc Natl Acad Sci USA 87:7280-7284

DePaoli-Roach AA 1989 Regulatory components of type 1 protein phosphatases. Adv Protein Phosphatases 5:479-500

Fischer EH, Charbonneau H, Tonks NK 1991 Protein tyrosine phosphatases: a diverse family of intracellular and transmembrane enzymes. Science (Wash DC) 253:401-406

Gould KL, Nurse P 1989 Tyrosine phosphorylation in the fission yeast *cdc2*$^{+}$ protein kinase regulates entry into mitosis. Nature (Lond) 342:39-45

Guan K, Dixon JE 1990 Protein tyrosine phosphatase activity of an essential virulence determinant in *Yersinia*. Science (Wash DC) 249:553-556

Haystead TAJ, Weiel JE, Litchfield DW, Tsukitani Y, Fischer EH, Krebs EG 1990 Okadaic acid mimics the action of insulin in stimulating protein kinase activity in isolated adipocytes: the role of protein phosphatase 2A in attenuation of the signal. J Biol Chem 265:16571-16580

Howell BW, Afar DE, Lew J et al 1991 STY, a tyrosine-phosphorylating enzyme with sequence homology to serine/threonine kinases. Mol Cell Biol 11:568-572

Hynes RO 1987 Integrins: a family of cell surface receptors. Cell 48:549-554

Ikebe M, Reardon S 1990 Phosphorylation of smooth muscle caldesmon by calmodulin-dependent protein kinase II. J Biol Chem 265:17607–17612
Kazlauskas A, Cooper JA 1988 Protein kinase C mediates platelet-derived growth factor-induced tyrosine phosphorylation of p42. J Cell Biol 106:1395–1402
Koretzky GA, Picus J, Thomas ML, Weiss A 1990 Tyrosine phosphatase CD45 is essential for coupling T-cell antigen receptor to the phosphatidyl inositol pathway. Nature (Lond) 346:66–68
Koretzky GA, Picus J, Schultz J, Weiss A 1991 Tyrosine phosphatase CD45 is required for T-cell antigen receptor and CD2-mediated activation of a protein tyrosine kinase and interleukin 2 production. Proc Natl Acad Sci USA 88:2037–2041
Krueger NK, Streuli M, Saito H 1990 Structural diversity and evolution of human receptor-like protein tyrosine phosphatases. EMBO (Eur Mol Biol Organ) J 9:3241–3252
Pasquale EB, Maher PA, Singer SJ 1986 Talin is phosphorylated on tyrosine in chicken embryo fibroblasts transformed by Rous sarcoma virus. Proc Natl Sci USA 83:5507–5511
Roach PJ 1990 Control of glycogen synthase by hierarchal protein phosphorylation. FASEB (Fed Am Soc Exp Biol) J 4:2961–2968
Ruoslahti E, Pierschbacher MD 1987 New perspectives in cell adhesion: RGD and integrins. Science (Wash DC) 238:491–497
Suchard SJ, Bourguignon LYW 1987 Further characterization of a fodrin-containing transmembrane complex from mouse T-lymphoma cells. Biochim Biophys Acta 896:35–46
Takeichi M 1990 Cadherins: a molecular family important in selective cell–cell adhesion. Annu Rev Biochem 59:237–252
Thomas ML 1989 The leukocyte common antigen family. Annu Rev Immunol 7:339–369
Tonks NK, Cicirelli MF, Diltz CD, Krebs EG, Fischer EH 1990 Effect of microinjection of a low $M_r$ human placenta protein tyrosine phosphatase on induction of meiotic cell division in *Xenopus* oocytes. Mol Cell Biol 10:458–463
Veillette A, Bookman MA, Horak EM, Samelson LE, Bolen JB 1989 Signal transduction through the CD4 receptor involves the activation of the internal membrane tyrosine-protein kinase $p56^{lck}$. Nature (Lond) 338:257–259
Villa Moruzzi E, Ballou LM, Fischer EH 1984 Phosphorylase phosphatase. Interconversion of active and inactive forms. J Biol Chem 259:5857–5863
Yarden Y, Ullrich A 1988 Molecular analysis of signal transduction by growth factors. Biochemistry 27:3113–3119

## DISCUSSION

*Yang:* The catalytic domain of tyrosine phosphatase contains three cysteine residues. Can it be inactivated by chemical modification of one or more of these cysteine residues?

*Fischer:* Yes, that has been done. There is one cysteine that is indispensable for catalytic activity. Site-directed mutagenesis of this residue renders the enzyme totally inactive. The catalytic domains of all the receptor and non-receptor phosphatases have this cysteine in a highly conserved region.

*Changeux:* What is the time scale of the hysteretic reaction that you described? Is this cycle of any physiological significance?

*Fischer:* In the case of phosphorylase phosphatase the half-life of the active catalytic subunit is 12.5 min. This is a fairly long time, although some of the hysteretic systems that have been described have half-lives of 20–25 min. If the system continues to turn over the enzyme could be maintained in its active, dephospho form.

*Changeux:* So you think a steady state will be maintained, with no physiologically significant oscillations.

*Fischer:* We have been unable to study whether there are fluctuations of activity in the cell; all the experiments have been done *in vitro*, and this is what happens with the cytoplasmic form of phosphorylase phosphatase.

*Cantrell:* In the experiments by Ledbetter et al (1988) the differing effects of CD45 cross-linkage on CD4 and CD3 activation suggesting that CD45 might be doing different things to these two activation pathways. CD4 associates with Lck (Viellette et al 1988) and CD3 with Fyn (Samelson et al 1990), so do you think that CD45 might differentially control Lck and Fyn?

*Fischer:* I am not absolutely convinced that Fyn is the tyrosine kinase associated with the T cell receptor. The question is, what controls the activity of CD45 and the other receptor phosphatases? With the growth factor receptors, for example, signal transduction results in autophosphorylation and appearance of tyrosine kinase activity. With CD45 we don't know what happens, whether binding of an external ligand (we don't know *any* of the ligands for these receptors) increases or decreases the activity, or targets the receptor towards different surface molecules. It is possible that one type of ligand targets the receptor towards surface molecule A, bringing about cell activation, whereas another ligand targets it towards surface molecule B, eliciting an inhibitory response.

*Cantrell:* There have been reports that CD45 can actually physically associate in the membrane with various molecules like CD2 and T cell antigen receptors.

*Fischer:* Jeff Ledbetter is continuing those experiments, trying to find specific targets for CD45. What impressed me about the results from Koretzky et al (1991) was that in $CD45^-$ cells the level of tyrosine phosphorylation was totally depressed whereas phosphorylation induced by TPA was not greatly affected. We have now confirmed this in $CD45^-$ cells.

*Cantrell:* In those $CD45^-$ mutant cells has a cytoplasmic domain-negative CD45 been put back in, rather than the full length molecule?

*Fischer:* That has not yet been done. In the $CD45^-$ cells we are expressing the full length and truncated T cell enzymes and Art Weiss is expressing the receptor PTP$\alpha$, to see whether signalling through the T cell receptor can be restored. Many tyrosine phosphatase isoforms are known; these may be differentially located within the cell. It is also possible that there is one phosphatase for each kinase. Experiments are being done to look at these alternatives.

*Fredholm:* Do you have any thoughts about the possible significance of the protein kinase C phosphorylation sites on CD45? We find that CD45 can be phosphorylated not only by phorbol ester treatment, but also when we activate CD3. Might this phosphorylation be involved in regulation of the physical association of CD45 and T cell receptor molecules such as CD3 or CD4?

*Fischer:* CD45 can be phosphorylated by protein kinase C, glycogen synthase kinase 3 and casein kinase 2, both *in vitro* and *in vivo*. I haven't seen any results indicating whether phosphorylation would affect localization or substrate specificity. The low molecular weight enzyme is phosphorylated on tyrosyl residues quite readily by the $p43^{v\text{-}abl}$ kinase, but undergoes immediate autodephosphorylation. It is also phosphorylated on serine/threonine residues. We don't know which of these phosphorylations, if any, are of physiological significance.

*Hunter:* Ostergaard & Trowbridge (1991) have found that ionomycin-induced calcium influx leads to a severe decrease in CD45 activity in T cells and simultaneous serine dephosphorylation. They think, therefore, that serine phosphorylation activates CD45, but they haven't yet shown that phosphatase treatment of CD45 actually reduces its activity. Also, TPA can induce redistribution of CD45, an effect which might be due to its phosphorylation.

*Fischer:* Ian Trowbridge has been expressing the external domain of CD45, searching for a potential ligand.

*Hunter:* I think nothing has come out of that yet.

*Michell:* In the 'original' tyrosine phosphatase there is one catalytic domain, but in all of the transmembrane homologues there are two catalytic-type domains. What is the significance of this? Do both catalytic domains work and do you have any idea why there should be this tandem arrangement?

*Fischer:* The second domain is said to have very low or no activity. Two or three protein kinases (S6 kinase, for example) have been reported to have two catalytic domains within the same chain. However, most regulatory enzymes have oligomeric structures. They can exist as homodimers or homotetramers, not for catalysis but for regulatory purposes. It is conceivable that the two catalytic domains of the receptor phosphatases form a homodimeric structure that would allow homotropic interactions involved in the regulation of the enzyme.

*Michell:* Is there evidence in any of these two-domain enzymes as to whether particular domains contribute to particular functions?

*Fischer:* Not as yet. Whether the second domain has a different kind of activity is a good question.

*Nishizuka:* Is there any evidence of cross-talk between calcium and tyrosine phosphatases?

*Fischer:* We don't have any evidence about that but I would guess that this is being investigated. At present, we have little indication of the state of activity of these enzymes *in vivo* or of what regulates their activity. In collaboration with Axel Ullrich's group in Münich, we have put the various phosphatase constructs and receptor kinases behind the cytomegalovirus promoter, and mix-match type experiments have been done to see if one enzyme is particularly active towards a specific receptor kinase.

*Thomas:* What are the effects on tyrosine phosphorylation when you overexpress the low $M_r$ T cell phosphatase in BHK cells?

*Fischer:* In control BHK cells that have been made quiescent for a few hours PDGF causes an immediate phosphorylation of a number of bands, including one that corresponds to phospholipase C. In cells overexpressing either the full length or the truncated phosphatase, after treatment with PDGF one sees *de*phosphorylation of these bands, not of the PDGF receptor itself. There is no difference in the extent of dephosphorylation caused by overexpressing the full length and truncated forms. BHK cells are transforming and can be grown on soft agar. Curiously, cells overexpressing the full length enzyme grow better on soft agar. Furthermore, their injection into nude mice causes formation of larger tumours than injection of control BHK cells. In contrast, cells overexpressing the truncated form grow poorly on soft agar and produce smaller tumours than control cells.

*Rasmussen:* Is there any evidence of a receptor-linked serine/threonine phosphatase?

*Fischer:* You would imagine that if there were receptor serine/threonine kinases, there ought to be serine/threonine phosphatases, but I don't know that any have been described.

*Hunter:* I don't know of any. There are serine/threonine *kinase* receptor-like molecules; the activin receptor is likely to be a serine kinase (Matthews & Vale 1991).

*Cantrell:* The results you have described, Dr Fischer, suggest that CD45 has a role in the early stages of T cell activation. Do you know anything about the role of CD45 in IL-2 regulation of T lymphocyte growth? IL-2 regulates tyrosine phosphorylation—is this response controlled by CD45?

*Fischer:* The β chain of the IL-2 receptor is associated with $p56^{lck}$. Lck becomes activated by IL-2; presumably, it can phosphorylate the β chain of the receptor and participate in signalling downstream (Hatakeyama et al 1991). CD45 could well be involved in these processes; it has been said to modulate IL-2 expression, but I don't know how it does that (Thomas 1989).

*Warner:* Is there any evidence for a role for CD45 in cells other than T cells? Does it have a role in neutrophil activation, for example?

*Fischer:* I imagine that it must have but I don't know. It is expressed only in haemopoeitic cells, in all except mature erythrocytes and their immediate progenitors. In non-myeloid or lymphoid cells, the external domains must be quite different because they are not recognized by antibodies directed against CD45.

*Warner:* I know it's expressed in all these cells but what does it *do*?

*Fischer:* I don't know. Now that we know it has enzymic activity it should be easier to define its exact function.

*Hunter:* There is some recent evidence that CD45 is needed for B cell activation as well.

*Fischer:* The 39 kDa and 48 kDa subunits of the B cell receptor are phosphorylated on tyrosyl residues. I suppose that, as in T cells, CD45 could act through the *src* family kinases that are presumably involved. Incidentally, even though our low molecular mass enzyme was cloned from a T cell and we call it a T cell enzyme, it's actually more abundant in B cells.

## References

Hatakeyama M, Kono T, Kobayashi N et al 1991 Interaction of the IL-2 receptor with the *src*-family kinase $p56^{lck}$: identification of novel intermolecular association. Science (Wash DC) 252:1523–1528

Koretzky GA, Picus J, Schultz J, Weiss A 1991 Tyrosine phosphatase CD45 is required for T-cell antigen receptor and CD2-mediated activation of a protein tyrosine kinase and interleukin 2 production. Proc Natl Acad Sci USA 88:2037–2041

Ledbetter JA, Tonks NK, Fischer EH, Clark EA 1988 CD45 regulates signal transduction and lymphocyte activation by specific association with receptor molecules on T cells or B cells. Proc Natl Acad Sci USA 85:8628–8632

Matthews LS, Vale WW 1991 Expression cloning of an activin receptor, a predicted transmembrane serine kinase. Cell 65:973–982

Ostergaard HL, Trowbridge IS 1991 Negative regulation of CD45 protein tyrosine phosphatase activity by ionomycin in T cells. Science (Wash DC) 253:1423–1425

Samelson LE, Phillips AF, Luong ET, Klausner RD 1990 Association of the fyn protein-tyrosine kinase with the T-cell antigen receptor. Proc Natl Acad Sci USA 87:4358–4362

Thomas ML 1989 The leukocyte common antigen family. Annu Rev Immunol 7:339–369

Veillette A, Bookman MA, Horak EM, Bolen JB 1988 The CD4 and CD8 T cell surface antigens are associated with the internal membrane tyrosine-protein kinase p56lck. Cell 55:301–308

# Calmodulin and protein kinase C cross-talk: the MARCKS protein is an actin filament and plasma membrane cross-linking protein regulated by protein kinase C phosphorylation and by calmodulin

Angus C. Nairn and Alan Aderem

*The Rockefeller University, 1230 York Avenue, New York, NY 10021, USA*

*Abstract.* The myristoylated, alanine-rich C kinase (PKC) substrate (MARCKS) is a major, specific substrate of PKC that is phosphorylated during macrophage and neutrophil activation, growth factor-dependent mitogenesis and neurosecretion. MARCKS is also a calmodulin-binding protein and binding of calmodulin inhibits phosphorylation of the protein by PKC. Several recent observations from our laboratories suggest a role for MARCKS in cellular morphology and motility. First, in macrophages MARCKS is located at points of cellular adherence where actin filaments insert at the plasma membrane and is released to the cytoplasm upon activation of PKC. Second, during neutrophil chemotaxis MARCKS undergoes a cycle of release from, and reassociation with, the plasma membrane. Third, *in vitro*, MARCKS is an F-actin cross-linking protein whose activity is inhibited by PKC-mediated phosphorylation and by binding to calmodulin. MARCKS therefore appears to be a regulated cross-bridge between actin and the plasma membrane. Regulation of the plasma membrane-binding and actin-binding properties of MARCKS represents a convergence of the PKC and calmodulin signal transduction pathways in the control of actin cytoskeleton–plasma membrane interactions.

*1992 Interactions among cell signalling systems. Wiley, Chichester (Ciba Foundation Symposium 164) p 145–161*

The protein kinases C (PKCs) are a family of diacylglycerol-activated, calcium-dependent protein kinases that regulate diverse cellular pathways in all eukaryotic cells. Although a number of protein substrates of known function have been shown to be phosphorylated and regulated by PKC, little is known about the functions of the major cellular substrates of PKC and their role in mediating biological responses. One of the most prominent substrates for PKC is an acidic protein with an apparent relative molecular mass of 68 000–87 000. The protein was first described as the '87 k protein' in rat brain synaptosomes that was found to be phosphorylated in response to potassium depolarization and treatment with phorbol esters (Wu et al 1982, Albert et al 1987, Patel & Kligman 1987). In fibroblasts, where it was known as the 80 k protein, it was found to be the major protein phosphorylated when the cells were treated with growth factors or with phorbol esters (Rozengurt et al 1983, Blackshear et al 1986). The protein was first shown to be myristoylated in macrophages (Aderem et al 1988), and this observation has since been confirmed in a variety of cell types (James & Olson 1989, Thelen et al 1990). Recently, cDNAs for bovine (Stumpo et al 1989), chicken (Graff et al 1989a) and murine (Seykora et al 1991) forms of the protein have been isolated. These encode proteins of $M_r$ 28 000–32 000, indicating that there is an extremely anomalous migration of the protein on SDS–PAGE. Because of these various properties, and its high proportion of alanine, the protein became known by the acronym MARCKS (for *m*yristoylated, *a*lanine-*r*ich *C* *k*inase *s*ubstrate) (Stumpo et al 1989).

This chapter describes analysis of the function of MARCKS and the role of its phosphorylation by PKC. MARCKS appears to cross-link actin filaments and the plasma membrane. Phosphorylation of the protein releases it from the plasma membrane and alters its actin cross-linking properties. MARCKS is also a calmodulin-binding protein (Graff et al 1989b) and the binding of calmodulin inhibits phosphorylation of the protein by PKC (Graff et al 1989b, Albert et al 1984). Calmodulin also inhibits its ability to cross-link actin. Therefore, MARCKS represents an important component in the cross-talk between calmodulin- and PKC-dependent pathways in the control of cellular morphology and motility at the level of actin cytoskeleton–plasma membrane interactions.

## Molecular characterization of MARCKS

MARCKS has been purified from bovine brain and characterized biochemically (Albert et al 1987). The amino acid composition of MARCKS is unusual: alanine accounts for about 29% of the residues, while glutamate and proline constitute 16% and 11% of the residues, respectively. Ten amino acids (Ala, Glu, Pro, Gly, Ser, Lys, Gln, Thr, Asp and Phe) represent over 95% of the residues. cDNAs for bovine, murine and chicken brain MARCKS encode proteins of 31.9, 29.6 and 28.7 kDa, respectively, which migrate on SDS–PAGE with apparent molecular masses of 87, 68 and 67 kDa (Stumpo et al 1989, Graff et al 1989a, Seykora et al 1991). The anomalous migration of MARCKS in SDS–PAGE is attributable to its large Stoke's radius (Albert et al 1987) and to its rod-like shape (dimensions 33 nm × 2.5 nm, Hartwig et al 1992). Analysis of the secondary structure of MARCKS suggests that it is approximately 76% helical, consistent with its rod-like shape.

Comparison of the primary sequences of MARCKS from mouse, cow and chicken reveals that the N-terminal domain and the phosphorylation domain are highly conserved, whereas the remainder of protein is not (Stumpo et al 1989, Graff et al 1989a, Seykora et al 1991). The N-terminal domain contains a myristoylation consensus sequence and the protein has been shown to be myristoylated in macrophages and other cell types. The second conserved domain, found in the centre of the molecule (for example, spanning amino acids 128–180 of murine MARCKS), contains all the known phosphorylation sites of the protein (serines 152, 156 and 163), as well as the calmodulin-binding site (Graff et al 1989c) and one or more actin-binding sites (see below).

## The association of MARCKS with the plasma membrane is regulated by cycles of phosphorylation and dephosphorylation

MARCKS has an unusual subcellular distribution in that the vast majority of the myristic acid-labelled, non-phosphorylated protein is associated with the plasma membrane in quiescent cells, whereas most of the phosphorylated protein is found in the cytosol of activated cells (Aderem et al 1988, Wang et al 1989). Activation of PKC in either macrophages or neutrophils results in the displacement of the myristoylated protein from the membrane, and this occurs without deacylation (Thelen et al 1991). Displacement also occurs in *in vitro* reconstituted systems where phosphorylation of MARCKS by PKC promotes its release from isolated membranes (Wang et al 1989, Thelen et al 1991). Furthermore, mutational analysis showed that the myristic acid moiety is required for the stable attachment of MARCKS to the membrane (Graff et al 1989d). These findings suggest that the myristic acid moiety targets MARCKS to the membrane, where it comes into close apposition with PKC. Upon activation, PKC phosphorylates MARCKS, which results in the release of the protein from the membrane.

Treatment of neutrophils with *N*-formyl-Met-Leu-Phe (f-Met-Leu-Phe), a chemotactic peptide that stimulates transient PKC-dependent activity via a receptor-mediated, G-protein coupled pathway (Dewald et al 1988), results in the rapid, but transient, phosphorylation of MARCKS. This is accompanied by its release from the plasma membrane and accumulation in the cytosol. Subsequently, the equilibrium between kinase and phosphatase activities changes, favouring the dephosphorylation of MARCKS. As dephosphorylation proceeds there is a concomitant reassociation of the protein with the membrane, such that most of the protein is membrane-bound when phosphorylation has returned to basal levels (Thelen et al 1991). Okadaic acid, a specific phosphatase inhibitor (Bialojan & Takai 1988, Haystead et al 1989), blocks both the dephosphorylation of MARCKS and its reassociation with the plasma membrane. Furthermore, when dephosphorylation is accelerated by addition of the receptor antagonist *tert*-butoxycarbonyl-Met-Leu-Phe the shift in the equilibrium towards the dephosphorylated species is accompanied by increased binding to the membrane (Thelen et al 1991). This cycle of membrane release and attachment is not influenced by cycloheximide, indicating that *de novo* synthesis of MARCKS does not account for the increase in the membrane-bound form of the protein observed on dephosphorylation.

## MARCKS is localized at points of focal adhesion in macrophage filopodia

The subcellular location of MARCKS in macrophages was analysed by immunofluorescence microscopy. The protein has a punctate distribution and is localized at the substrate-adherent surface of macrophage pseudopodia and filopodia (Rosen et al 1990). Many of the structures containing MARCKS also stain for vinculin and talin (Rosen et al 1990). A punctate distribution of vinculin and talin is in contrast to the typical plaque-like distribution seen in adherent fibroblasts (Burridge et al 1987), but has previously been observed in macrophages (Marchisio et al 1987) and at the active cell edge of fibroblasts during the early stages of fibroblast attachment (Bershadsky et al 1985). It seems likely, therefore, that MARCKS is located in the initial, more transient adhesion complex that is formed at the leading edge of the motile macrophage.

Immuno-electron microscopy has further defined the cellular localization of MARCKS. The protein is found in clusters near the plasma membrane in Lowicryl-embedded sections of fixed macrophages. The distribution of MARCKS was also examined in cells mechanically unroofed by a technique that preserves both the membrane and its associated cytoskeletal elements, thereby allowing connections between filaments and the cytoplasmic side of the plasma membrane to be visualized (Hartwig et al 1989). Anti-MARCKS gold labelled antibodies are found in small, widely spaced clusters at points where actin filaments interact with the cytoplasmic surface of the plasma membrane (Rosen et al 1991).

Immunofluorescence studies in macrophages have indicated that PKC is a component of the substrate-adherent punctate structures which contain MARCKS, vinculin and talin (Rosen et al 1990). Previously, PKC has been found to be a component of focal adhesions in fibroblasts (Jaken et al 1989), and it is known that PKC rapidly phosphorylates MARCKS when activated (Rosen et al 1989). Activation of PKC with phorbol esters causes marked cell spreading and rounding, and the almost complete disappearance of filopodia. This is accompanied by disappearance of punctate staining of MARCKS and by a modest increase in the level of diffuse staining. The kinetics of the morphological changes and the disappearance of punctate staining of MARCKS (Rosen et al 1990) mirror the kinetics of phosphorylation of MARCKS by PKC in intact macrophages (Rosen et al 1989) and those of the translocation of MARCKS from the membrane to the cytosol (Rosen et al 1990). The effect of PMA (phorbol 12-myristate 13-acetate; also known as TPA, 12-*O*-tetradecanoyl phorbol-13-acetate) on the distribution of vinculin and talin is quite different. After PKC activation by PMA, vinculin and talin still stain prominently, with a patchy, punctate organization, and are particularly obvious at the phase-dense cell edge (Rosen et al 1990).

Immuno-electron microscopy confirms that the amount of plasma membrane-associated MARCKS noticeably diminishes upon activation of PKC, while cytoplasmic levels of the protein increase proportionally. Experiments using mechanically unroofed macrophages show that MARCKS remains associated with the sides of actin filaments in PMA-treated cells, but that the filaments to which the gold label is attached are now spatially displaced from the membrane surface.

## Regulation of MARCKS by lipopolysaccharide and tumour necrosis factor

Bacterial lipopolysaccharide (LPS) promotes the synthesis and myristoylation of MARCKS, and concomitantly alters its profile of phosphorylation (Aderem et al 1988). The steady-state level of MARCKS mRNA in murine macrophages is increased 20–50-fold by exposure of the cells to LPS (Seykora et al 1991). MARCKS synthesis is also induced by tumour necrosis factor $\alpha$ in human neutrophils, where it constitutes approximately 90% of the total protein synthesized in response to this cytokine (Thelen et al 1990).

LPS greatly increases the adherence of cultured mouse macrophages to surfaces. This is accompanied by a striking increase in the number and prominence of lamellipodia, filopodia and membrane veils (Rosen et al 1992). Stimulation of macrophages with LPS also results in a dramatic increase in the number of punctate structures containing MARCKS, which are seen over the entire substrate-adherent surface of lamellipodia as well as along filopodia and veils of membrane. Electron microscopy confirms the results from immunofluorescence microscopy, and shows that the total amount of membrane-associated MARCKS, as well as the concentration of MARCKS in the clusters, increases. The close correlation between synthesis and acylation of MARCKS and its enrichment in pseudopodia and filopodia suggests that MARCKS might serve a function in these plastic structures. The pronounced staining of MARCKS in filopodia and the presence in filopodia of a conspicuous, central actin bundle (Mitchison & Kirschner 1988) prompted us to examine the distribution of MARCKS and actin in the same cell. Double staining and electron microscopy experiments show MARCKS spotted alongside filopodial actin bundles in LPS-treated macrophages (Rosen et al 1992). Furthermore, clusters of MARCKS are intimately associated with the cytoplasmic surface of the plasma membrane in unroofed cells at points where large numbers of actin filaments contact the cytoplasmic surface of the plasma membrane.

## MARCKS is an actin-binding protein

The results described above suggested a role for MARCKS in actin-based motility. The MARCKS protein also appears to interact with actin *in situ*, because actin co-purifies with MARCKS through numerous chromatography and centrifugation steps, and the two proteins can be separated only by reverse-phase chromatography. It was therefore important to determine whether MARCKS binds actin *in vitro* and whether this binding was influenced by phosphorylation. Both non-phosphorylated MARCKS and MARCKS phosphorylated *in vitro* with PKC bind to F-actin in a co-sedimentation assay (Hartwig et al 1992). The interaction of MARCKS with actin filaments was also studied by electron microscopy using negative staining. When incubated at a ratio of 1 molecule of dephospho-MARCKS to 10 actin subunits, the actin filaments become aggregated and decorated with rod-shaped MARCKS molecules, which are attached to the actin fibres near the mid-points of the rods. At molar excess of MARCKS over actin subunits, the actin filaments become highly entangled and are periodically decorated, every 20 nm, with dephospho-MARCKS molecules. Actin filaments incubated in the presence of phospho-MARCKS appear bare and unaggregated. Thus, although phosphorylation does not diminish the binding of MARCKS to the actin filament, it does change the manner of its binding (Hartwig et al 1992).

The effect of MARCKS binding on the functioning of F-actin, and alterations mediated by phosphorylation, were further evaluated by morphological, optical and hydrodynamic methods. Non-phosphorylated MARCKS cross-linked actin filaments into loose aggregates, which were easily discernible by electron microscopy. The addition of molar ratios of $\leqslant 1$ MARCKS to 20 actin filament subunits also increased the viscosity of actin, consistent with filament cross-linking. Increasing the ratio of non-phosphorylated MARCKS to actin, however, decreased the viscosity of actin solutions relative to actin alone. These results suggest that at low ratios of MARCKS to actin MARCKS links actin into a network and that at higher concentrations it aggregates filaments into bundles. Dynamic light-scattering confirms that non-phosphorylated MARCKS causes the lateral alignment of actin filaments into larger structures. In contrast, phosphorylated MARCKS has minimal effects on the viscosity of or light-scatter intensity from F-actin solutions. These studies demonstrate that MARCKS binds to the sides of actin filaments and cross-links them. High concentrations of the phosphorylated or non-phosphorylated protein do not affect the extent of linear actin assembly, making it unlikely that MARCKS interacts with the ends of actin polymers.

Non-phosphorylated MARCKS binds calmodulin in the presence of calcium whereas the phosphorylated protein does not (Graff et al 1989b). In addition, calmodulin prevents PKC-dependent phosphorylation of MARCKS (Albert et al 1984, Graff et al 1989b). Taken together, these findings suggest that the calmodulin-binding site and the phosphorylation sites of MARCKS are closely linked. Because the phosphorylation of MARCKS inhibits its actin cross-linking activity, it was of interest to determine whether calmodulin prevented the actin cross-linking activity of non-phosphorylated MARCKS. This proved to be the case: stoichiometric amounts of calmodulin completely inhibit the actin cross-linking activity of non-phosphorylated MARCKS (Hartwig et al 1992).

As discussed above, the MARCKS protein contains two highly conserved domains—an N-terminal membrane-binding domain, which is myristoylated, and a highly charged phosphorylation domain, which also contains the calmodulin binding-site (Graff et al 1989b). A synthetic peptide corresponding to residues 146–163 of murine MARCKS, which includes all the serine residues known to be phosphorylated (Seykora et al 1991), both increases the light-scatter intensity of F-actin in solution and aggregates actin filaments into tight bundles that are observable by electron microscopy. Phosphorylation of the serine residues in this peptide, however, prevents the peptide from cross-linking actin filaments, as shown by light-scattering and viscosity measurements and by electron microscopy (Hartwig et al 1992). Phosphorylation, therefore, modulates the ability of this peptide to cross-link actin. Because calmodulin inhibits the actin-bundling activity of non-phosphorylated MARCKS, and because the synthetic peptide contains the calmodulin-binding site (Graff et al 1989b), the effect of calmodulin on the actin-bundling activity of the peptide was also determined. Calmodulin, at a 1:1 molar ratio to the peptide, completely inhibits the actin-bundling activity of the non-phosphorylated peptide.

The synthetic peptide containing the phosphorylation sites, the calmodulin-binding site and the actin-binding site is predicted to form an $\alpha$-helix in aqueous solution. A helical wheel analysis positions the five lysine residues in the peptide on one side of the helix, such that a consensus calmodulin-binding site is formed (O'Neil & Degrado 1990). This analysis also suggests that the lysine residues are positioned similarly to those found in known actin-binding domains (Vandekerckhove 1989). The phosphorylatable serines are positioned on the opposite side of the helix. It is likely, therefore, that the lysine residues comprise both the actin- and calmodulin-binding sites of MARCKS; this hypothesis is supported by the observation that calmodulin binding prevents actin-bundling activity. The cross-linking of actin filaments requires monomeric molecules either to express two actin-binding sites, or to dimerize. It is not yet clear whether the synthetic peptide has a second actin-binding site or whether it forms multimers.

## Conclusions

During neurosecretion, leukocyte activation and growth factor-dependent mitogenesis cells undergo cytoskeletal rearrangements while MARCKS is both rapidly phosphorylated and redistributed to the cytoplasm from the cytoplasmic surface of the plasma membrane (Aderem et al 1988, Rosen et al 1990, Wang et al 1989). Phosphorylation of MARCKS results in its translocation from the plasma membrane into the cytosol, where it still appears to be closely associated with actin filaments. This phosphorylation-regulated translocation from the plasma membrane is accompanied by the marked reorganization of the actin cytoskeleton that occurs on activation of PKC (Phaire-Washington et al 1980, Schliwa et al 1984). The phosphorylation-dependent regulation of binding of

MARCKS to the membrane might therefore serve to modify the attachment of actin filaments to the membrane, thereby influencing cytoskeletal organization and morphology in response to signals that activate PKC.

*Acknowledgements*

We thank Antony Rosen, Marcus Thelen, John Seykora, Jeffrey Ravitch, John Hartwig and Paul Janmey, who all made major contributions to the work described here. This work was supported by National Institutes of Health AI 25032 and by grants-in-Aid from the Squibb Institute for Medical Research and the Cancer Research Institute. Alan Aderem is an Established Investigator of the American Heart Association.

## References

Aderem AA 1988 Protein myristoylation as an intermediate step during signal transduction in macrophages: its role in arachidonic acid metabolism and in responses to interferon. J Cell Sci Suppl 9:151–167

Aderem AA, Albert KA, Keum MM, Wang JKT, Greengard P, Cohn ZA 1988 Stimulus-dependent myristoylation of a major substrate for protein kinase C. Nature (Lond) 332:362–364

Albert KA, Wu WS, Nairn AC, Greengard P 1984 Inhibition by calmodulin of calcium/phospholipid-dependent protein phosphorylation. Proc Natl Acad Sci USA 81:3622–3625

Albert KA, Nairn AC, Greengard P 1987 The 87k protein, a major specific substrate for protein kinase C. Purification from bovine brain and characterization. Proc Natl Acad Sci USA 84:7046–7050

Bershadsky AD, Tint IS, Ney Fakh AA Jr, Vasilier JM 1985 Focal contacts of normal and RSV-transformed quail cells. Exp Cell Res 158:433–444

Bialojan C, Takai A 1988 Inhibitory effect of a marine-sponge toxin, okadaic acid, on protein phosphatases. Biochem J 256:283–290

Blackshear PJ, Wen L, Glynn BP, Witters LA 1986 Protein kinase C-stimulated phosphorylation in vitro of a Mr 80,000 protein phosphorylated in response to phorbol esters and growth factors in intact fibroblasts. Distinction from protein kinase C and prominence in brain. J Biol Chem 261:1459–1469

Burridge K, Molony L, Kelly T 1987 Adhesion plaques: sites of transmembrane interaction between the extracellular matrix and the actin cytoskeleton. J Cell Sci 8:211–229

Dewald B, Thelen M, Baggiolini M 1988 Two transduction sequences are necessary for neutrophil activation by receptor agonists. J Biol Chem 263:16179–16184

Graff JM, Stumpo DJ, Blackshear PJ 1989a Molecular cloning, sequence, and expression of a cDNA encoding the chicken *m*yristoylated *a*lanine-*r*ich *C* *k*inase *s*ubstrate (MARCKS). Mol Endocrinol 3:1903–1906

Graff JM, Young TN, Johnson JD, Blackshear PJ 1989b Phosphorylation-regulated calmodulin binding to a prominent cellular substrate for protein kinase C. J Biol Chem 264:21818–21823

Graff JM, Stumpo DJ, Blackshear PJ 1989c Characterization of the phosphorylation sites in the chicken and bovine myristoylated alanine-rich C kinase substrate protein, a prominent cellular substrate for protein kinase C. J Biol Chem 264:11912–11919

Graff JM, Gordon JI, Blackshear PJ 1989d Myristoylated and nonmyristoylated forms of a protein are phosphorylated by protein kinase C. Science (Wash DC) 246:503–506

Hartwig JH, Chambers KA, Stossel TP 1989 Association of gelsolin with actin filaments and cell membranes of macrophages and platelets. J Cell Biol 109:467–479

Hartwig JH, Rosen A, Janmey PA, Thelen M, Nairn AC, Aderem A 1992 The MARCKS protein is an actin filament crosslinking protein regulated by protein kinase C-mediated phosphorylation and by calcium/calmodulin. Submitted

Haystead TAJ, Sim ATR, Carling D et al 1989 Effects of tumor promoter okadaic acid on intracellular protein phosphorylation and metabolism. Nature (Lond) 337:78–81

Jaken S, Leach K, Klauck T 1989 Association of type 3 protein kinase C with focal contacts in rat embryo fibroblasts. J Cell Biol 109:697–704

James G, Olson EN 1989 Myristoylation, phosphorylation, and subcellular distribution of the 80-kDa protein kinase C substrate in BC3H1 myocytes. J Biol Chem 264:20928–20933

Marchisio PC, Cirillo D, Teti A, Zambonin-Zallone A, Tarone G 1987 Rous sarcoma virus-transformed fibroblasts and cells of monocytic cell origin display a peculiar dot-like organization of cytoskeletal proteins involved in microfilament-membrane interactions. Exp Cell Res 169:202–214

Mitchison T, Kirschner M 1988 Cytoskeletal dynamics and nerve growth. Neuron 1:761–772

O'Neil KT, Degrado WF 1990 How calmodulin binds its targets: sequence independent recognition of amphiphilic alpha-helices. Trends Biochem Sci 15:59–64

Patel J, Kligman D 1987 Purification and characterization of an Mr 87,000 protein kinase C substrate from rat brain. J Biol Chem 262:16686–16691

Phaire-Washington L, Silverstein SC, Wang E 1980 Phorbol myristate acetate stimulates microtubule and 10-nm filament extension and lysosome redistribution in mouse macrophages. J Cell Biol 86:641–655

Rosen A, Nairn AC, Greengard P, Cohn ZA, Aderem AA 1989 Bacterial lipopolysaccharide regulates the phosphorylation of the 68K protein kinase C substrate in macrophages. J Biol Chem 264:9118–9121

Rosen A, Keenan KF, Thelen M, Nairn AC, Aderem AA 1990 Activation of protein kinase C results in the displacement of its myristoylated, alanine-rich substrate from punctate structures in macrophage filopodia. J Exp Med 172:1211–1215

Rosen A, Hartwig J, Thelen M, Keenan KF, Nairn AC, Aderem A 1992 Regulated clustering of the MARCKS protein at points of actin attachment to the substrate adherent surface of the plasma membrane. Submitted

Rozengurt E, Rodriguez-Pena M, Smith KA 1983 Phorbol esters, phospholipase C, and growth factors rapidly stimulate the phosphorylation of a Mr 80,000 protein in intact quiescent 3T3 cells. Proc Natl Acad Sci USA 80:7244–7248

Schliwa M, Nakamura T, Porter KR, Euteneuer U 1984 A tumor promoter induces rapid and coordinated reorganization of actin and vinculin in cultured cells. J Cell Biol 99:1045–1059

Seykora JT, Ravetch JV, Aderem A 1991 Cloning and molecular characterization of the murine macrophage '68-kDa' protein kinase C substrate, and its regulation by bacterial lipopolysaccharide. Proc Natl Acad Sci USA 88:2505–2509

Stumpo DJ, Graff JM, Albert KA, Greengard P, Blackshear PJ 1989 Molecular cloning, characterization, and expression of a cDNA encoding the '80- to 87-kDa' myristoylated alanine-rich C kinase substrate: a major cellular substrate for protein kinase C. Proc Natl Acad Sci USA 86:4012–4016

Thelen M, Rosen A, Nairn AC, Aderem A 1990 Tumor necrosis factor α modifies agonist-dependent responses in human neutrophils by inducing the synthesis and myristoylation of a specific protein kinase C substrate. Proc Natl Acad Sci USA 87:5603–5607

Thelen M, Rosen A, Nairn AC, Aderem A 1991 Phosphorylation regulates the reversible association of a myristoylated protein kinase C substrate with the plasma membrane. Nature (Lond) 351:320–322

Vandekerckhove J 1989 Structural principles of actin-binding proteins. Curr Opin Cell Biol 1:15–22

Wang JKT, Walaas SI, Sihra TS, Aderem AA, Greengard P 1989 Phosphorylation and associated translocation of the 87-kDa protein, a major protein kinase C substrate, in isolated nerve terminals. Proc Natl Acad Sci USA 86:2253–2256
Wu WS, Walaas, SI, Nairn AC, Greengard P 1982 Calcium/phospholipid regulates phosphorylation of a Mr "87k" substrate protein in brain synaptosomes. Proc Natl Acad Sci USA 79:5249–5253

## DISCUSSION

*Cantrell:* Is MARCKS a substrate for all protein kinase C isotypes?

*Nairn:* I haven't tested that directly. The synthetic peptide containing all the phosphorylation sites is a substrate for the $\alpha$, $\beta$ and $\gamma$ isoforms and is phosphorylated by these with apparently similar kinetics. I haven't tested the other isotypes.

*Warner:* Do different doses of f-Met-Leu-Phe have different effects? For example, a high dose, such as 1 μM, might cause degranulation whereas a lower dose, such as 10 nM, would cause chemotaxis, enabling you to dissociate the two effects.

*Nairn:* The experiments were done with 500 nM f-Met-Leu-Phe. We had already demonstrated the effects of phorbol esters and wanted to look at a physiological agonist. At this concentration, f-Met-Leu-Phe would cause chemotaxis and degranulation.

*Fischer:* Does MARCKS bind to or affect microtubules?

*Nairn:* We have no evidence for that. We purified microtubules to see whether there was a MARCKS protein associated with them during purification but at no stage was there any enrichment in microtubule preparations, nor was there any association of the purified MARCKS and purified microtubules.

*Fischer:* Does it affect microtubule polymerization or cold stability?

*Nairn:* No.

*Fischer:* Have you looked at the localization of MARCKS in transformed cells?

*Nairn:* We have started to do those experiments. Work done in Ian Macara's (Wolfman et al 1987) and Nancy Colburn's (Simek et al 1989) laboratories suggest that the level of MARCKS in certain transformed cells is very low. In addition, MARCKS does not appear to be present in every cell type (Wolfman et al 1987, Harlan et al 1991). It is possible, therefore, that the absence of MARCKS might be associated with changes in the interaction of the cytoskeleton with the plasma membrane. For example, the low levels of MARCKS in transformed cells might explain some of the alterations in cell shape in these cells.

*Rasmussen:* What does calcium do to the binding of calmodulin to MARCKS?

*Nairn:* From the experiments I have done the binding of calmodulin–calcium to MARCKS seems typical of other calcium-dependent calmodulin-binding proteins.

*Rasmussen:* Is there a difference in the affinity in the absence and presence of calcium?

*Nairn:* MARCKS binds calmodulin only in the presence of μM calcium, in contrast to GAP-43 (growth-associated protein-43), which apparently binds calmodulin in the absence of calcium, and binding of calcium causes calmodulin to be released. MARCKS seems to behave as a normal calmodulin-binding protein.

*Mikoshiba:* What kind of a molecule do you imagine is attached to MARCKS in the plasma membrane?

*Nairn:* Experiments with v-*src* protein suggest that there may be a receptor for the myristoylated protein (Resh & Ling 1990). The punctate distribution of the MARCKS protein leads us to believe that there might also be some sort of MARCKS receptor. The interaction would involve the myristoylated part of the protein in some way, but would possibly include a protein–protein interaction. We have no evidence as yet that such a receptor exists.

*Krebs:* People were interested in MARCKS originally because of its use as an indicator of PKC activation. How good is this method?

*Nairn:* Studies in Blackshear's and Rozengurt's laboratories in a number of different cell types with different stimuli suggest that where PKC is activated MARCKS is phosphorylated (Blackshear et al 1986, Rozengurt et al 1983). Because of its ubiquitous distribution MARCKS is probably one of the best indicators of PKC activity that we have.

*Cantrell:* In the T cell interleukin 2, which is thought not to activate protein kinase C because it doesn't cause any changes in diacylglycerol or phosphorylation of other PKC substrates, does in fact cause phosphorylation of a protein that appears, from its pI, to be MARCKS, which is a PKC substrate in that cell. The question then is, what kinase is involved in that pathway?

*Nairn:* We have tested many kinases other than PKC. Cyclic AMP-dependent protein kinase can phosphorylate the protein *in vitro* at low rates, but this is not believed to be physiologically important.

*Cantrell:* Have you tried Raf?

*Nairn:* No, we haven't. Sue Jaken has some results suggesting that there is a subpopulation of PKC which is associated with focal adhesions (Jaken et al 1989). There may be a particular population of PKC which is intimately associated with MARCKS or other membrane-associated substrates and bulk PKC may not always be involved in the phosphorylation of MARCKS.

*Parker:* When PKC is associated with cytoskeletal elements, at least in our hands, it does not show activity either *in situ* or after any method of extraction that we have tried. Neither is there phosphorylation of endogenous proteins in such preparations.

*Nairn:* We haven't tried to actually measure enzyme activity because the amount of PKC associated with the cytoskeleton is a small fraction of the total PKC. The results convince us that PKC is associated with the cytoskeleton, so presumably it's possible to activate it.

*Tsien:* It seems possible that in a system such as a nerve terminal, where calcium levels increase and PKC is activated, calcium-driven calmodulin binding might result. This might temporarily occlude phosphorylation by PKC. One implication is that someone might underestimate how rapidly PKC is turned on if they were using MARCKS as a substrate. One possible experiment is to see whether the presence of BAPTA (1,2-bis[2-aminophenoxy]ethane *N,N,N′,N′*-tetraacetic acid, a $Ca^{2+}$ chelator) in a synaptosome affects the time course of phosphorylation. If you clamp the calcium concentration at a low level you should prevent calmodulin binding and actually hasten the phosphorylation.

*Nairn:* One of the difficult aspects of this work is dissecting the two signals, the calcium/calmodulin signal and the diacylglycerol/PKC signal. Results from studies in synaptosomes indicate that depolarization with high $K^+$, which would result in a higher $Ca^{2+}$ concentration, does lead to increased phosphorylation of MARCKS (Wu et al 1982, Wang et al 1989). These results suggest that the diacylglycerol signal predominates over the calcium/calmodulin signal. However, the experiments you suggest would help to clarify the relative contributions and the time course of the two signals.

*Tsien:* Many of the systems that we are concerned with in this meeting are 'AND' gates, where two different types of information need to act in a combinatorial manner. You seem to be dealing with an 'OR' gate, where either the calmodulin or the PKC pathway will suffice. Do you have evidence for such an 'OR' function in a more physiological setting?

*Nairn:* I presented only *in vitro* data. I need to do the types of experiments that you suggested in intact cells.

*Michell:* The usual idea is that when PKC is turned on diacylglycerol is the primary trigger and that calcium has a rather permissive role. As a complement to the experiment that Professor Tsien suggested, you could first treat cells with an ionophore and then add diacylglycerol or a phorbol ester after a minute or two, to see whether the phosphorylation is ablated or slowed.

*Nairn:* We have thought of doing such experiments. However, $Ca^{2+}$ ionophores are not physiological and it is difficult to provide a physiological calcium stimulation and a physiological PKC activation independently and actually distinguish between them.

*Cantrell:* Martin Gullberg has actually done that experiment in T cells. Treatment with ionomycin and phorbol dibutyrate doesn't seem to affect phosphorylation of MARCKS (Cantrell et al 1989).

*Daly:* Instead of using an ionophore you could use maitotoxin, which will cause a sustained calcium influx and will also activate translocation of protein kinase C. It would be very interesting to see what effect that would have on MARCKS phosphorylation.

*Michell:* Better still would be to use a caged compound that can provide 'instantaneous' calcium release.

*Nairn:* That would follow on from the experiment with BAPTA that Professor Tsien suggested.

*Daly:* With maitotoxin and a sustained input of calcium the calmodulin site should be occupied and this would presumably prevent the phosphorylation by activated PKC.

*Carpenter:* Graff et al (1989) have suggested that MARCKS could serve to inactivate calmodulin by forming a complex with it. Do you favour that idea?

*Nairn:* I doubt that a protein with such an interesting structure would have been designed simply to bind calmodulin. I think that the results I presented here suggest that MARCKS acts as an actin-binding protein and that the calmodulin sink idea is probably simplistic. I do believe that binding of calmodulin to MARCKS is important in the regulation of its actin-binding properties.

*Tsien:* This idea might seem simplistic for MARCKS, but might it apply to neuromodulin (GAP-43)?

*Nairn:* My bias for GAP-43 is that it too does something in addition to binding calmodulin. In fact, the MARCKS protein and GAP-43 have similar amino acid compositions and similar predicted secondary structures. GAP-43 is palmitoylated at the N-terminus and its site of phosphorylation by PKC is in the calmodulin-binding domain. It wouldn't surprise me if GAP-43 were a specialized form of MARCKS.

*Krebs:* Do you know what type of serine phosphatases are most active against MARCKS?

*Nairn:* We have tested phosphatases 1, 2A and 2B. Both 1 and 2A worked fairly well, but 2A was slightly better than 1 (K. A. Albert & A. C. Nairn, unpublished results). I don't know what is responsible for dephosphorylation in intact cells, though the results I described on the effects of okadaic acid suggest that phosphatase 2A may be involved.

*Kanoh:* Do the kinetics of phosphorylation and dephosphorylation of MARCKS correlate with altered interaction of actin filaments with the plasma membrane, as you proposed in your model?

*Nairn:* If you stimulate neutrophils with f-Met-Leu-Phe, phosphorylation peaks at about 40 seconds and after about 2–4 min the level of phosphorylation returns to base-line levels. Similar stimulation of neutrophils causes transient polymerization and depolymerization of actin filaments which occurs within the same time period. In the experiments with macrophages, stimulation with phorbol esters increased phosphorylation of MARCKS within seconds, and a peak was reached in several minutes. This correlated quite well with the loss of MARCKS immunoreactivity associated with the macrophage filopodia and with the changes in cell shape.

*Nishizuka:* Do you know the binding constant for actin?

*Nairn:* We are working on that at the moment. The initial experiments suggest that it's submicromolar, and there are other indications that it must be a high affinity interaction. One of the problems that we encountered in purifying

MARCKS is that it associates with actin throughout purification. In several cases MARCKS has been purified by heat denaturation, which we obviously do not want to use when looking for functional activity. Even so, we have to prepare the protein using reverse-phase HPLC because of the problem of its tight association with actin.

*Nishizuka:* I believe MARCKS binds to calmodulin with very high affinity also.

*Nairn:* That's correct. The binding constant is 3 nM (Graff et al 1989).

*Hunter:* For there to be bundling of F-actin the protein either has a single binding site and self-associates, or it must have many binding sites. Therefore, either the peptide that you made dimerizes or oligomerizes, or it must have two binding sites.

*Nairn:* That's correct. We don't favour one or the other at the moment. Initially, we favoured the latter possibility, and we felt that one of the actin-binding sites would be in the phosphorylation site domain. However, the fact that the peptide is able to bundle actin suggests that the peptide is either aggregating, or has two binding sites.

*Hunter:* You said that MARCKS decorates the F-actin regularly.

*Nairn:* With a large excess of dephosphorylated MARCKS, using negative staining you can see decoration of actin with a 20 nm spacing on the actin filaments. Phosphorylated MARCKS protein seems to decorate actin differently but we can't really get quantitative data from the negative staining images.

*Hunter:* Does it cap actin during polymerization?

*Nairn:* We have no evidence of capping. MARCKS seems to be a side-binding actin-binding protein.

*Hunter:* It's distribution in the cell doesn't look like that of tropomyosin, or anything that decorates actin filaments, so there must be something else that localizes MARCKS more specifically than binding to F-actin.

*Nairn:* Presumably there is some other protein component of focal adhesions, or at least a subpopulation of focal adhesions, that is involved in binding MARCKS. In addition, MARCKS is probably only associated with the small amount of actin filaments that are associated with filopodia.

*Hunter:* MARCKS would need to have a higher affinity for that other protein than it does for F-actin, otherwise when it is dephosphorylated and it has to return to where it belongs it would simply bind to the nearest F-actin and would become distributed across the F-actin network.

*Nairn:* The results suggest that there may be a binding site for the plasma membrane which is distinct from the actin-binding domain; that would be the region that would be released on dephosphorylation.

*Hunter:* But the binding of MARCKS to whatever attaches it to the membrane is also lost on phosphorylation.

*Nairn:* The MARCKS protein is released, but whatever binds MARCKS to the membrane would not be released.

*Hunter:* When MARCKS is phosphorylated it is lost from the focal contacts and it becomes soluble, so it's not bound to microfilaments.

*Nairn:* After the protein has been released from the membrane there is still association of the phosphorylated protein with the microfilaments.

*Hunter:* That's what the immuno-electron microscopy shows, but fractionation experiments suggest that it is not associated with microfilaments.

*Nairn:* The immuno-electron microscopy suggest that there is still some MARCKS associated with the microfilaments. Biochemical analysis suggests that while phospho-MARCKS still interacts with actin filaments, the interaction is different from that with dephospho-MARCKS in that the actin filaments are not bundled. In our fractionation experiments the cells are lysed by nitrogen cavitation and the lysates are separated by discontinuous sucrose gradient centrifugation. The conditions of lysis may cause dissociation of phospho-MARCKS from the microfilaments or their depolymerization. Though a large amount of filamentous actin remains associated with other components of the cytoskeleton, there are significant levels of cytosolic actin. We did not find any changes in the amount of cytosolic actin after treatment of cells with phorbol esters. However, given that MARCKS interacts with a small percentage of the total cellular actin, we might not expect to be able to measure this.

*Carpenter:* You said that calmodulin inhibits binding of MARCKS to actin. Can calmodulin disassemble the actin bundles that are formed by MARCKS?

*Nairn:* We have not done that experiment, but one would predict that it could.

*Exton:* How much MARCKS is there in the cell in relation to calmodulin and actin, which are abundant proteins?

*Nairn:* My estimates of the concentrations of these proteins in the brain are; MARCKS 1–5 μM, calmodulin 50–100 μM and actin 1mM.

*Exton:* So MARCKS influences only a relatively small proportion of these proteins.

*Nairn:* When purified, MARCKS is in association with actin, but there's a lot of actin that's not associated with MARCKS. In fact, there could be a specific population of actin molecules that interact with MARCKS.

*Tsien:* Is it possible to design an experiment that will tell you whether the molecules tend to return to the same place on the surface of the cell during the cycle of release and reattachment?

*Nairn:* One of the problems with working with this protein is that most of the antibodies that have been raised recognize the carboxy terminal and don't cross-react between species. We might be able to exploit that to answer your question, using MARCKS proteins from different sources and specific antibodies.

*Yang:* What is the pI of the MARCKS protein?

*Nairn:* It is predicted from the sequence to be 4.1.

*Yang:* The protein has a high content of glutamic acid. Do the anionic groups have a particular role in the protein's biological activity?

*Nairn:* The fact that there is a fairly non-conserved array of glutamate, alanine, proline and glycine residues throughout the molecule suggests that these residues are important for the secondary structure, and, by implication, for function.

*Kanoh:* Is there a requirement for phosphatidylserine for binding of MARCKS to actin? Can MARCKS bind phosphatidylserine?

*Nairn:* In some conditions the purified protein binds to phospholipid vesicles but that is presumably because of the myristic acid portion.

*Nishizuka:* TPA closes gap junctions in some cell types. Is there any localization of MARCKS at gap junctions?

*Nairn:* No.

*Daly:* Has anyone investigated whether there is a fraction of the MARCKS protein that doesn't have an available binding site in the membrane, by overexpressing the protein?

*Nairn:* Perry Blackshear has overexpressed certain forms of the MARCKS protein and John Seykora in Alan Aderem's lab is presently doing that type of experiment.

*Daly:* So you don't know as yet whether there are a limited number of sites in the membrane with which it can associate.

*Nairn:* That's correct—we don't know.

*Tanaka:* Have you done any immunohistochemical studies of MARCKS in the central nervous system?

*Nairn:* Charlie Ouimet has done detailed studies at the light and electron microscopy level of the distribution of MARCKS in the CNS (Ouimet et al 1990). The MARCKS protein was initially identified in synaptosomes, and it's clear from light microscopy that the protein is present in neurons. However, a large amount of MARCKS in the CNS is associated with microglia, which are the nervous system's equivalent of macrophages.

*Carpenter:* It's interesting that phosphorylation causes relocalization of MARCKS to the cytoplasm, but it's not clear to me how that might occur. Do you think that localization in the membrane requires both myristoylation and interaction of non-phosphorylated MARCKS with actin bundles and that phosphorylation releases it from the actin bundles?

*Nairn:* There are various models that can be proposed but we have no information to support any of them. Whatever you could think of would be just as good as what I could think up! There are a number of possibilities. The simplest interpretation is that there is a receptor that requires both the myristic acid and part of the amino terminus of the protein, and that phosphorylation somehow changes the interaction with the protein component so that the interaction of myristic acid with membrane lipids is no longer strong enough to keep the protein attached to the membrane. Is is also possible that there is a protein receptor for MARCKS which can be phosphorylated and that phosphorylation of the receptor causes release of MARCKS from the plasma membrane.

## References

Blackshear PJ, Wen L, Glynn BP, Witters LA 1986 Protein kinase C-stimulated phosphorylation in vitro of a Mr 80,000 protein phosphorylated in response to phorbol esters and growth factors in intact fibroblasts. Distinction from protein kinase C and prominence in brain. J Biol Chem 261:1459–1469

Cantrell DA, Friedrich B, Davies AA, Gullberg M, Crumpton MJ 1989 Evidence that a kinase distinct from protein kinase C induces CD3 gamma-subunit phosphorylation without a concomitant down-regulation in CD3 antigen expression. J Immunol 142:1626–1630

Graff JM, Young TN, Johnson JD, Blackshear PJ 1989 Phosphorylation-regulated calmodulin binding to a prominent cellular substrate for protein kinase C. J Biol Chem 264:21818–21823

Harlan DM, Graff JM, Stumpo DJ et al 1991 The human myristoylated alanine-rich C kinase substrate (MARCKS) gene (MACS): analysis of its gene product, promotor and chromosomal localization. J Biol Chem 266:14399–14406

Jaken S, Leach K, Klauck T 1989 Association of type 3 protein kinase C with focal contacts in rat embryo fibroblasts. J Cell Biol 109:697–704

Ouimet CC, Wang JKT, Walaas SI, Albert KA, Greengard P 1990 Localization of the MARCKS (87 kDa) protein, a major specific substrate for protein kinase C, in rat brain. J Neurosci 10:1683–1698

Resh MD, Ling H-p 1990 Identification of a 32k plasma membrane protein that binds to the myristylated amino-terminal sequence of p60$^{v\text{-}src}$. Nature (Lond) 346:84–86

Rozengurt E, Rodriguez-Pena M, Smith KA 1983 Phorbol esters, phospholipase C, and growth factors rapidly stimulate the phosphorylation of a Mr 80,000 protein in intact quiescent 3T3 cells. Proc Natl Acad Sci USA 80:7244–7248

Simek SL, Kligman D, Patel J, Colburn NH 1989 Differential expression of an 80-kDa protein kinase C substrate in preneoplastic and neoplastic mouse JB6 cells. Proc Natl Acad Sci USA 86:7410–7414

Wang JKT, Walaas SI, Sihra TS, Aderem AA, Greengard P 1989 Phosphorylation and associated translocation of the 87-kDa protein, a major protein kinase C substrate in isolated nerve terminals. Proc Natl Acad Sci USA 86:2253–2256

Wolfman A, Wingrove TG, Blackshear PJ, Macara IG 1987 Down-regulation of protein kinase C and of an endogenous 80-kDa substrate in transformed fibroblasts. J Biol Chem 262:16546–16552

Wu WCS, Walaas SI, Nairn AC, Greengard P 1982 Calcium/phospholipid regulates phosphorylation of an 87 kilodalton substrate protein in brain synaptosomes. Proc Natl Acad Sci USA 79:5249–5253

# The synaptic activation of NMDA receptors and $Ca^{2+}$ signalling in neurons

G. L. Collingridge*†, A. D. Randall*‡, C. H. Davies*† and S. Alford*†

*Department of Pharmacology*, The University of Birmingham, The Medical School, Edgbaston, Birmingham B15 2TT, UK and the Departments of Pharmacology† and Biochemistry‡, School of Medical Sciences, University of Bristol BS8 1TD, UK*

*Abstract.* Long-term potentiation (LTP) in the hippocampus is a model system for understanding the synaptic basis of learning and memory. We have studied the mechanism of induction of LTP using voltage-clamp techniques and confocal imaging of $Ca^{2+}$ in rat hippocampal slices. In the Schaffer collateral-commissural pathway the neurotransmitter L-glutamate activates two classes of ionotropic receptor, named after the selective ligands AMPA (α-amino-3-hydroxy-5-methyl-4-isoxazole propionate) and NMDA (*N*-methyl-D-aspartate). During low frequency transmission the excitatory postsynaptic potential (EPSP) is mediated predominantly by AMPA receptors. NMDA receptors play a minor role because their ion channels are substantially blocked by $Mg^{2+}$, and this block is intensified by GABA-mediated synaptic inhibition. During high frequency transmission the GABA-mediated inhibition is depressed, by mechanisms initiated by $GABA_B$ autoreceptors. This allows a greater contribution from the NMDA receptors, through which $Ca^{2+}$ enters the dendrites of the postsynaptic neurons to initiate a cascade of biochemical processes which ultimately result in enhanced synaptic efficiency.

*1992 Interactions among cell signalling systems. Wiley, Chichester (Ciba Foundation Symposium 164) p 162–175*

L-Glutamate is the major excitatory neurotransmitter in the brain of vertebrates. Its effects are mediated by a variety of receptors which either directly gate ion channels (ionotropic receptors) or affect second messenger systems (metabotropic receptors). One type of ionotropic receptor, named after the selective agonist *N*-methyl-D-aspartate (NMDA), has been the focus of considerable attention because it is involved in synaptic plasticity, neurotrophic mechanisms and a variety of pathological processes (Lodge & Collingridge 1990). We have been studying how this receptor system is activated by synaptically released L-glutamate in the hippocampus of the rat. We are interested in how its activation leads to the induction of long-term potentiation (LTP), a model system for understanding the synaptic basis of learning and memory (Bliss & Lynch 1988). Here, we describe how the synaptic activation of the NMDA receptor system is controlled by dynamic changes in GABA-mediated synaptic inhibition, through an interaction at the level of the membrane potential. We also summarize our attempts to detect the $Ca^{2+}$ entering through synaptically activated NMDA receptor-gated ion channels, because this signal is believed to initiate the cascade of biochemical processes that mediate the enhancement of synaptic efficiency in LTP.

## Low frequency synaptic transmission

In our studies we have used the CA1 region of the hippocampus, which contains the highest density of NMDA receptors in the adult rat brain. We believe that the synaptic mechanisms that operate here are found throughout the vertebrate central nervous system and are involved in one of the fundamental mechanisms of synaptic plasticity. The experiments have been performed on 200–400 μm thick slices of rat hippocampus maintained *in vitro* using standard techniques. A stimulating electrode, placed in the stratum radiatum, is used to activate excitatory fibres (which originate from CA3 pyramidal cells and form the glutamatergic Schaffer collateral-commissural pathway) and inhibitory fibres (of $\gamma$-aminobutyric acid [GABA]-releasing interneurons). The electrophysiological recordings described here have been obtained using a patch pipette (or conventional microelectrode) placed on (or in) a neuron in the CA1 cell body layer (Fig. 1).

Low frequency stimulation (for example, one stimulus every minute) evokes a characteristic excitatory postsynaptic potential (EPSP)–inhibitory postsynaptic potential (IPSP) sequence (Fig. 2A). The EPSP is made up of two components—a faster component mediated by AMPA receptors (named after the selective agonist $\alpha$-amino-3-hydroxy-5-methyl-4-isoxazole propionate) and a slower component mediated by NMDA receptors. The IPSP is also composed of two kinetically distinct phases—a faster component mediated by $GABA_A$ receptors and a slower component mediated by $GABA_B$ receptors (see Figs. 1 and 2).

The NMDA receptor-mediated conductance is subject to a highly voltage-dependent block by the $Mg^{2+}$ present in the synaptic cleft (Ault et al 1980, Nowak et al 1984, Mayer et al 1984, Sah et al 1990, Randall et al 1990, Konnerth et al 1990). The voltage dependence of this block is shown in Fig. 2. It can be seen that the block intensifies as the cell is hyperpolarized below $-35$ mV, so that at membrane potentials near resting ($-70$ mV) the conductance is 80–90% blocked. Significantly, the block is further intensified as the cell becomes hyperpolarized by GABA-mediated synaptic inhibition. The combination of this voltage-dependent block and the relative time courses of the four synaptic components explains why during low frequency stimulation there is little

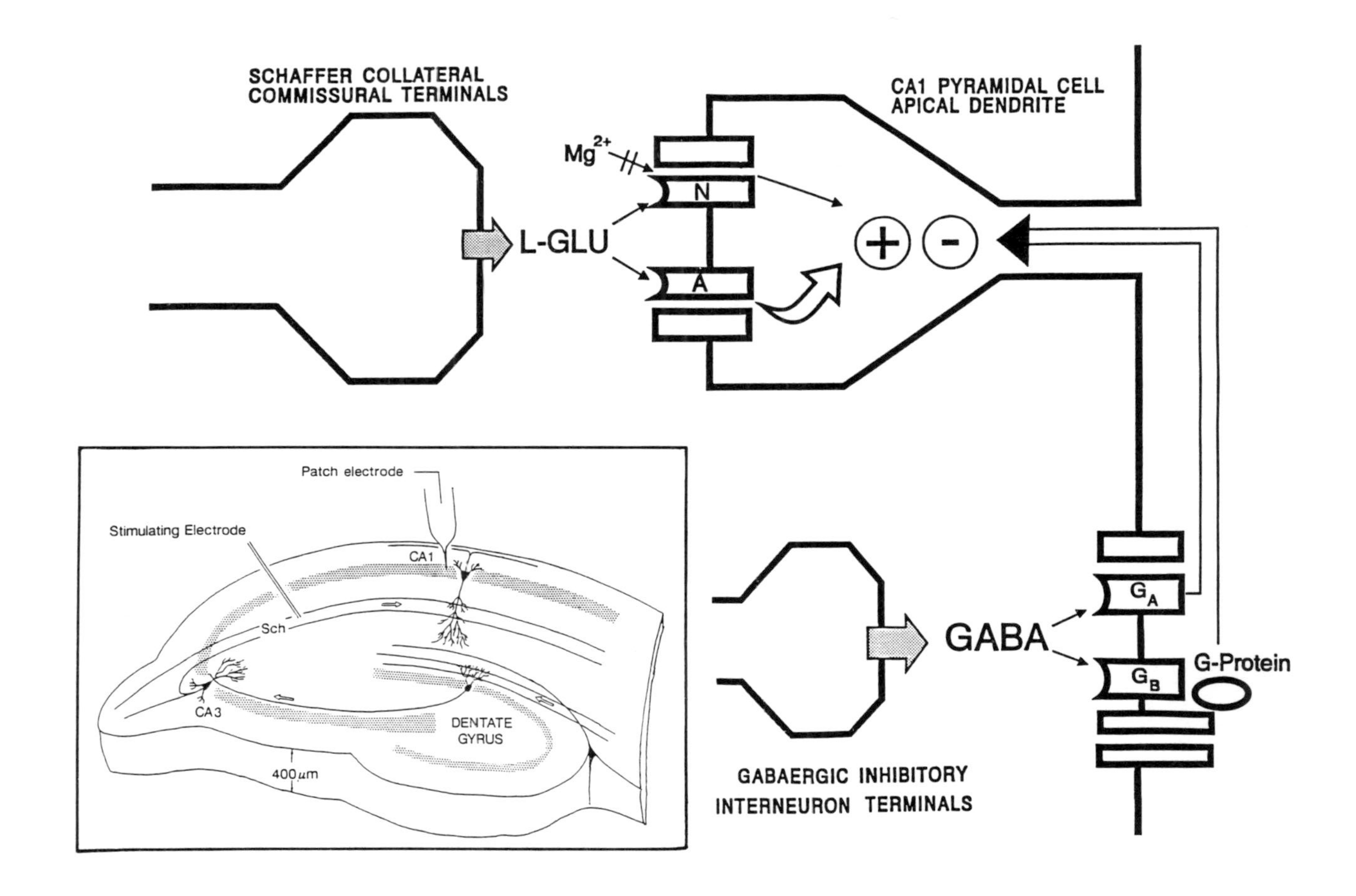
SCHAFFER COLLATERAL
COMMISSURAL TERMINALS
L-GLU
Mg2+
N
A
CA1 PYRAMIDAL CELL
APICAL DENDRITE
GABA
GA
GB
G-Protein
GABAERGIC INHIBITORY
INTERNEURON TERMINALS
Patch electrode
Stimulating Electrode
CA1
Sch
CA3
DENTATE
GYRUS
400μm

contribution from NMDA receptors to the synaptic response (Collingridge et al 1988). Thus, by the time that the sluggish NMDA receptor-mediated conductance has been fully activated, the GABA-mediated IPSP has hyperpolarized the cell into a region where NMDA channels are substantially (>90%) blocked by $Mg^{2+}$. This ability of IPSPs to limit the synaptic activation of NMDA receptors has been shown experimentally in many different ways (e.g. Herron et al 1985, Dingledine et al 1986, Collingridge et al 1988).

## High frequency synaptic transmission

Although there is a finite conductance contributed by NMDA receptors during low frequency transmission, this does not result in detectable alterations in synaptic efficiency and makes, at most, a minor contribution to the input–output relationship of a cell. However, the role of NMDA receptors during high frequency transmission is very different; they contribute greatly to the synaptic response, promoting repetitive firing (Herron et al 1986). They can also initiate the biochemical processes which lead to LTP (Collingridge et al 1983). Experimentally, a period of high frequency stimulation (termed a tetanus) is usually used to induce LTP. This is often a train of many impulses (for example, 100 shocks delivered at 100 Hz), although much more modest tetanus paradigms are effective (Rose & Dunwiddie 1986, Larson & Lynch 1986). We have used a 'primed-burst' paradigm (a single 'priming' pulse followed 200 ms later by a 'burst' of four shocks at 100 Hz) to demonstrate how dynamic changes in GABA-mediated synaptic inhibition permit sufficient activation of the NMDA receptor system for the induction of LTP. The 'priming' pulse depresses both $GABA_A$ and $GABA_B$ receptor-mediated synaptic inhibition such that a subsequent brief burst of appropriately timed shocks can provide EPSPs which summate and allow greatly enhanced expression of the NMDA receptor-mediated component. Because of the slow time course of the NMDA receptor-mediated conductance, the NMDA receptor-mediated EPSPs (once released from their inhibitory hold) summate very effectively. The depression of GABA-mediated synaptic inhibition

---

FIG. 1. Amino acid receptors involved in synaptic transmission in the Schaffer collateral-commissural pathway (Sch) of the rat. The excitatory neurotransmitter, probably L-glutamate (L-GLU), acts on two types of ionotropic receptor. During low frequency transmission the AMPA ($\alpha$-amino-3-hydroxy-5-methyl-4-isoxazole propionate) receptors (A) provide most of the current, because the ion channels gated by the NMDA receptors (N) are blocked for most of the time by $Mg^{2+}$, present in the synaptic cleft. The inhibitory neurotransmitter (GABA, $\gamma$-aminobutyric acid) acts on $GABA_A$ and $GABA_B$ receptors ($G_A$ and $G_B$) which act to hyperpolarize the cell and thereby intensify the $Mg^{2+}$ block. The inset shows a schematic of the slice preparation. (See Collingridge et al 1988, Randall et al 1990 for details of methods.)

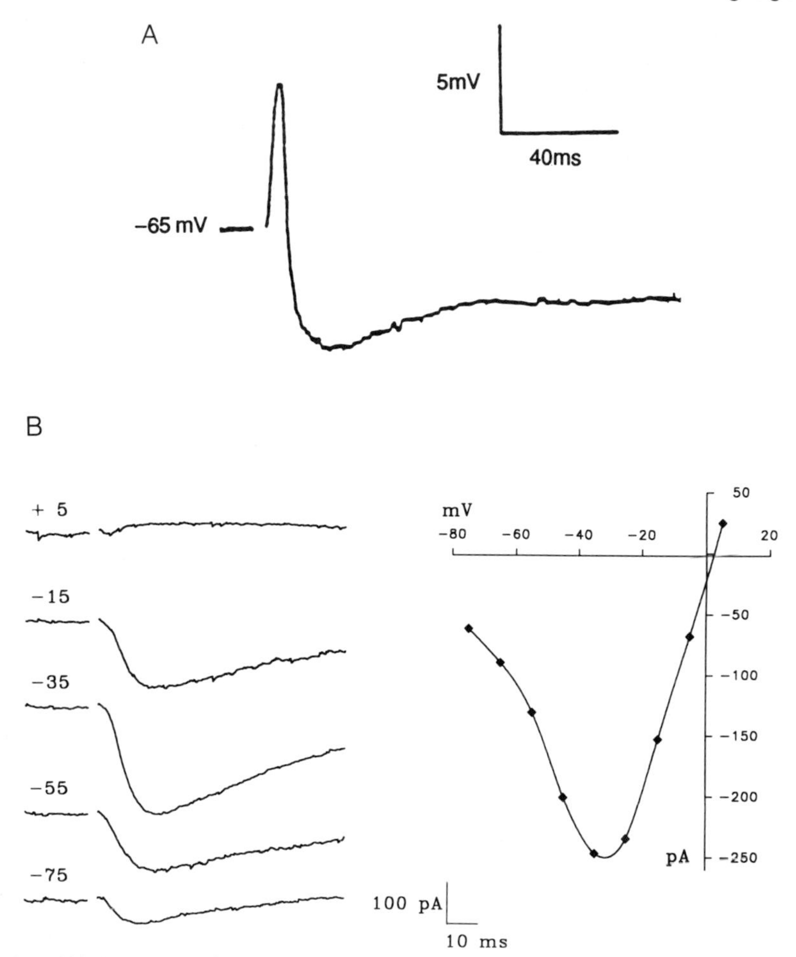


FIG. 2. (A) A characteristic EPSP–IPSP (excitatory postsynaptic potential–inhibitory postsynaptic potential) complex recorded intracellularly from a rat CA1 pyramidal cell after low frequency stimulation of the Schaffer collateral-commissural pathway. The upward deflection is an AMPA receptor-mediated EPSP which is curtailed by a biphasic IPSP mediated by $GABA_A$ and $GABA_B$ receptors, respectively. (B) A pharmacologically isolated NMDA receptor-mediated excitatory postsynaptic current (EPSC) recorded at a variety of different membrane potentials. The graph plots the peak amplitude of this synaptic current versus the voltage-clamped membrane potential. Note the region of negative slope conductance (at potentials more negative than −35 mV). See Randall et al 1990 for further details.

is an active process, caused by GABA feeding back and inhibiting its own release via a mechanism initiated by the activation of $GABA_B$ autoreceptors (Davies et al 1991). This mechanism is summarized in Fig. 3 (p 168).

In summary, frequency-dependent changes in the GABA system directly regulate the synaptic activation of the NMDA receptor system through an interaction at the level of the membrane potential. Once activated, the NMDA receptor system contributes directly to synaptic transmission by mediating a slow inward current that is mainly carried by $Na^+$ ions. More significant for contemporary theories of LTP is that these channels also provide a direct route of $Ca^{2+}$ entry into neurons (MacDermott et al 1986, Jahr & Stevens 1987, Ascher & Nowak 1988). It is known that high frequency stimulation leads to accumulation of $Ca^{2+}$ in CA1 cell dendrites (Regehr & Tank 1990) and that postsynaptic $Ca^{2+}$ is required for the induction of LTP (Lynch et al 1983, Malenka et al 1988). Because of these and similar observations it is often assumed that $Ca^{2+}$ entry directly through NMDA channels is necessary for the initiation of the biochemical processes that lead to LTP. In this way, assuming that the $Ca^{2+}$ entering through the channels of the NMDA receptors is localized (for example within spines), one can account for the specificity of LTP to tetanized fibres. In the next section we describe our first attempts to visualize this $Ca^{2+}$ signal which has such a central function in LTP.

## $Ca^{2+}$ entry through synaptically activated NMDA receptor channels

In order to distinguish $Ca^{2+}$ entry through channels gated by NMDA receptors from that entering via voltage-gated channels we performed experiments in voltage-clamped neurons. The patch pipette, while being used to control membrane potential and to measure membrane current, was also used to load the $Ca^{2+}$ indicator fluo-3 (Tsien 1988) into the cell. The change in fluorescence of the dye when excited by an argon ion laser was taken as an indication of an alteration in cytosolic $Ca^{2+}$ concentration and was visualized using a scanning confocal microscope (Fine et al 1988). As shown in Fig. 4 (see colour plate facing p 170), tetanic stimulation resulted in a transient increase in $Ca^{2+}$ concentration in the dendrites of voltage-clamped neurons. In contrast, there was no increase in $Ca^{2+}$ in the soma of voltage-clamped cells, indicating the specificity of the change to the region of the cell where the synaptic inputs are located. (The $Ca^{2+}$ level was increased in the soma in cells that were not voltage-clamped, presumably because the tetanus-induced synaptic depolarization was able to activate voltage-gated $Ca^{2+}$ channels.) The dendritic transient calcium influx was blocked by the selective NMDA antagonist D-2-amino-5-phosphonopentanoate (D-AP5) and, in a manner expected for an NMDA receptor-mediated conductance in the presence of $Mg^{2+}$, increased with depolarization (between $-90$ and $-35$ mV). Analysis of small segments of dendrites (about 4 μm lengths) showed that the $Ca^{2+}$ transient decayed in a heterogeneous manner, with some regions showing pronounced oscillations indicative of $Ca^{2+}$-induced $Ca^{2+}$ release (Fig. 5). These oscillations were prevented by D-AP5, suggesting that they may be initiated by the $Ca^{2+}$ entering through NMDA receptor-gated channels. In contrast, the rise in $Ca^{2+}$ levels during the tetanus was smooth and is, we believe, caused by $Ca^{2+}$ entry directly through NMDA receptor-gated channels.

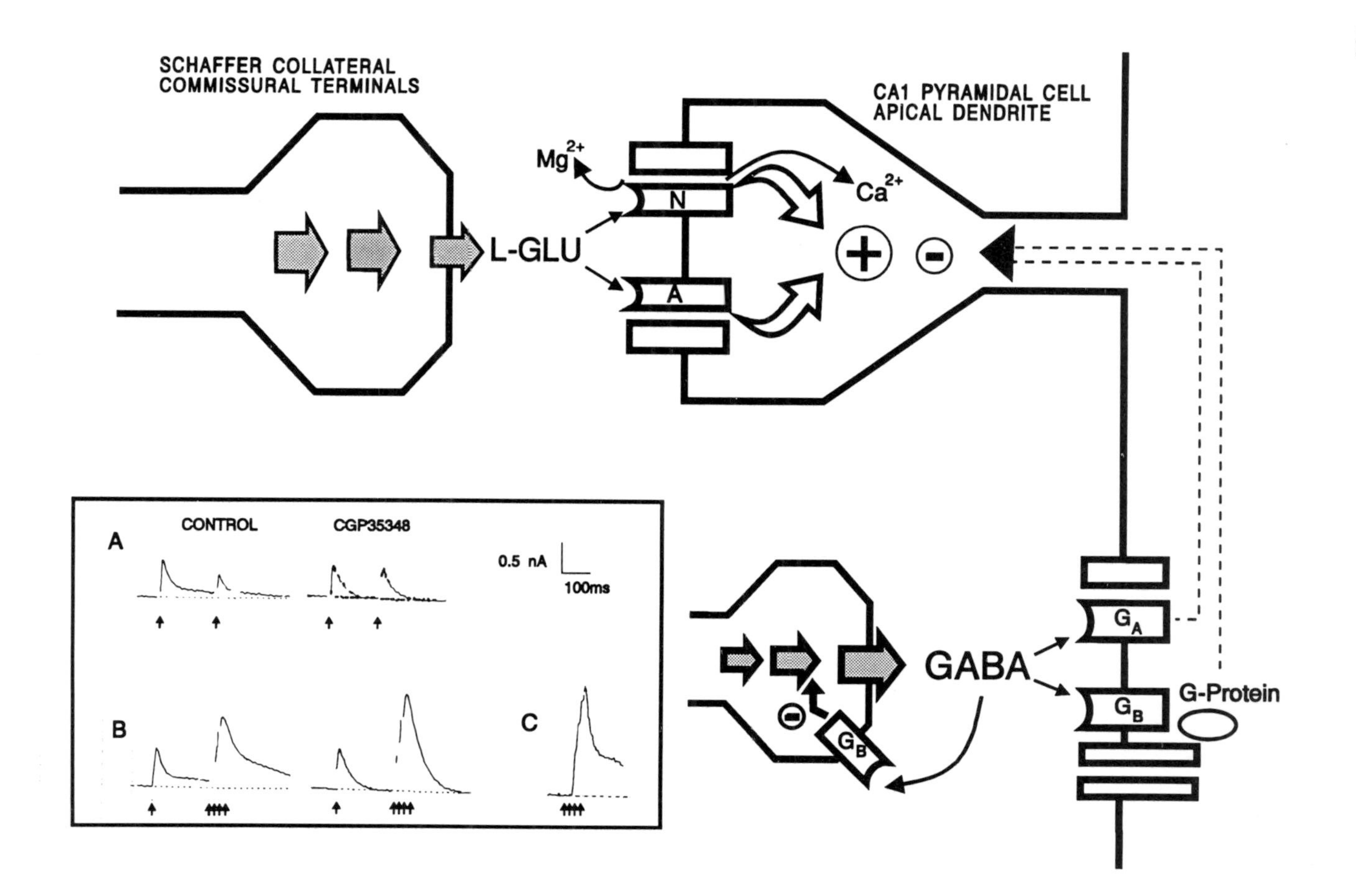

SCHAFFER COLLATERAL
COMMISSURAL TERMINALS
CA1 PYRAMIDAL CELL
APICAL DENDRITE
Mg2+
Ca2+
L-GLU
N
A
CONTROL
CGP35348
A
0.5 nA
100ms
B
C
GABA
GA
GB
G-Protein
GB

## Concluding remarks

In summary, the induction of LTP involves interaction between the two major neurotransmitter systems in the brain; this interaction is primarily brought about by the GABA-mediated IPSP affecting the level of the $Mg^{2+}$ block of NMDA receptor-gated ion channels through its hyperpolarizing action. During synaptic activity of a low frequency nature this block is intensified by GABAergic IPSPs. However, during high frequency transmission GABA-mediated IPSPs are autoinhibited, allowing NMDA receptors to contribute to the synaptic response sufficiently to induce LTP. $Ca^{2+}$ enters directly into the dendrites of CA1 neurons through channels in NMDA receptors located at the activated synapses. This localized $Ca^{2+}$ entry probably confers the specificity of LTP and initiates a sequence of biochemical processes that lead to the persistent increase in synaptic strength (see Malgaroli et al 1992, this volume). Because LTP involves complex interactions (i) between populations of neurons, (ii) at the receptor level, both pre- and postsynaptically, (iii) between intracellular signalling pathways and (iv) at the level of gene expression, the possibility for modulation via other receptor systems is substantial. Interest has arisen in interactions between muscarinic acetylcholine receptor activation and LTP because of their mutual involvement in learning processes and pathological states such as Alzheimer's disease. The interactions with the LTP process of other systems, for example, monoamine and peptide neurotransmitters, growth factors and hormones, are being increasingly investigated.

In conclusion, interactions between cell signalling systems are of prime importance in the processes that are likely to constitute the principal synaptic

---

FIG. 3. Dynamic changes in synaptic inhibition during high frequency transmission. During high frequency transmission GABA depresses its own release through an action on $GABA_B$ autoreceptors. With the hyperpolarizing influence of the IPSPs greatly reduced, the postsynaptic cell is more readily depolarized by L-glutamate, allowing much greater activation of the NMDA receptor system. The inset shows some of the experimental results from which this mechanism was deduced. (A) Paired pulse depression of inhibitory postsynaptic currents (IPSCs), recorded in the presence of excitatory amino acid antagonists. The $GABA_B$ antagonist CGP35348 (1 mM) blocks this depression and the postsynaptic $GABA_B$ receptor-mediated component of the response (late phase of the IPSC). (B) The effects of a 'primed-burst' protocol (four shocks at 100 Hz, preceded by a single 'priming' shock 200 ms earlier) on IPSCs. CGP35348, by blocking the 'primed' depression, allows much greater summation of the $GABA_A$ receptor-mediated component of the response. (C) shows how four IPSCs, evoked at 100 Hz, would summate if no frequency-dependent changes occurred (computed from the 'priming' IPSC under control conditions). Note that CGP35348 almost completely prevents the priming depression of the $GABA_A$ receptor-mediated response. This indirect potentiation of $GABA_A$ receptor-mediated inhibition can block the induction of LTP (see Davies et al 1990, 1991 for more details).

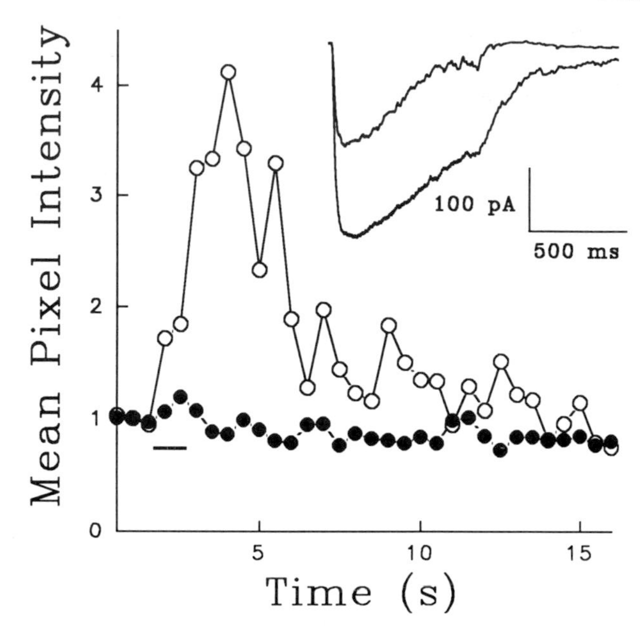


FIG. 5. $Ca^{2+}$ oscillations induced by $Ca^{2+}$ entry through NMDA receptor-gated channels. The graph plots the change in fluorescence caused by a tetanus (as in Fig. 4), measured from a 4 μm length of dendrite, before addition of (open circles) and in the presence of the selective NMDA antagonist D-2-amino-5-phosphonopentanoate (100 μM; filled circles). Note the pronounced oscillations in the decaying phase of the response. The synaptic current was greatly depressed by the NMDA antagonist (the residual component is mediated by AMPA receptors). The cell was voltage-clamped at −35 mV.

mechanisms underlying learning (and other forms of neuronal plasticity) in the vertebrate brain.

*Acknowledgements*

We thank Professor J. C. Watkins for gifts of compounds, J. George Schofield for facilities and the MRC and the Wellcome Trust for financial support.

## References

Ascher P, Nowak L 1988 The role of divalent cations in the N-methyl-D-aspartate responses of mouse central neurones in culture. J Physiol (Lond) 399:247–266

Ault B, Evans RH, Francis AA, Oakes DJ, Watkins JC 1980 Selective depression of excitatory amino acid induced depolarizations by magnesium ions in isolated spinal cord preparations. J Physiol (Lond) 307:413–428

Bliss TVP, Lynch MA 1988 Long-term potentiation of synaptic transmission in the hippocampus—properties and mechanisms. Neurol Neurobiol (NY) 35:3–72

Collingridge GL, Kehl SJ, McLennan H 1983 Excitatory amino acids in synaptic transmission in the Schaffer collateral-commissural pathway of the rat hippocampus. J Physiol (Lond) 334:33–46

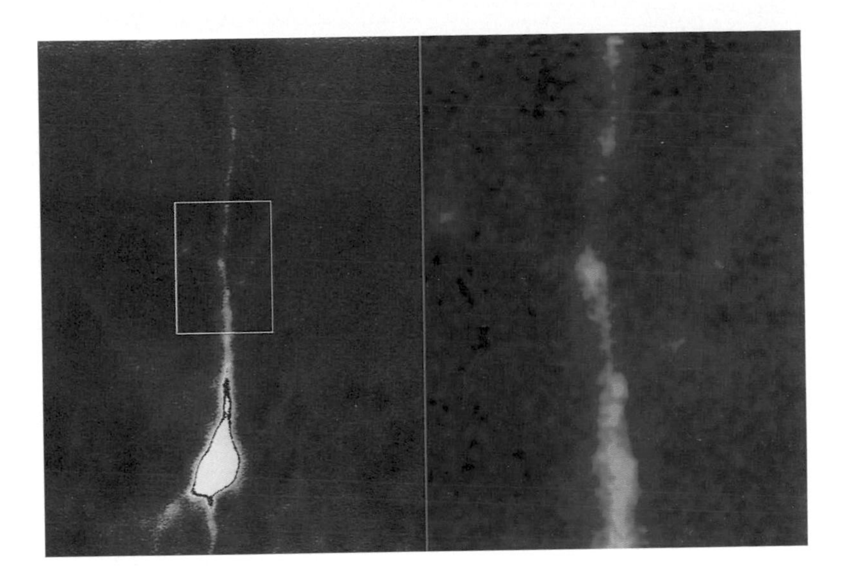

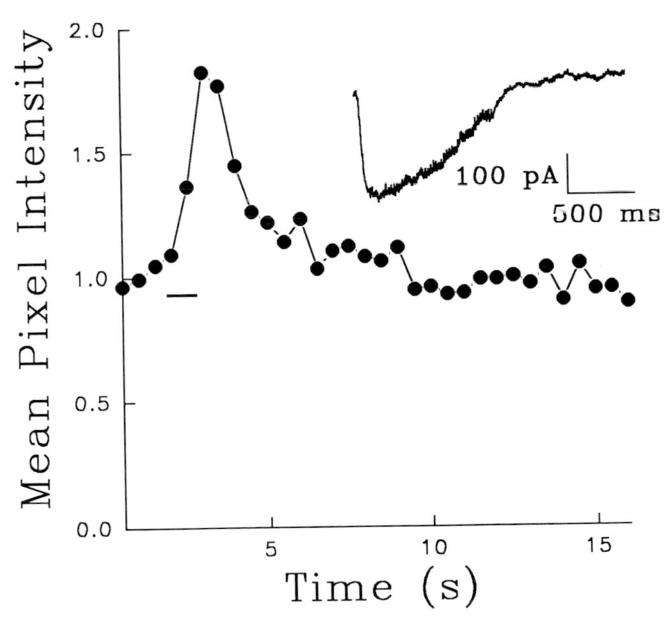


FIG. 4. Tetanic stimulation increases $Ca^{2+}$ concentration in the dendrites of a voltage-clamped CA1 neuron. *Left, top:* An optical section of a cell and *right, top* an expanded view of the boxed dendritic segment from which measurements were obtained. The box is 40 $\mu$m across. The graph plots the mean fluorescence in the dendrite, normalized to the three pre-tetanus values. The time of the tetanus (100 Hz, 1 s) is indicated by the bar and the synaptic current during and shortly following the tetanus is shown above the graph. The cell was voltage-clamped at −35 mV.

Collingridge GL, Herron CE, Lester RAJ 1988 Synaptic activation of N-methyl-D-aspartate receptors in the Schaffer collateral-commissural pathway of rat hippocampus. J Physiol (Lond) 399:283–300
Davies CH, Davies SN, Collingridge GL 1990 Paired-pulse depression of monosynaptic GABA-mediated inhibitory postsynaptic responses in rat hippocampus. J Physiol (Lond) 424:513–531
Davies CH, Starkey SJ, Pozza MF, Collingridge GL 1991 $GABA_B$ autoreceptors regulate the induction of LTP. Nature (Lond) 349:609–611
Dingledine R, Hynes MA, King GL 1986 Involvement of N-methyl-D-aspartate receptors in epileptiform bursting in the rat hippocampal slice. J Physiol (Lond) 380:175–189
Fine A, Amos WB, Durbin RM, McNaughton PA 1988 Confocal microscopy: applications in neurobiology. Trends Neurosci 11:346–351
Herron CE, Williamson R, Collingridge GL 1985 A selective N-methyl-D-aspartate antagonist depresses epileptiform activity in rat hippocampal slices. Neurosci Lett 61:255–260
Herron CE, Lester RAJ, Coan EJ, Collingridge GL 1986 Frequency-dependent involvement of NMDA receptors in the hippocampus: a novel synaptic mechanism. Nature (Lond) 322:265–268
Jahr CE, Stevens CF 1987 Glutamate activates multiple single channel conductances in hippocampal neurons. Nature (Lond) 325:522–525
Konnerth A, Keller BU, Ballanyi K, Yaari Y 1990 Voltage sensitivity of NMDA-receptor mediated postsynaptic currents. Exp Brain Res 81:209–212
Larson J, Lynch G 1986 Induction of synaptic potentiation in hippocampus by patterned stimulation involves two events. Science (Wash DC) 232:985–990
Lodge D, Collingridge GL (eds) 1990 The pharmacology of excitatory amino acids. Trends Pharmacol Sci (Spec Rep) 1–89
Lynch G, Larson J, Kelso S, Barrionuevo G, Schottler F 1983 Intracellular injections of EGTA block induction of hippocampal long-term potentiation. Nature (Lond) 305:719–721
MacDermott AB, Mayer ML, Westbrook GL, Smith SJ, Barker JL 1986 NMDA-receptor activation increases cytoplasmic calcium concentration in cultured spinal cord neurones. Nature (Lond) 321:519–522
Malenka RC, Kauer JA, Zucker RS, Nicoll RA 1988 Postsynaptic calcium is sufficient for potentiation of hippocampal synaptic transmission. Science (Wash DC) 242:81–84
Malgaroli A, Malinow R, Schulman H, Tsien RW 1992 Persistent signalling and changes in presynaptic function in long-term potentiation. In: Interactions among cell signalling systems (Ciba Found Symp 164) p 176–196
Mayer ML, Westbrook GL, Guthrie PB 1984 Voltage-dependent block by $Mg^{2+}$ of NMDA responses in spinal cord neurones. Nature (Lond) 309:261–263
Nowak L, Bregestovski P, Ascher P, Herbet A, Prochiantz A 1984 Magnesium gates glutamate-activated channels in mouse central neurones. Nature (Lond) 307:462–465
Randall AD, Schofield JG, Collingridge GL 1990 Whole-cell patch-clamp recordings of an NMDA receptor-mediated synaptic current in rat hippocampal slices. Neurosci Lett 114:191–196
Regehr WG, Tank DW 1990 Postsynaptic NMDA receptor-mediated calcium accumulation in hippocampal CA1 pyramidal cell dendrites. Nature (Lond) 345:807–810
Rose GM, Dunwiddie TV 1986 Induction of hippocampal long-term potentiation using physiologically patterned stimulation. Neurosci Lett 69:244–248
Sah P, Hestrin S, Nicoll RA 1990 Properties of excitatory postsynaptic currents recorded in vitro from rat hippocampal interneurones. J Physiol (Lond) 430:605–616
Tsien RY 1988 Fluorescence measurement and photochemical manipulation of cytosolic free calcium. Trends Neurosci 11:419–424

## DISCUSSION

*Changeux:* You record from pyramidal cells. How many synapses are modified under long-term potentiation? You showed one nice synapse in your scheme, but reality might be very different from that.

*Collingridge:* Of course, our scheme is simplified. In reality we stimulate a large band of fibres in our experiments. The size of the synaptic currents that we normally record suggests that we are activating about 20–25 synapses.

*Changeux:* In one cell could you generate LTP on one part of the dendrite and check whether the other synapses are modified?

*Collingridge:* Yes; this has been shown many times before by other groups. The $Ca^{2+}$-imaging techniques may enable us to investigate the mechanism that underlies the specificity of the LTP process.

*Tsien:* Synaptic potentiation is of course geographically specific.

*Collingridge:* That's correct—it's geographically specific on a given dendrite. The theory is that the specificity arises because only those synapses which are activated appropriately provide the necessary calcium (through the NMDA receptor channels) and, furthermore, that the calcium signal is highly localized, perhaps within the dendritic spines.

*Changeux:* Could there be diffusion of calcium from activated synapses to synapses which have not been activated by LTP?

*Collingridge:* That is the sort of question we now want to address. The working hypothesis is that it's localized and that this explains the specificity.

*Tsien:* The calcium oscillations bring to mind Professor Mikoshiba's talk about $InsP_3$ receptors and the related organelle $Ca^{2+}$ channel, the ryanodine receptor. Have you looked at the spatial distribution of calcium during that period of time? You showed time plots in one 4 μm zone and calcium was oscillating every second or so. If you plot $Ca^{2+}$ concentration as a function of space at a particular time, what is the spatial periodicity, if any?

*Collingridge:* Well we haven't analysed this in great detail, but there were no obvious waves.

*Tsien:* Even so, you could still ask, on a scale of micrometres, if calcium is high at a particular point, how far away from that point can you go before it's not likely that the calcium concentration will still be high.

*Collingridge:* We arbitrarily selected 4 μm squares, and within those squares the oscillations, whatever they are, are actually smaller than this. We need to re-analyse our data before we can answer your question fully.

*Changeux:* You assigned the depression of GABA-mediated inhibition to a presynaptic $GABA_B$ receptor effect. However, could it not be explained by desensitization of the postsynaptic $GABA_B$ receptor?

*Collingridge:* We have a lot of results, that I didn't have time to go into, which indicate the effect is presynaptic and is mediated by $GABA_B$ receptors (see Davies et al 1991). For example, the $GABA_A$ and $GABA_B$ receptor-mediated components of the IPSP are depressed by high frequency stimulation in an identical manner. This frequency-dependent depression is prevented by the $GABA_B$ antagonist CGP35348. If the phenomenon were postsynaptic, there would have to be a postsynaptic process affecting a $GABA_A$ receptor–chloride channel-linked process *and* a $GABA_B$ receptor–potassium-linked process identically in terms of the time course and magnitude of the effect. I suppose that this is possible, but I think the results are more easily explained by there being less GABA released.

*Changeux:* You are recording from the soma. Do you think that the GABAergic synapse is on the same dendrite as the excitatory synapse?

*Collingridge:* It's difficult to know exactly where the synapses are anatomically, but what we do know is that the GABA synapses interact functionally with the NMDA channels, which we postulate are on the dendritic spines, because the activation of the NMDA receptor system is very sensitive to the level of GABA-mediated synaptic inhibition.

*Akaike:* We have dissociated rat nucleus basalis of Meynert (nBM) neurons with the presynaptic terminals, termed synaptic boutons, still attached. Spontaneous inhibitory postsynaptic currents (IPSCs) mediated by GABA can be recorded from the nBM neurons in whole-cell mode using the patch-clamp technique. In $Ca^{2+}$-free external solution containing EGTA or BAPTA (1,2-bis[2-aminophenoxy]ethane $N,N,N',N'$-tetraacetic acid) the IPSCs decrease. The IPSCs are also reduced by the $GABA_B$ receptor agonist baclofen, and enhanced by the antagonist phaclofen. The results indicate that the $GABA_B$ receptor regulates the release of GABA from the presynaptic nerve terminals and acts as a negative autoreceptor (Tateishi et al 1990).

The postsynaptic $GABA_A$ receptor of dissociated rat hippocampal pyramidal neurons is regulated by the intracellular ATP concentration. The response to GABA decreases in a concentration-dependent manner when the ATP concentration decreases below 2.5 mM (Akaike & Shirasaki 1991). When the intracellular $Ca^{2+}$ concentration of the postsynaptic nerve cell body increases markedly during tetanic stimulation, the intracellular ATP concentration might become depleted within a few minutes. In that situation, exogenous GABA cannot elicit any response.

*Collingridge:* I think that under the conditions of our experiments the results can be best explained purely by a presynaptic $GABA_B$ receptor-mediated event, for some of the reasons that I have given and many more that I could go into. I am not saying that there are not other mechanisms which come into play under different circumstances. The more intense the tetanus, the more likely it is that the GABA system will change through other mechanisms, including shifts in the $Cl^-$ and $K^+$ equilibrium potentials. Under our experimental conditions the results can all be explained by autoinhibition through $GABA_B$ receptors.

*Fredholm:* Is the decrease in the inhibitory influence of GABA that you talked about relevant to the action of NMDA receptors in ischaemic cell damage? Or, is there another reason why the inhibitory GABAergic influence appears to be decreased in ischaemic hypoxia?

*Collingridge:* The classic view of ischaemia is that glutamate levels are increased because of the failure of re-uptake mechanisms. Clearly, that will cause depolarization of the cell through the NMDA and AMPA receptors, irrespective of the functioning of the GABAergic inhibitory system. I don't think it's *necessary* to postulate that this mechanism is involved, but it could be an important component of ischaemia. I tend to think of ischaemia, epilepsy and LTP as being the same process taken to different extremes.

*Hunter:* Ligand-binding studies suggest that there are multiple AMPA-type receptors that probably have different properties (Hollmann et al 1991). How does that affect the interpretation of your results?

*Collingridge:* I said little about what expresses the potentiation of a synaptic response, partly because Professor Tsien will be talking about it and partly because it's very controversial. One idea is that the process is maintained, in part, by modification of the AMPA receptor response, through a postsynaptic mechanism (see, for example, Davies et al 1989). Sackmann and Seeburg's work on the conductance properties of alternatively spliced AMPA receptors suggests that in LTP there might be changes in the expression of AMPA receptors (Sommer et al 1990). That's only a theory though.

*Hunter:* I believe that some of the alternatively spliced forms of AMPA receptors have different functional properties.

*Collingridge:* Yes; those termed 'flip' have a fast desensitizing component whereas those termed 'flop' generate small sustained currents.

*Akaike:* Tetanic stimulation of mossy fibres also induces LTP in guinea pig CA3 pyramidal neurons. However, NMDA receptor blockers have no effect on mossy fibre LTP (Harris & Cotman 1986). So, involvement in the induction of LTP of either the kainic acid receptor or the metabotropic glutamate receptor has been proposed, because these receptors also increase the intracellular calcium level (Goh & Pennefather 1989, Ito et al 1990, Masu et al 1991).

*Collingridge:* I have not done any experiments on that myself. Clearly, as you say, the NMDA receptor is not involved in that LTP process, nor are there the Hebbian-type synaptic properties that the NMDA receptor confers upon synapses.

*Mikoshiba:* Do you think $K^+$ channels have any role in LTP? I ask the question because we are working on a bee peptide called MCD peptide (mast cell-degranulating peptide) which induces long-term potentiation in hippocampal slices (Kondo et al 1990). It's effects and those of tetanic stimulation are additive. This peptide actually blocks potassium channels.

*Collingridge:* In our system a direct role for $K^+$ channels is not necessary. There have been reports (for example, Aniksztejn & Ben-Ari 1991) that there is a form of LTP that can be induced by blocking $K^+$ channels, but that appears to be a different process. It is not yet known whether it has any physiological role. Potassium channels are probably important in the *modulation* of the type of LTP that we study. To activate the NMDA receptor system both glutamate and depolarization are needed; thus, neurotransmitter systems could easily interact with the LTP process via potassium channels. For example, acetylcholine, which is clearly implicated in memory function, could inhibit potassium channel function, facilitating the synaptic activation of NMDA receptors by allowing the cell to become depolarized more easily.

## References

Akaike N, Shirasaki T 1991 Intracellular ATP regulates GABA responses in rat dissociated hippocampal neurons. In: Imachi S, Nakazawa M (eds) Role of adenosine and adenine nucleotides in the biological system. Elsevier Science Publishers, p 645–651

Aniksztejn L, Ben-Ari Y 1991 Novel form of long-term potentiation produced by a $K^+$ channel blocker in the hippocampus. Nature (Lond) 349:67–69

Davies SN, Lester RAJ, Reymann KG, Collingridge GL 1989 Temporally distinct pre- and post-synaptic mechanisms maintain long-term potentiation. Nature (Lond) 338:500–503

Davies CH, Starkey SJ, Pozza MF, Collingridge GL 1991 $GABA_B$ autoreceptors regulate the induction of LTP. Nature (Lond) 349:609– 611

Goh JW, Pennefather PS 1989 A pertussis toxin-sensitive G protein in hippocampal long-term potentiation. Science (Wash DC) 244:980–983

Harris EW, Cotman CW 1986 Long-term potentiation of guinea pig mossy fiber responses is not blocked by N-methyl-D-aspartate antagonists. Neurosci Lett 70:132–137

Hollmann M, Hartley M, Heinemann S 1991 $Ca^{2+}$ permeability of KA-AMPA- gated glutamate receptor channels depends on subunit composition. Science (Wash DC) 252:851–853

Ito I, Tanabe S, Kohda A, Sugiyama H 1990 Allosteric potentiation of quisqualate receptors by a nootropic drug aniracetam. J Physiol (Lond) 424:533–543

Kondo T, Ikenaka K, Kato H et al 1990 Long-term enhancement of synaptic transmission by synthetic mast cell degranulating peptide and its localization of binding sites in hippocampus. Neurosci Res 8:147–157

Masu M, Tanabe Y, Tsuchida K, Shigemoto R, Nakanishi S 1991 Sequence and expression of a metabotropic glutamate receptor. Nature (Lond) 349:760–765

Sommer B, Keinänen K, Verdoorn KA et al 1990 Flip and flop: a cell- specific functional switch in glutamate-operated channels of the CNS. Science (Wash DC) 249:1580–1585

Tateishi N, Uneo S, Akaike N 1990 Synaptic transmission between isolated rat nucleus basalis of Meynert (NbM) neurons and attached synaptic boutons. Japn J Pharmacol (suppl) 52:97

# Persistent signalling and changes in presynaptic function in long-term potentiation

Antonio Malgaroli*, Roberto Malinow[†]*, Howard Schulman‡ and Richard W. Tsien*

*Departments of Molecular and Cellular Physiology* and Pharmacology‡, Stanford University Medical Center, Stanford CA 94305-5425, USA*

*Abstract*. Long-term potentiation (LTP) is an example of a persistent change in synaptic function in the mammalian brain, thought to be essential for learning and memory. At the synapse between hippocampal CA3 and CA1 neurons LTP is induced by a $Ca^{2+}$ influx through glutamate receptors of the NMDA (*N*-methyl-D-aspartate) type (see Collingridge et al 1992, this volume). How does a rise in $[Ca^{2+}]_i$ lead to enhancement of synaptic function? We have tested the popular hypothesis that $Ca^{2+}$ acts via a $Ca^{2+}$-dependent protein kinase. We found that long-lasting synaptic enhancement was prevented by prior intracellular injection of potent and selective inhibitory peptide blockers of either protein kinase C (PKC) or $Ca^{2+}$/calmodulin-dependent protein kinase II (CaMKII), such as PKC(19–31) or CaMKII(273–302), but not by control peptides. Evidently, activity of both PKC and CaMKII is somehow necessary for the postsynaptic *induction* of LTP. To determine if these kinases are also involved in the *expression* of LTP, we impaled cells with microelectrodes containing protein kinase inhibitors after LTP had already been induced. Strikingly, established LTP was not suppressed by a combination of PKC and CaMKII blocking peptides, or by intracellular postsynaptic H-7. However, established LTP remained sensitive to bath application of H-7. Thus, the persistent signal may be a persistent kinase, but if so, the kinase cannot be accessed within the postsynaptic cell. Evidence for a presynaptic locus of expression comes from our studies of quantal synaptic transmission under whole-cell voltage clamp. We find changes in synaptic variability expected to result from enhanced presynaptic transmitter release, but little or no increase in quantal size. Furthermore, miniature synaptic currents in hippocampal cultures are increased in frequency but not amplitude as a result of a glutamate-driven postsynaptic induction. The combination of postsynaptic induction and presynaptic expression necessitates a retrograde signal from the postsynaptic cell to the presynaptic terminal.

*1992 Interactions among cell signalling systems. Wiley, Chichester (Ciba Foundation Symposium 164) p 176–196*

---

[†]*Present address*: Department of Physiology and Biophysics, University of Iowa, Iowa City IA 52242, USA

Long-lasting changes in synaptic function are thought to provide a cellular basis for learning and memory in both vertebrates and invertebrates (Hebb 1949, Goelet et al 1986). The most thoroughly characterized example of such synaptic plasticity in the mammalian nervous system is long-term potentiation (LTP). A remarkable aspect of LTP is that a short burst of synaptic activity can trigger persistent enhancement of synaptic transmission lasting at least several hours, and possibly weeks or longer (see Bliss et al 1990 for review). This phenomenon was first found in the hippocampus but has recently been described in other areas of the mammalian CNS, including the visual cortex and the motor cortex. There is intense interest in understanding the cellular and molecular basis of this form of synaptic plasticity (Collingridge & Bliss 1987, Brown et al 1988, Kennedy 1989, Bliss & Lynch 1988, Malenka et al 1989a, Bliss et al 1990, Madison et al 1991).

Most studies on mechanisms of LTP have focused on synaptic transmission in the CA1 field of the hippocampus, at synapses between Schaffer collaterals and CA1 pyramidal cells; this area is rich in NMDA (*N*-methyl-D-aspartate) receptors (Cotman et al 1989) and $Ca^{2+}$-dependent protein kinases (Nishizuka 1988, Kennedy 1989). Induction of LTP requires a temporal conjunction of presynaptic transmitter release and postsynaptic depolarization (see Collingridge & Bliss 1987), a combination of events resembling that envisioned by Hebb (1949). It is generally agreed that these factors work together, glutamate activating the NMDA receptor channel and postsynaptic depolarization freeing the channel from its extracellular $Mg^{2+}$ block (reviewed by Ascher & Nowak 1988). The NMDA receptor allows a significant $Ca^{2+}$ influx and increases $[Ca^{2+}]_i$ in postsynaptic spines (Regehr & Tank 1990); the rise in $[Ca^{2+}]_i$ is necessary for LTP (Lynch et al 1983, Malenka et al 1988).

Many questions remain about what happens next. What is the mechanism of the long-lasting synaptic modification and its relationship to $Ca^{2+}$ signalling? Our work has been directed toward understanding the role of $Ca^{2+}$-dependent protein kinases and elucidating the nature of the persistent change. We have also addressed the question of whether synaptic transmission is strengthened by increased transmitter release or enhanced postsynaptic receptivity. Our experiments, along with those of others, raise the possibility of a two-way communication between signalling mechanisms in presynaptic terminals and postsynaptic spines, in what might be called transsynaptic cross-talk.

## Experimental strategy and methods

We used three complementary approaches to study plastic changes in synaptic transmission between hippocampal neurons. Microelectrode recordings were made in CA1 neurons in transverse hippocampal slices, to monitor intracellular synaptic potentials and to introduce agents that modify intracellular signalling (Malinow et al 1989). Whole-cell recordings were made from CA1 neurons two

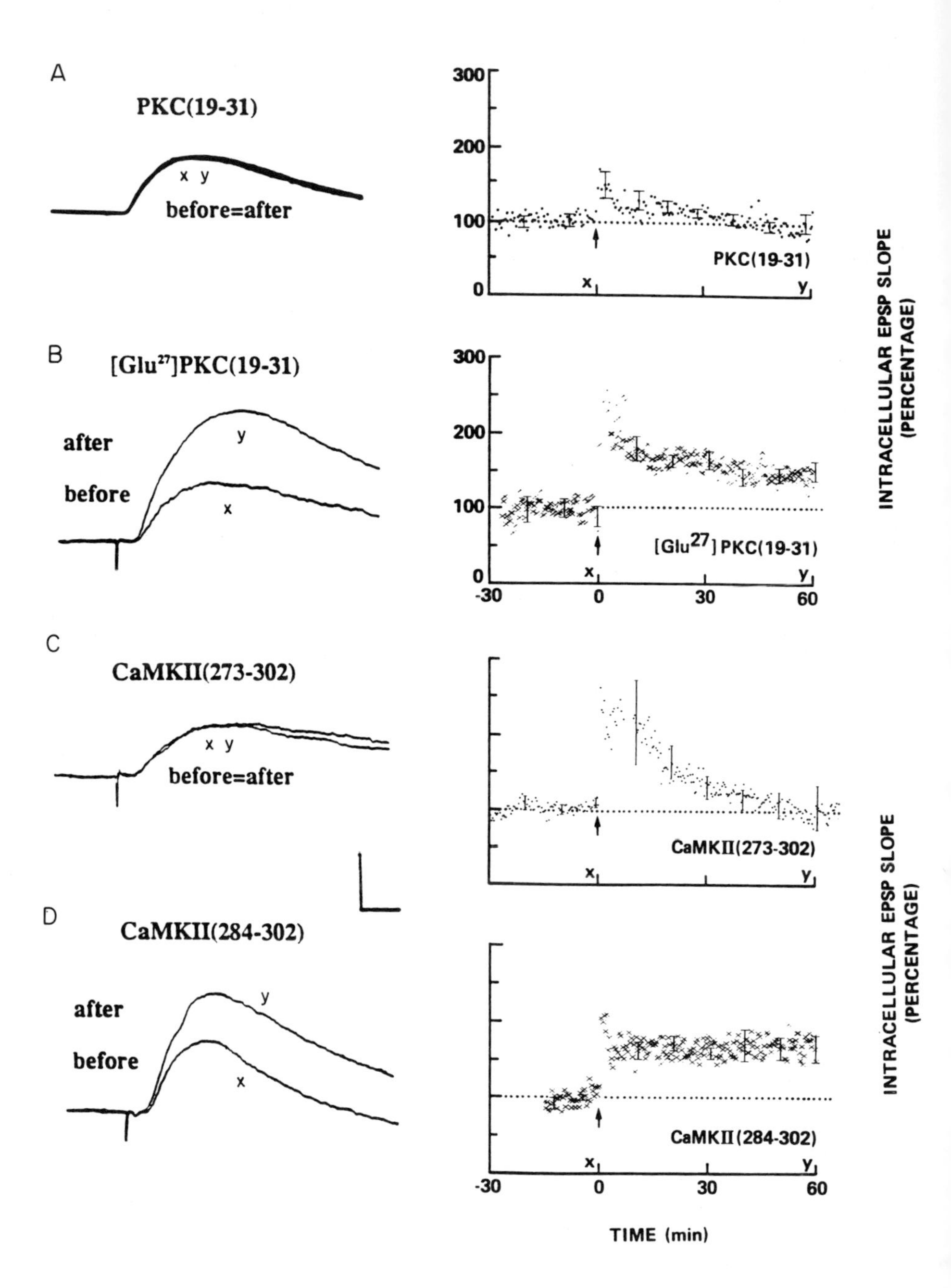
A
PKC(19-31)
x y
before=after
300
200
100
0
PKC(19-31)
x
y
B
[Glu27]PKC(19-31)
after
y
before
x
300
200
100
0
[Glu27] PKC(19-31)
x
y
-30
0
30
60
INTRACELLULAR EPSP SLOPE (PERCENTAGE)
C
CaMKII(273-302)
x y
before=after
CaMKII(273-302)
x
y
D
CaMKII(284-302)
after
y
before
x
CaMKII(284-302)
x
y
-30
0
30
60
TIME (min)
INTRACELLULAR EPSP SLOPE (PERCENTAGE)

or three cell layers below the surface of the slices to monitor excitatory postsynaptic currents with relatively little background noise (Malinow & Tsien 1990, Tsien & Malinow 1990). We also studied evoked and spontaneous synaptic currents at synapses between hippocampal neurons in culture (Malgaroli & Tsien 1991).

## $Ca^{2+}$-dependent protein kinases and postsynaptic induction

A key question in LTP is how a rise in postsynaptic $[Ca^{2+}]_i$ can lead to a long-lasting enhancement of presynaptic function. A generic working hypothesis, favoured by many investigators, is that $Ca^{2+}$ acts through a signal transduction pathway involving a $Ca^{2+}$-dependent protein kinase such as protein kinase C (PKC) or multifunctional $Ca^{2+}$/calmodulin-dependent protein kinase II (CaMKII) (see Malinow et al 1988a, Cotman et al 1989, Kennedy 1989, Malenka et al 1989b).

We have tested these ideas by postsynaptic intracellular injection of peptides that are potent and selective inhibitors of either PKC or CaMKII (Malinow et al 1989, Tsien et al 1990). PKC(19–31) and PKC(19–36) are peptide fragments constituting the pseudosubstrate region of the PKC regulatory domain (House & Kemp 1987). They are over 600-fold more potent as blockers of PKC ($IC_{50} \approx 0.1\ \mu M$) than as blockers of CaMKII. In contrast to the parent Arg-27 peptide, the Glu-27 derivative of PKC(19–31) is relatively inactive against either kinase and serves as a useful control against non-specific effects.

The results of intracellular injection of peptides are illustrated in Fig. 1 (*left*). PKC(19–31) and PKC(19–36) blocked LTP when delivered to the postsynaptic

FIG. 1. Selective postsynaptic block of PKC or CaMKII prevents LTP. (A) Synaptic potentials monitored with an intracellular microelectrode, the tip of which is filled with 3 mM PKC(19–31) ($n = 8$). Error bars indicate SEM values for representative time points. Traces (*left*) are averages of 10 consecutive potentials obtained at the times designated on time axis (*right*). Scale bars, 5 mV and 12.5 ms. After the tetanus (arrow) there is no persistent potentiation. Transmission in a non-tetanized pathway, monitored through the same PKC(19–31)-containing electrode, is constant throughout the experiment, indicating there is not a non-specific depressant effect on basal synaptic transmission; extracellular monitoring shows there is LTP after tetanic stimulation from synapses on neighbouring cells (data not shown). (B) Transmission monitored in a different set of slices with 3 mM [Glu-27]PKC(19–31) in the intracellular electrode shows LTP after a conditioning tetanus ($n = 6$ pathways from 3 slices). (C) Synaptic potentials monitored with an intracellular electrode containing 1.1 mM CaMKII(273–302) show no persistent potentiation after tetanic conditioning. Transmission in a non-tetanized pathway, monitored with the same CaMKII(273–302)-containing electrode, is constant throughout the experiment; extracellular monitoring shows there is LTP after tetanic stimulation from synapses on neighbouring cells (data not shown). (D) Transmission monitored in a different set of slices, with 1.1 mM CaMKII(284–302) in the intracellular electrode, shows LTP after a tetanus ($n = 5$ pathways from 3 slices). From Malinow et al 1989.

cell with the recording intracellular microelectrode; they had no significant effect on transmission to a group of neighbouring cells, monitored extracellularly; [Glu-27] PKC(19–31), the control peptide, failed to block LTP. These results are compatible with the effects of injecting PKC (Hu et al 1987) or PKC inhibitory peptides (Andersen et al 1990).

To investigate the role of postsynaptic CaMKII, we used the peptide fragment CaMKII(273–302), which inhibits CaMKII at concentrations much lower than those at which it inhibits PKC. CaMKII(273–302) was designed to be selective for catalytic activity of CaMKII, as opposed to calmodulin binding. The N-terminal region of the peptide includes several residues on either side of the RQET sequence that is thought to lie at the heart of CaMKII's autoinhibitory activity. Calmodulin-blocking action was avoided by terminating the peptide with the C-terminal sequence -RKLKGA, this sequence being only partway along a domain crucial for calmodulin binding (RKLKGAILTTMLA). That the peptide did not block calmodulin binding was confirmed experimentally by showing that the inhibitory action of CaMKII(273–302) on CaMKII is not overcome by raising the calmodulin concentration. Note that the sequence and action of CaMKII(273–302) are considerably different from those of the CaMKII(281–309) peptide studied by Malenka et al (1989b) and Smith et al (1990), which acts largely as a calmodulin inhibitor (Malenka et al 1989b). A control was provided by CaMKII(284–302), a shorter peptide that is much less effective at blocking CaMKII.

When delivered to postsynaptic cells through the recording microelectrode, CaMKII(273–302) blocks LTP (Fig. 1C, *right*), while CaMKII(284–302) does not (Fig. 1D, *right*). We obtained similar results with a calmodulin-blocking peptide (not shown; see Malenka et al 1989b). The effects of the CaMKII fragments are almost certainly not explained by inhibition of PKC. If this were the case, one would expect CaMKII(284–302) to be more effective than CaMKII(273–302) in blocking LTP, because it is more potent as a PKC blocker. This result was not found (Malinow et al 1989), suggesting that activity of postsynaptic CaMKII is also required in some way for the establishment of LTP.

## What is the nature of the persistent signal?

Another question concerns the nature of the persistent signal that maintains LTP. Keeping in mind the evidence for involvement of protein phosphorylation, one may consider a generic signal transduction cascade, involving the activation and effect of a single protein kinase (depicted in Fig. 2, *top*). The persistent signal in LTP could be a long-lived kinase activator (such as $Ca^{2+}$ or diacylglycerol), a persistently active protein kinase (for example, PKC translocated to the membrane) or an enduring phosphoester bond (such as a persistently phosphorylated ion channel). To help distinguish between these possible signals, we used sphingosine and H-7, agents that are thought to

<u>Persistent signal in LTP</u>

activator ⇒ active kinase ⇒ phosphoprotein
↑ *sphingosine* ↑ *H-7*

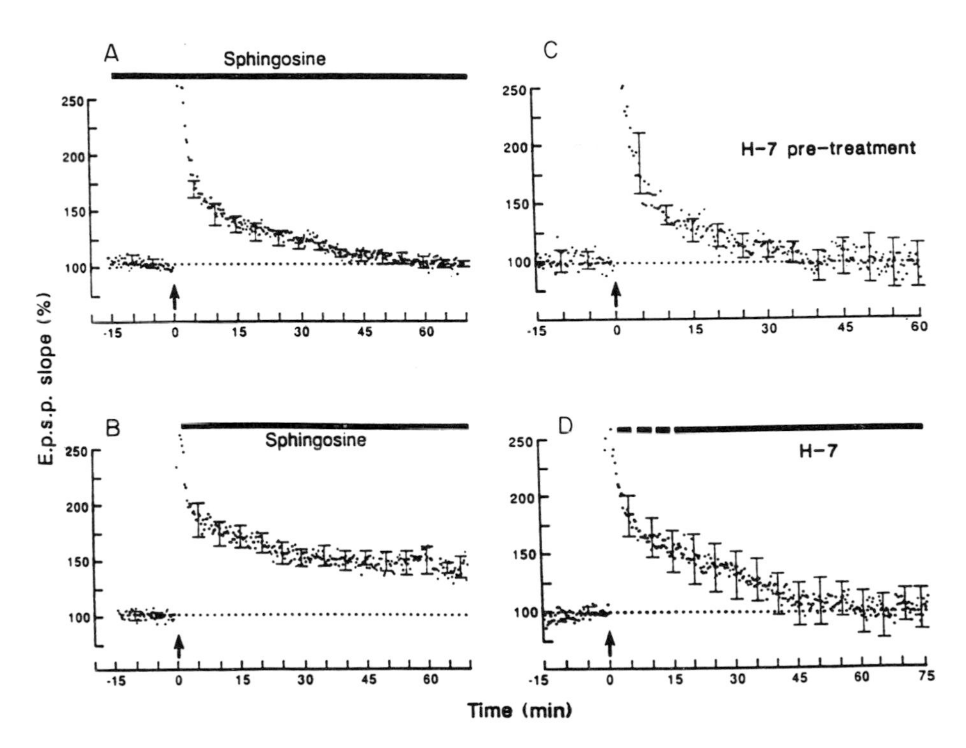


FIG. 2. Use of sphingosine and H-7 to distinguish between possible candidates for the persistent signal in LTP; the scheme (*top*) gives the rationale behind the experiment. Results show that sphingosine blocks LTP if applied before tetanic stimulation, but not if applied after, whereas LTP is blocked by H-7 applied either before or after tetanic stimulation. (A) Tetanic stimulation produced no significant LTP when sphingosine was delivered by bath application 15–30 min before the tetanus (arrow). (B) Sphingosine applied immediately after the tetanus did not block LTP, the EPSP remaining significantly potentiated ($n = 8$, $P < 0.01$). (C) Responses collected from slices incubated with 300 μM H-7 for 1–8 h before recording (pre-treatment, $n = 7$). (D) Data collected ($n = 7$) from slices superfused with 300 μM H-7, beginning 0–15 min (dashed bar) after tetanic stimulation (arrow) and continuing to the end of the experiment (solid bar). No significant LTP was produced, regardless of whether H-7 was applied before or after the tetanic stimulation.

interfere with kinase action at the steps of activation and catalytic activity, respectively (Fig. 2, inset). Sphingosine competes with activators of PKC (Hannun & Bell 1987) and CaMKII (Jefferson & Schulman 1988). H-7 resembles ATP in its chemical structure, and is known to compete with ATP at the catalytic domains of protein kinases (Hidaka et al 1984) in its action as a general protein kinase inhibitor. In collaboration with Daniel Madison, we found that both agents were effective in abolishing sustained potentiation if present at the time of the tetanus. However, the actions of these agents differed when delivered after the tetanus (Fig. 2). H-7 suppressed established LTP, suggesting that the synaptic potentiation is not sustained by a stable phosphorylated substrate (Malinow et al 1988a). In contrast, sphingosine failed to inhibit established LTP, suggesting that such kinase activity is not maintained by a long-lived activator. Inactive congeners of sphingosine and H-7, *N*-acetylsphingosine and HA1004, served as useful controls for non-specific effects, and were found to be ineffective. This combination of results leads to the tentative conclusion that the persistent signal may be a persistently active protein kinase.

If kinase activity is persistent enough, one might be able to pharmacologically interrupt the catalytic activity of the kinase on a temporary basis, without disrupting the persistent signal. We have tested this hypothesis by applying H-7 and looking for recovery of LTP after its removal. We found that the H-7 block of expressed LTP was reversible: LTP returned after washout of the drug (Fig. 3). Evidently, the kinase immediately underlying LTP remains activated, even though its catalytic activity is interrupted, arguing against such kinase activity sustaining itself simply through continual autophosphorylation. Note that the effect of bath-applied H-7 appears selective for the potentiated pathway under our experimental conditions (Fig. 3).

## Does the persistent signal reside within the postsynaptic cell?

To determine if the postsynaptic kinases are involved in the *expression* of LTP, we impaled cells with microelectrodes containing protein kinase inhibitors *after* the high frequency stimulation (arrow in Fig. 4) had already established potentiated transmission (insets, Fig. 4). To our surprise, we found that established LTP was *not* suppressed by intracellular postsynaptic H-7. Delayed introduction of a combination of PKC(19–31) and CaMKII(273–302) was similarly ineffective (Malinow & Tsien 1990). Control procedures showed that the agents were successfully delivered (D) and that LTP had indeed been established before the impalement (E,F). Thus, once established, the persistent signal is inaccessible to postsynaptic injection of H-7 (and to the kinase-blocking peptides). However, the established LTP remains sensitive to bath application of H-7 (Fig. 4E). These results suggest that LTP is maintained by a signalling pathway not entirely contained within the postsynaptic cell, that may even extend to the presynaptic terminal.

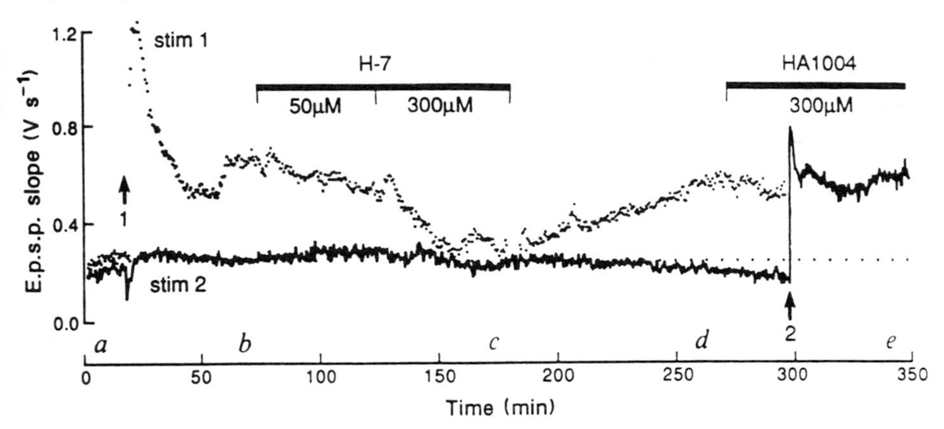


FIG. 3. H-7 applied after tetanus reversibly blocks LTP, but has no effect on basal synaptic transmission, whereas HA1004 is not effective. The graph shows EPSP slopes from a single recording site produced in response to stimuli alternating between two separate inputs. After a tetanic stimulation at stim 1 (arrow 1), the slice was exposed to H-7, which had little effect at 50 μM and inhibited LTP at 300 μM, but had no effect on the untetanized pathway (stim 2). When H-7 was washed from the slice, synaptic transmission in the tetanized pathway recovered to its potentiated level. HA1004 did not reverse LTP in the stim 1 pathway, nor did it prevent potentiation when stim 2 was subsequently tetanized (arrow 2).

## Statistical analysis of synaptic currents suggests presynaptic expression

There is general agreement that LTP is triggered postsynaptically in the CA1 region of the hippocampus (see Collingridge et al 1992, this volume), but the locus of persistent modification of synaptic transmission remains controversial. To test for possible changes in presynaptic function, we have analysed the variability in synaptic currents elicited by minimal stimulation before and during LTP (Malinow & Tsien 1990, Tsien & Malinow 1990). Whole-cell recordings were obtained from CA1 neurons 2–3 cell diameters below the surface of rat hippocampal slices, cut 400–500 μm thick. This method (Blanton et al 1989) allows recordings from cells with largely intact dendritic structures, while also providing much better signal resolution and biochemical access than conventional microelectrode recording methods. We found that robust LTP could be evoked by a steady postsynaptic depolarization to ≈0 mV during continued minimal activation of the test pathway at 2 Hz for 20 s (pairing). Typically, mean synaptic current was increased three-fold after pairing. Prolonged dialysis of the postsynaptic cell blocked the triggering of LTP, but had no effect on established LTP. Synaptic currents displayed a large trial-to-trial variability, reflecting the probabilistic nature of transmitter release. Block of postsynaptic receptors with CNQX (6-cyano-7-nitroquinoxaline-2,3-dione, a non-NMDA glutamate receptor blocker) attenuated the responses but left their relative variability unchanged.

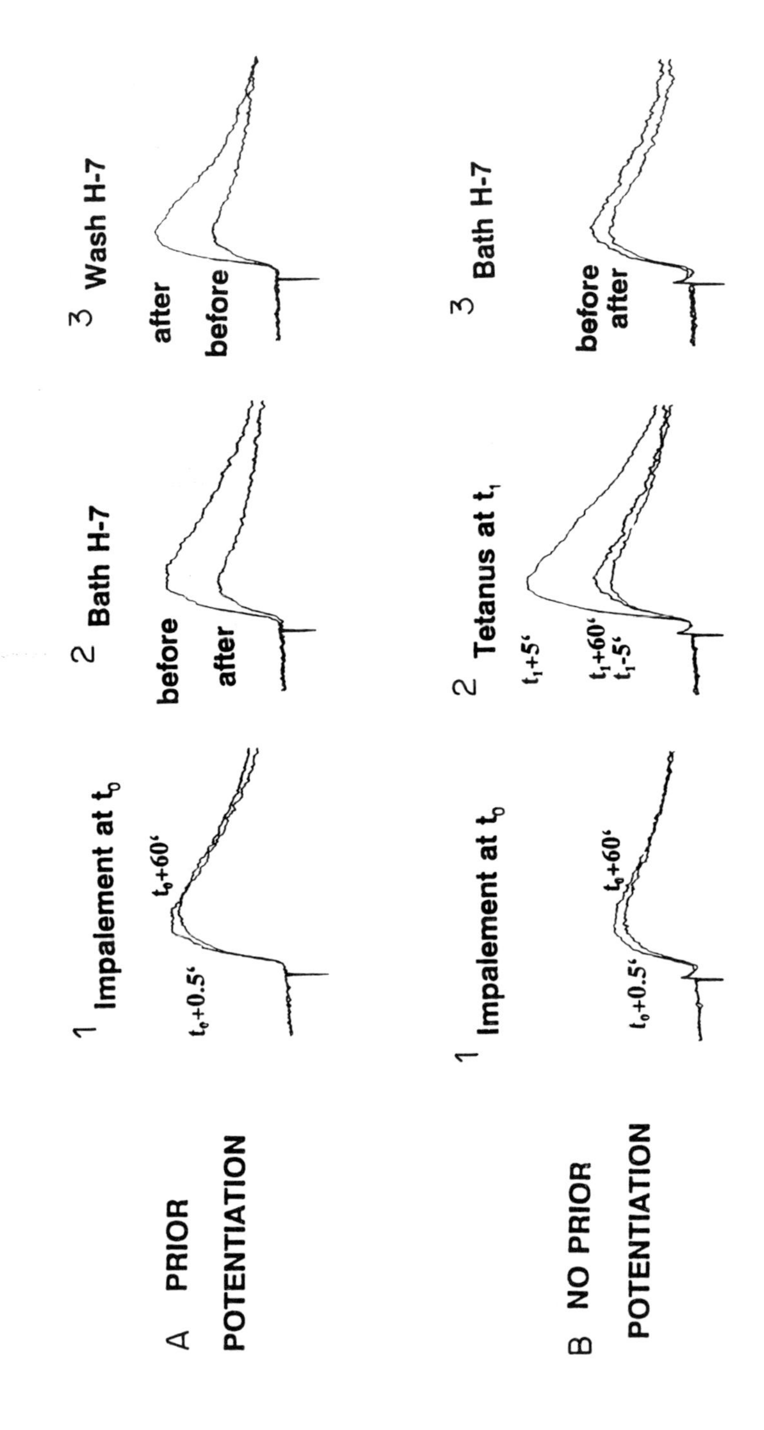
A PRIOR POTENTIATION
1 Impalement at $t_0$
$t_0$+0.5'
$t_0$+60'
2 Bath H-7
before
after
3 Wash H-7
after
before
B NO PRIOR POTENTIATION
1 Impalement at $t_0$
$t_0$+0.5'
$t_0$+60'
2 Tetanus at $t_1$
$t_1$+5'
$t_1$+60'
$t_1$-5'
3 Bath H-7
before
after

In contrast, manoeuvres that alter presynaptic release (elevating $[Ca^{2+}]_o$, lowering $[Mg^{2+}]_o$, adding 4-aminopyridine or increasing stimulus strength to recruit more fibres) produced the expected alterations in synaptic variability: variance of the mean synaptic current increased, and the histogram of synaptic current amplitudes shifted from a distribution skewed to small currents toward a symmetrical bell shape. We found consistent and qualitatively similar changes during LTP (14 experiments). In addition, the proportion of synaptic failures sharply decreased during LTP, when failures could be clearly resolved. Although not excluding the possibility of some change in postsynaptic responsivity (cf. Davies et al 1989), increases in variance of the mean synaptic current and decreases in failures both support an enhanced likelihood of transmitter release during LTP.

## Potentiation of frequency, not amplitude, of spontaneous miniature synaptic currents

Another argument in favour of presynaptic involvement in LTP expression is provided by our recent studies of glutamate-induced long-term potentiation of miniature excitatory postsynaptic current (EPSC) frequency (Malgaroli & Tsien 1991). Analysis of spontaneous miniature events is a classical approach to synaptic phenomena, but is difficult to use to study LTP in brain slices where it is unclear whether individual 'minis' arise from potentiated or unpotentiated inputs. To circumvent this problem, we recorded from hippocampal pyramidal cells in culture and used direct application of glutamate to induce a global synaptic enhancement, presumably encompassing all synaptic inputs to a neuron.

---

FIG. 4. Expression of LTP is not inhibited by postsynaptic delivery of PKC(19–31) and CaMKII(273–302) but is reversibly blocked by subsequent bath application of 300 μM H-7. Traces show results of intracellular monitoring with a microelectrode containing 3 mM PKC(19–31) plus 1 mM CaMKII(273–302). One pathway (A1–3) was potentiated with a tetanus prior to intracellular impalement. If PKC or CaMKII activity is required for the expression of LTP, as these inhibitors diffuse into the postsynaptic cell one should see a decrement of transmission through the previously potentiated pathway. However, this was not the case, as transmission through both pathways stayed constant between 0.5 and 60 min after the impalement (A1). To test if the inhibitors had arrived at postsynaptic targets, we delivered a tetanus to the pathway which had not already been potentiated (B1–3). We saw a block of LTP through the intracellular electrode (B2), despite there being LTP measured with the extracellular electrode (data not shown), demonstrating that the peptides were effective in reaching the postsynaptic targets and blocking the induction of LTP. We subsequently applied H-7 in the bath and saw a decay of transmission through the previously potentiated pathway (A2). This suggested that transmission through this pathway was in fact potentiated. After washout of the drug, the potentiated transmission returned (A3), consistent with H-7 blocking the expression, and not the maintenance, of LTP. Transmission through the non-potentiated pathway was not significantly affected by bath H-7 (B3), as previously reported (see Fig. 3).

Dissociated hippocampal pyramidal neurons were obtained from 2–3-day-old rat pups and kept in culture. The neurons form synapses in culture, as detected by evoked and spontaneous synaptic currents and synapsin I immunostaining. Large neurons with CA1-like morphology were studied with whole-cell recordings with pipettes filled with a Cs-gluconate-based internal solution containing 0.6 mM EGTA. We found that in the presence of 1 μM tetrodotoxin (TTX), which blocks electrical activity, mini frequency was strongly increased, 2–10-fold, by a 30 s exposure to a solution containing 25–50 μM glutamate and no added $Mg^{2+}$ (Fig. 5). As with LTP in slices, the glutamate-induced enhancement of mini frequency was: (1) sustained up to the end of the recording period (as long as 80 min); (2) mediated by NMDA receptors, as judged by reversible inhibition by the NMDA receptor channel blocker MK-801 (10 μM); (3) induced postsynaptically, as indicated by the finding that the potentiation could be suppressed simply by hyperpolarizing the postsynaptic cell. In almost all cases, the distribution of mini amplitudes was not significantly different before and after glutamate application. Postsynaptic responsiveness was also tested directly with brief puffer applications of glutamate, and remained unchanged during the period of enhanced mini frequency. This argues against the idea that mini frequency increased because of an all-or-nothing recruitment of previously latent spines. Our results indicate that enhancement of presynaptic transmitter release can be achieved by glutamate receptor activation even when presynaptic action potentials are blocked. Furthermore, the elevation in mini frequency was not affected by removal of external $Ca^{2+}$ (see Fig. 5), ruling out a continued increase in presynaptic $Ca^{2+}$ influx as the mechanism of potentiation. Instead, we suggest that the synaptic enhancement might result from increased efficiency of the internal machinery controlling neurosecretion, such as a change in vesicle availability or the efficiency of vesicle fusion.

## A working hypothesis for signalling in LTP

We are only beginning to unravel the web of signalling events that underlies long-term potentiation. The following scheme summarizes our current working hypothesis:

$Ca^{2+}$ entry via NMDA receptors
↓
rise in spine $[Ca^{2+}]_i$
↓
activation of postsynaptic kinase(s)
↓
retrograde signalling
↓
persistent presynaptic kinase activity
↓
increased vesicular release

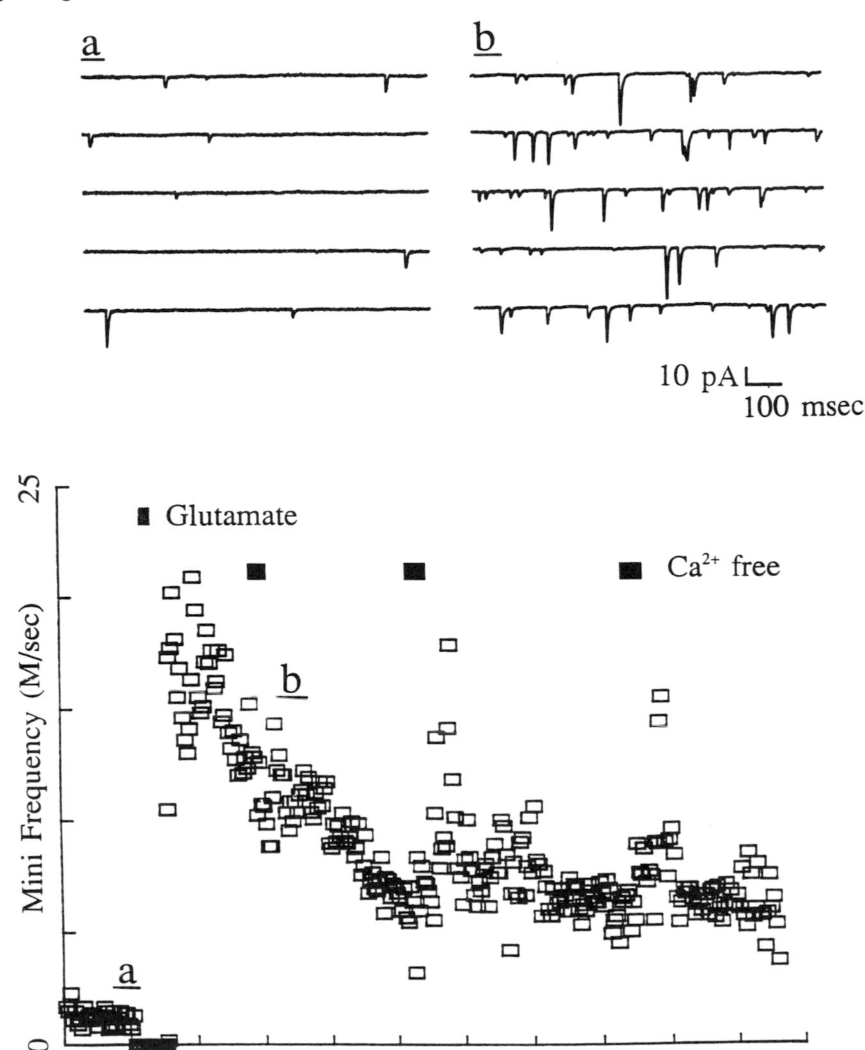


FIG. 5. Glutamate-induced increase in frequency of spontaneous miniature EPSCs. a,b; Whole-cell recordings of membrane current in a cultured hippocampal pyramidal cell, before (a) and about 10 min after a 30 s local application of 50 μM glutamate, $Mg^{2+}$-free solution (b). 1 μM tetrodotoxin was present throughout. The graph (*bottom*) shows miniature EPSC frequency in consecutive 10 s epochs. Mini frequency (M/sec) settles to a steady level of about 7/s, more than five-fold higher than the basal frequency. Dark bars indicate 2 min periods of superfusion with 0 mM $Ca^{2+}$/ 4 mM $Mg^{2+}$ external solution. Minis are not abolished during removal of $Ca^{2+}$ and show a rebound increase in frequency after restoration of 2 mM $Ca^{2+}$/2 mM $Mg^{2+}$ solution.

This scheme provides a framework for more focused discussion and for further experiments.

## Are postsynaptic PKC and CaMKII elements of an enzyme cascade?

Our experiments support the involvement of both PKC and CaMKII in the establishment of LTP. Further work is needed to clarify the cross-talk between these kinase activities. A number of different scenarios could accommodate the evidence from intracellular injection of the various peptide inhibitors.

(1) One or both of the protein kinases might provide basal activity necessary to keep other postsynaptic signalling mechanisms ready for the actual process of induction (cf. Malenka 1991). As a specific and rather extreme example, one might imagine that tonic activity of CaMKII modulates the function of NMDA receptors to allow the entry of $Ca^{2+}$ that triggers LTP, but that an increase in CaMKII activity is not involved in transducing the $Ca^{2+}$ signal.

Other alternatives give both kinases an active role in some sort of enzyme cascade.

(2) PKC and CaMKII might act at different steps along a series of sequential steps. For example, PKC might phosphorylate a cytoskeletal protein such as GAP-43 (neuromodulin) or MARCKS (Nairn & Aderem 1992, this volume), thereby freeing calmodulin, allowing it to activate CaMKII (Alexander et al 1987) to somehow cause LTP expression. Another possibility, suggested by James Schwartz, is that PKC might phosphorylate a cytoskeletal protein that forms a complex with CaMKII, thereby liberating CaMKII.

(3) PKC and CaMKII might work in parallel, not in series, to jointly phosphorylate a substrate protein. It is conceivable that a common substrate for PKC and CaMKII might itself be a kinase. There is precedence for such a scenario in platelets, where latent protein kinases can be activated by the concerted actions of a phorbol ester and a $Ca^{2+}$ inophore, but not by either stimulus alone (Ferrell & Martin 1989).

## Is the persistent signal sustained kinase activity?

Although sphingosine and H-7 are less suitable than peptide inhibitors for identification of protein kinases that contribute to LTP, because they are less specific, they do offer some advantages: they are sufficiently membrane permeable to have actions when applied on the outside of cells, and they act by fundamentally different mechanisms in inhibiting PKC and CaMKII. On a purely operational level, H-7 provides a means for reversibly suppressing the expression of LTP, without abolishing the underlying process that maintains it, and with little or no effect on basal transmission (Malinow et al 1988a).

On a more mechanistic level, the identification of the persistent signal as sustained protein kinase activity rests on the assumption that H-7 acts by interfering with protein kinase activity, rather than by some as yet unknown pharmacological action. Thus, it is important to look for more direct biochemical evidence for persistent protein kinase activity. Klann & Sweatt (1990) have reported a roughly three-fold increase in persistent protein kinase activity in hippocampal slices that correlates with the establishment of LTP. After electrophysiological recording, homogenates of the slices were tested for their ability to phosphorylate an exogenous substrate, myosin light chain. There was a large and persistent elevation of kinase activity that could be suppressed by PKC(19–36), suggesting the existence of persistent PKC activity. Their experiments do not rule out an accompanying persistent change in CaMKII activity.

## Possible involvement of a presynaptic protein kinase

Clearly, much work remains to be done to identify the retrograde messenger. Arachidonic acid or nitric oxide are the most often-mentioned candidates (see Bliss et al 1990 for review). The presynaptic mechanisms responsible for enhanced transmitter release from presynaptic terminals also remain to be identified. PKC is a logical candidate for a presynaptic effector system (Routtenberg 1985). Phorbol esters (presumably acting by stimulation of PKC) are able to greatly increase the frequency of spontaneous minis recorded from hippocampal neurons in culture (Finch & Jackson 1990). In slices, delivery of phorbol esters (e.g. phorbol diacetate or dibutyrate) to small groups of presynaptic afferent fibres greatly increases synaptic responses mediated by those inputs, but not by other afferents that form synapses on the same postsynaptic cell (Malinow et al 1988b). Stimulation of PKC by phorbol esters not only enhances various types of high voltage-activated $Ca^{2+}$ channels in hippocampal neurons (Madison 1989), but also facilitates $Ca^{2+}$-independent steps in exocytosis in various secretory systems. Thus, it is reasonable to hypothesize that presynaptic PKC might be a link in the chain of events leading up to enhanced transmitter release. However, presynaptic CaMKII is another good candidate; CaMKII phosphorylation of synapsin I is thought to promote the uncaging of synaptic vesicles and to increase transmitter release at the squid giant synapse and in mammalian synaptosomes (Llinas et al 1990, Nichols et al 1990).

## References

Alexander KA, Cimler BM, Meier KE, Storm DR 1987 Regulation of calmodulin binding to P-57. Neurospecific calmodulin binding protein. J Biol Chem 262:6108–6113.

Andersen P, Godfraind JM, Greengard P et al 1990 Injection of a peptide inhibitor of protein kinase C blocks the induction of long-term potentiation in rat hippocampal cells *in vitro*. J Physiol 429:25P

Ascher P, Nowak L 1988 Electrophysiological studies of NMDA receptors. Trends Neurosci 10:284–288
Blanton MG, LoTurco JJ, Kriegstein AR 1989 Whole cell recording from neurons in slices of reptilian and mammalian cerebral cortex. J Neurosci Methods 30:203–210
Bliss TVP, Lynch M 1988 Long-term potentiation of synaptic transmission in the hippocampus: properties and mechanisms. In: Landfield, PW, Deadwyler SA (eds) Long term potentiation: mechanisms and key issues. From biophysics to behavior. Alan R Liss, New York p 3–72
Bliss TVP, Clements MP, Errington ML, Lynch MA, Williams JH 1990 Presynaptic changes associated with long-term potentiation in the dentate gyrus. Semin Neurosci 2:345–354
Brown TH, Chapman PF, Kairiss EW, Keenan CL 1988 Long-term synaptic potentiation. Science (Wash DC) 242:724–728
Collingridge GL, Bliss TVP 1987 NMDA receptors—their role in long-term potentiation. Trends Neurosci 10:288–293
Collingridge GL, Randall AD, Davies CH, Alford S 1992 The synaptic activation of NMDA receptors and $Ca^{2+}$ signalling in neurons. In: Interactions among signalling systems. Wiley, Chichester (Ciba Found Symp 164) p 162–175
Cotman CW, Bridges RJ, Taube JS, Clark AS, Geddes JW, Monaghan DT 1989 The role of the NMDA receptor in central nervous system plasticity and pathology. J Natl Inst Health Research 1:65–74
Davies SN, Lester RAJ, Reymann KG, Collingridge GL 1989 Temporally distinct pre- and post-synaptic mechanisms maintain long-term potentiation. Nature (Lond) 338:500–503
Ferrell JE Jr, Martin GS 1989 Thrombin stimulates the activities of multiple previously unidentified protein kinases in platelets. J Biol Chem 264:20723–20729
Finch DM, Jackson MB 1990 Presynaptic enhancement of synaptic transmission in hippocampal cell cultures by phorbol esters. Brain Res 518:269–273
Goelet P, Castellucci VF, Schacher S, Kandel ER 1986 The long and short of long-term memory—a molecular framework. Nature (Lond) 322:419–422
Hannun YA, Bell RM 1987 Lysosphingolipids inhibit protein kinase C: implications for the sphingolipidoses. Science (Wash DC) 235:670–674
Hebb DO 1949 The organization of behavior. Wiley, New York
Hidaka H, Inagaki M, Kawamoto S, Sasaki Y 1984 Isoquinolinesulfonamides, novel and potent inhibitors of cyclic nucleotide dependent protein kinase and protein kinase C. Biochemistry 23:5036–5041
House C, Kemp BE 1987 Protein kinase C contains a pseudosubstrate prototype in its regulatory domain. Science (Wash DC) 238:1726–1728
Hu GY, Hvalby O, Walaas SI et al 1987 Protein kinase C injection into hippocampal pyramidal cells elicits features of long-term potentiation. Nature (Lond) 328:426–429
Jefferson AB, Schulman H 1988 Sphingosine inhibits calmodulin-dependent enzymes. J Biol Chem 263:15241–15244
Kennedy MB 1989 Regulation of synaptic transmission in the central nervous system: long-term potentiation. Cell 59:777–787
Klann E, Sweatt JD 1990 Persistent alteration of protein kinase activity during the maintenance phase of long-term potentiation. Soc Neurosci Abstr 16:144
Llinas R, McGuinness TL, Leonard CS, Sugimori M, Greengard P 1985 Intraterminal injection of synapsin I or calcium/calmodulin dependent protein kinase II alters neurotransmitter release at the squid giant synapse. Proc Natl Acad Sci USA 82:3035–3039
Lynch G, Larson J, Kelso S, Barrionuevo G, Schottler F 1983 Intracellular injections of EGTA block induction of hippocampal long-term potentiation. Nature (Lond) 305:719–721

Madison D 1989 Phorbol esters increase unitary calcium channel activity in cultured hippocampal neurons. Soc Neurosci Abstr 15:16
Madison DV, Malenka RC, Nicoll RA 1991 Mechanisms underlying long-term potentiation of synaptic transmission. Annu Rev Neurosci 14:379–397
Malenka RC 1991 Postsynaptic factors control the duration of synaptic enhancement in area CA1 of the hippocampus. Neuron 6:53–60
Malenka RC, Kauer JA, Zucker RS, Nicoll RA 1988 Postsynaptic calcium is sufficient for potentiation of hippocampal synaptic transmission. Science (Wash DC) 242:81–84
Malenka RC, Kauer JA, Perkel DJ, Nicoll RA 1989a The impact of postsynaptic calcium on synaptic transmission—its role in long term potentiation. Trends Neurosci 12:444–450
Malenka RC, Kauer JA, Perkel DJ et al 1989b An essential role for postsynaptic calmodulin and protein kinase activity in long-term potentiation. Nature (Lond) 340:554–557
Malgaroli A, Tsien RW 1991 Glutamate-induced long-term potentiation of miniature EPSP frequency in cultured hippocampal neurons. Biophysical J Abstr 59: 19a
Malinow R, Tsien RW 1990 Presynaptic enhancement revealed by whole cell recordings of long-term potentiation in rat hippocampal slices. Nature (Lond) 346:177–180
Malinow R, Madison DV, Tsien RW 1988a Persistent protein kinase activity underlying long-term potentiation. Nature (Lond) 335:820–824
Malinow R, Madison DV, Tsien RW 1988b Selective activation of pre-synaptic protein kinase C (PKC) enhances synaptic transmission in rat hippocampal slices. Soc Neurosci Abstr 14:18
Malinow R, Schulman H, Tsien RW 1989 Inhibition of postsynaptic PKC or CaMKII blocks induction but not expression of LTP. Science (Wash DC) 245:862–866
Nairn AC, Aderem A 1992 Calmodulin and protein kinase C cross-talk: the MARCKS protein is an actin filament and plasma membrane cross-linking protein regulated by protein kinase C phosphorylation and by calmodulin. In: Interactions among cell signalling systems. Wiley, Chichester (Ciba Found Symp 164) p 145–161
Nichols RA, Sihra TS, Czernik AJ, Nairn AC, Greengard P 1990 Calcium/calmodulin-dependent protein kinase II increases glutamate and noradrenaline release from synaptosomes. Nature (Lond) 343:647–651
Nishizuka Y 1988 The molecular heterogeneity of protein kinase C and its implications for cellular regulation. Nature (Lond) 334:661–665
Regehr WG, Tank DW 1990 Postsynaptic NMDA-receptor mediated calcium accumulation in hippocampal CA1 pyramidal cell dendrites. Nature (Lond) 345:807–810
Routtenberg A 1985 Protein kinase C activation leading to protein F1 phosphorylation may regulate synaptic plasticity by presynaptic terminal growth. Behav Neural Biol 44:186–200
Smith MK, Colbran RJ, Soderling TR 1990 Specificities of autoinhibitory domain peptides for four protein kinases. J Biol Chem 265:1837–1840
Tsien RW, Malinow R 1990 Long-term potentiation: presynaptic enhancement following postsynaptic activation of $Ca^{2+}$-dependent protein kinases. Cold Spring Harbor Symp Quant Biol 55:147–159
Tsien RW, Schulman H, Malinow R 1990 Peptide inhibitors of PKC and CaMK block induction but not expression of long-term potentiation. In: Nishizuka Y, Endo M, Tanaka C (eds) The biology and medicine of signal transduction. Raven Press, New York (Adv Second Messenger Phosphoprotein Res 24) p 101–107

## DISCUSSION

*Changeux:* In your experiments on cultured cells you use a suspension of cells from the hippocampus, so you have a mixture of different categories of cells.

*Tsien:* The dentate gyrus is removed before the dissociation, leaving areas CA3 and CA1. We also restrict the growth of glial cells.

*Changeux:* How do you control the connectivity of the neurons? For example, how do you eliminate connections between pyramidal cells?

*Tsien:* We are happy for there to be lots of synapses between pyramidal cells.

*Changeux:* But if there are interconnecting pyramidal cells, what is postsynaptic for one cell becomes presynaptic for the one to which it is connected.

*Tsien:* We record in the presence of tetrodotoxin, which eliminates action potentials and minimizes the possibility of reverberating circuits.

*Changeux:* When you apply glutamate depolarization may occur all along the cell, axonal dendrites included, and thus directly affect the presynaptic release mechanism. How can you exclude any direct effect of the applied glutamate on the nerve endings themselves?

*Tsien:* We don't claim that there is no action of glutamate on the nerve endings. Even in the process of tetanic stimulation in slices, glutamate is released, so it's not unphysiological for the presynaptic terminal to 'see' glutamate. The key question is whether postsynaptic events are a necessary step for induction, as is characteristic of the kind of long-term potentiation that we are trying to study. The crucial experiments are the demonstrations of the hyperpolarization block and of the MK-801 block; the effect of MK-801 shows that the NMDA receptor is involved and the effect of postsynaptic hyperpolarization shows that the potentiation is postsynaptically induced.

*Changeux:* There is no inhibition by GABA.

*Tsien:* That's fine for studying mechanisms of LTP *per se*. In experiments, both in slices and in hippocampal cultures, we use picrotoxin to block GABAergic IPSPs, to try to work with LTP in its most elemental form. This is a different issue from the role of GABAergic inhibition that Professor Collingridge is interested in.

*Collingridge:* Chuck Stevens had to use a lot of tricks to see LTP in culture (Bekkers & Stevens 1990). Which of those tricks do you find are necessary; that is, what conditions are required to get LTP in cultured cells?

*Tsien:* Bekkers & Stevens found that they could not produce LTP under standard culture conditions and suspected that the synapses had already experienced it in culture. They cultured the cells in 50 μM D-2-amino-5-phosphonopentanoate (AP5), a blocker of NMDA receptor channels, to try to reduce any background LTP. We found that this was not necessary in our case. The density of the cultures may be important here. Fortunately, we see a very consistent mini frequency increase in about two thirds of the cells challenged briefly with glutamate—it's a very robust, long-lasting, large phenomenon. Evoked responses are also potentiated.

*Thomas:* One of the arguments that can be levelled at the experiments with the PKC inhibitory peptides is that they might inhibit a kinase that you don't know about. Have you tried to do a 'rescue' experiment, injecting increasing amounts of the kinase to titrate out the peptide?

*Tsien:* We have not yet done that. We would have to perfuse the pipette and this is difficult to do with a sharp microelectrode. Paul Greengard and Per Andersen are doing experiments with various kinase inhibitors. The effect we find with the PKC-inhibiting peptide fits with their results, and our findings with the CAMKII-inhibiting peptide are in accord with those of Malenka et al (1989) with calmodulin inhibitors. None of these results excludes other kinases from being relevant. Remember, though, that we cannot abolish established LTP even by delivering H-7 through the micropipette; even this biochemically non-selective method of kinase inhibition doesn't abolish established LTP.

*Thomas:* That assumes a lot about H-7.

*Tsien:* It assumes that it can block the action of most, if not all, kinases; this seems reasonable, given what is known about the strong family resemblance among the ATP-binding sites of various protein kinases.

*Fischer:* Is it possible that another kinase is involved? Eric Kandel saw an effect of cyclic AMP on short-term and long-term potentiation in *Aplysia*; is that difference because he works with *Aplysia* rather than the mammalian brain?

*Tsien:* The idea of the involvement of a third kinase is an intriguing one. Ferrell & Martin (1989) looked for protein kinases in platelets that were activated by the conjunction of ionomycin and TPA. They separated protein kinases by SDS–PAGE and then renatured them. They found several kinases whose activity depended jointly on TPA and calcium delivery by ionomycin, suggesting that there might be a whole set of kinases in platelets that need both PKC and calmodulin-dependent kinases for their activation. That kind of system might be involved in allowing the convergence of signals from separate pathways.

I doubt there's a general rule that *Aplysia* uses cyclic AMP and the mammalian brain doesn't. Cyclic AMP is so important everywhere that it would be no surprise were it to be involved in some way; but in the systems where LTP is most intensely studied there doesn't seem to be any direct evidence for the involvement of cyclic AMP. We should probably inject PKI, a peptide inhibitor of cAMP-dependent protein kinase, to test this more rigorously.

*Nairn:* Per Andersen's group has done this experiment. Injection of PKI into postsynaptic hippocampal neurons had no effect on LTP (P. Anderson, A. C. Nairn & P. Greengard, unpublished results).

*Fredholm:* The presynaptic enhancement of amino acid release by cyclic AMP-stimulating agents could have the same background as the proposed 'memory' mechanisms in *Aplysia*; the same inhibition of $Ca^{2+}$-dependent $K^+$ conductance may occur in the hippocampus, even though it might not be terribly important in

LTP. β-Adrenergic enhancement of noradrenaline release from sympathetic nerve endings could similarly be brought about by the same mechanism as that studied by Kandel and co-workers for presynaptic facilitation in *Aplysia*.

*Krebs:* Professor Tsien, how long would the peptides have to stay intact and active in order to produce the effect that you want to see?

*Tsien:* They would have to remain active long enough for them to diffuse from the cell body out to dendritic spines—that is, for several minutes at least.

*Krebs:* The synthetic peptides that you use contain several lysine and arginine residues, so they would be readily attacked by proteases. Small peptides are often more vulnerable to protease attack than larger proteins.

*Tsien:* That would be most critical in the experiments in which we failed to see an effect of the peptide. When we see a block we know that some peptide has survived. In experiments where delivery of a peptide fails to block established LTP, we use controls to check that the same peptide can block a subsequent attempt to induce LTP by an anatomically separate pathway, which means that it has remained biologically active.

*Krebs:* Some time ago, Jim Maller showed that when kemptide was injected into $^{32}$P-labelled *Xenopus* oocytes it became phosphorylated within a few seconds, but within a couple of minutes the oocyte contained three or four degradation products of the peptide (Maller et al 1978). We wanted to do some inhibition studies using various peptide analogues, but gave up because of their short life within the oocytes.

*Tsien:* Clearly, one has to be careful. Our experiments with control peptides certainly help in the interpretation.

*Hunter:* I believe that Mike Greenberg has evidence that stimulation of the NMDA receptor increases tyrosine phosphorylation of MAP kinase in hippocampal neuron cultures (Bading & Greenberg 1991). According to Ed Krebs (Ahn et al 1992, this volume), that could be due to autophosphorylation. Also, Eric Kandel has some evidence that tyrosine kinase inhibitors block LTP in hippocampal slices (O'Dell et al 1991). Those are some other types of protein kinase that might be involved.

*Nishizuka:* The kinase theory of LTP is fashionable now. What is the current view about Lynch's calcium-dependent protease hypothesis (Lynch & Baudry 1984)?

*Tsien:* Lynch & Baudry have proposed that LTP is entirely postsynaptic. They originally suggested that calpain might be activated and change the structure of the synapse, but I think they are modifying this hypothesis to consider several postsynaptic or presynaptic scenarios (Lynch & Baudry 1991).

*Nishizuka:* What are your thoughts about retrograde signals?

*Tsien:* The two leading candidates are arachidonic acid and nitric oxide, which are both known to carry signals from one cell to another. This is in line with cell–cell signalling in other systems, such as between smooth muscle cells and endothelial cells, and possibly between B and T cells. The evidence is not very convincing for arachidonic acid because the effects of exogenous arachidonate take a long time to develop. Our results point to changes in the amount of transmitter released within the first minute after the inducing stimulus, so even though the effect lasts a long time it starts relatively early.

Nitric oxide is the other possible candidate (Schumann & Madison 1991) and I would be interested to know if nitric oxide has effects on PKC or any signalling molecule other than guanylate cyclase (which is probably not involved).

*Michell:* Has anyone tested inhibitors of nitric oxide synthesis?

*Tsien:* Schumann & Madison (1991) have done that. They see block of LTP with $N^G$-monomethylarginine.

*Collingridge:* Tim Bliss and colleagues have found that the nitric oxide synthase inhibitors $N^G$-monomethyl-L-arginine and $N^W$-nitro-L-arginine do not affect LTP in the CA1 region, but reduce the magnitude of LTP in the dentate gyrus (personal communication).

*Exton:* What are the results that lead you to argue against the involvement of guanylate cyclase?

*Tsien:* If nitric oxide stimulates guanylate cyclase in the mechanism of LTP something would need to happen downstream of cyclic GMP formation, but, to my knowledge, no potentiating effect of dibutyryl cyclic GMP has been reported.

*Daly:* Has nitroprusside been used?

*Tsien:* Böhme et al (1991) have found potentiation and occlusion of LTP with sodium nitroprusside.

*Collingridge:* There are problems with H-7. Have you tried any other inhibitors that act in the same manner, such as the PKC inhibitor K-252B?

*Tsien:* No, but that might be a good idea.

*Kanoh:* The same is true for sphingosine. Many other unknown mechanisms might be activated or deactivated, because sphingosine can affect a wide variety of biological activities.

*Tsien:* We are not using sphingosine to identify a kinase, so 90% of the criticism does not apply. We know that it *is* effective in blocking activation of CaMKII and PKC. The broadness of its action as an inhibitor makes it all the more interesting that it fails to suppress already established LTP.

## References

Ahn NG, Seger RL, Bratlien RL, Krebs EG 1992 Growth factor-stimulated phosphorylation cascades: activation of growth factor-stimulated MAP kinase. In: Interactions among cell signalling systems. Wiley, Chichester (Ciba Found Symp 164) p 113–131

Bading H, Greenberg ME 1991 Stimulation of protein tyrosine phosphorylation by NMDA receptor activation. Science (Wash DC) 253:912–914

Bekkers JM, Stevens CF 1990 Presynaptic mechanism for long-term potentiation in the hippocampus. Nature (Lond) 346:724–729

Böhme GA, Bon C, Stutzman J-M, Doble A, Blanchard J-C 1991 Possible involvement of nitric oxide in long-term potentiation. Eur J Pharmacol 199:379–381

Ferrell JE Jr, Martin GS 1989 Thrombin stimulates that activities of multiple previously identified protein kinases in platelets. J Biol Chem 264:20723–20729
Lynch G, Baudry M 1984 The biochemistry of memory—a new and specific hypothesis. Science (Wash DC) 224:1057–1063
Lynch G, Baudry M 1991 Reevaluating the constraints on hypotheses regarding LTP expression. Hippocampus 1:9–14
Malenka RC, Kauer JA, Perkel DJ 1989 An essential role for postsynaptic calmodulin and protein kinase activity in long-term potentiation. Nature (Lond) 340:554–557
Maller JL, Kemp BE, Krebs EG 1978 *In vivo* phosphorylation of a synthetic peptide substrate of cyclic AMP-dependent protein kinase. Proc Natl Acad Sci USA 75:248–251
O'Dell TJ, Kandel ER, Grant SGN 1991 Long-term potentiation in the hippocampus is blocked by tyrosine kinase inhibitors. Nature (Lond) 353:558–560
Schumann EW, Madison DV 1991 A requirement for the intercellular messenger nitric oxide in long-term potentiation. Science (Wash DC) 254:1503–1506

# General discussion I

## The metabotropic glutamate receptor

*Akaike:* There are two types of glutamate receptor—ionotropic (iGluR) and metabotropic (mGluR). The NMDA (*N*-methyl-D-aspartate), kainate and AMPA (α-amino-3-hydroxy-5-methyl-4-isoxalone propionic acid) receptors, with their cation-specific ion channels, are ionotropic glutamate receptors. The mGluR is coupled to a pertussis toxin-sensitive G protein and stimulates the production of $InsP_3$, which results in an increase in the intracellular calcium concentration. The mGluR expressed in *Xenopus* oocytes after injection of rat brain mRNA (I. Ito et al 1990) or mGluR mRNA (Masu et al 1991) responded to glutamate and to *t*ACPD (*t*-1-aminocyclopentane dicarboxylic acid), an agonist of the mGluR, increasing $Cl^-$ conductance. In addition, the sequence of the mGluR has been reported (Masu et al 1991). However, there is no information about the functional role of the mGluR in native mammalian neurons. We therefore studied mGluR-mediated responses, using the nystatin-perforated patch-clamp technique, which leaves the inositol phosphate/$Ca^{2+}$ second messenger intracellular signalling system intact.

Hippocampal pyramidal neurons were dissociated from the brains of 1–4-week-old Wistar rats by mechanical dissociation after enzyme treatment. Drugs were applied by the 'Y-tube' technique, which allows rapid exchange of external solution, within 10–20 ms. Application of glutamate or quisqualate induced an inward current that was followed by an outward current. The outward current was resistant to the agonists AP5, CNQX and DNQX (amino-5-phosphonopentanoate, 6-cyano-7-nitroquinoxaline-2,3-dione, 6,7-dinitroquinoxaline-2,3-dione; Fig. 1B), and it disappeared when we switched to the whole-cell configuration in which the interior of the cell was perfused with patch-pipette solution containing 5 mM BAPTA, 1,2-bis(2-aminophonoxy)ethane *N,N,N′,N′*-tetraacetic acid (Fig. 1A). Because this outward current was (1) abolished by pretreatment with caffeine, ryanodine and acetylcholine, (2) activated in $Ca^{2+}$-free external solution and (3) had a reversal potential close to $E_K$, we concluded that the mGluR in native rat hippocampal CA1 neurons activated a $Ca^{2+}$-dependent $K^+$ channel via a second messenger system (probably $InsP_3$). The dose–response relationship of the quisqualate-induced potassium current had the following characteristics. The threshold, the $K_a$ value (the apparent dissociation constant), and the Hill coefficient were $10^{-8}$ M, $1.1 \times 10^{-7}$ M and 2.28, respectively. This $K_a$ value was 100-fold lower than that of the inward current induced by quisqualate. Figure 1C shows the mGluR response induced by *t*ACPD. The transient outward mGluR response

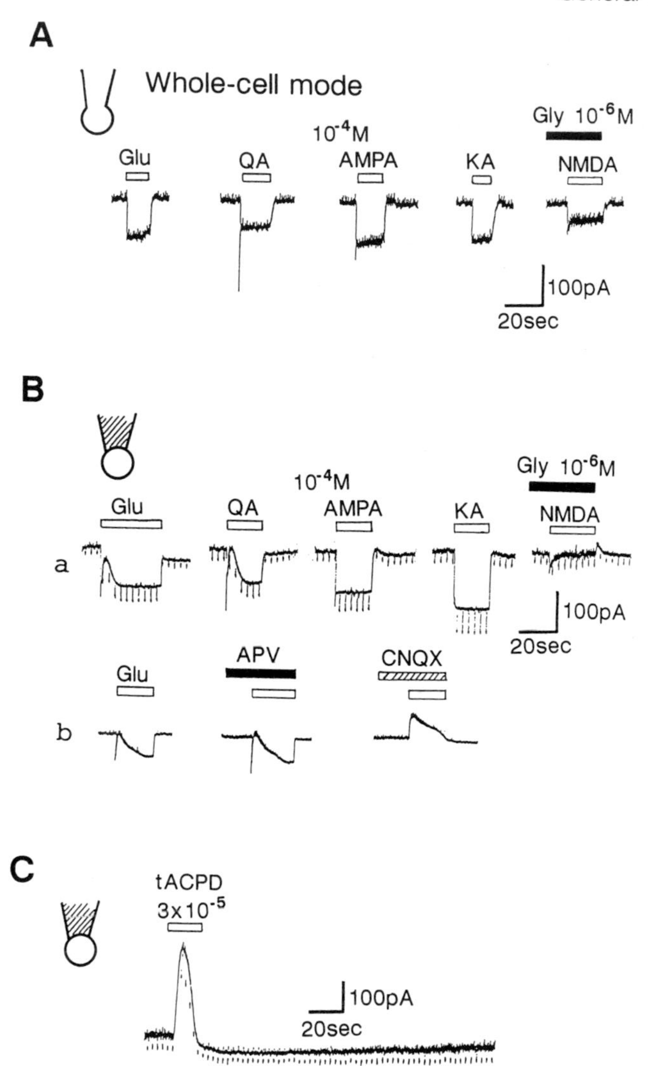


FIG. 1. (*Akaike*) Amino acid-induced responses in rat hippocampal pyramidal neurons. Ionic currents in (A) were recorded using a conventional patch pipette, whereas the nystatin-perforated patch-pipette technique was used in (B) and (C). For observation of changes in membrane conductance during responses to glutamate (Glu), quisqualate (QA), kainate (KA) and *N*-methyl-D-aspartate (NMDA) in B(a) hyperpolarized step command pulses of 8 mV, of duration 300 ms, were applied every 2 s; 10 mV pulses of the same duration and frequency were used in (C) to observe the response to *t*ACPD (*t*-1-aminocyclopentane dicarboxylic acid). The holding potential was −40 mV. All recordings were taken from four neurons. AP5, amino-5-phosphonopentanoate; Gly, glycine; AMPA, α-amino-3-hydroxy-5-methyl-4-isoxalone propionic acid; CNQX, 6-cyano-7-nitroquinoxaline-2,3-dione.

was followed by a long-lasting inward current that was maintained even after the removal of *t*ACPD from the medium. The former and latter currents were followed by the increase and decrease, respectively, of $K^+$ conductance. The results suggest that the mGluR may play an important role in neuronal excitability.

## Distribution of protein kinase C subspecies in the hippocampal formation

*Tanaka:* As Professor Tsien described, there is increasing evidence for the involvement of protein kinase C (PKC) at certain stages of LTP in the hippocampal formation (hippocampus and dentate gyrus). Protein kinase C is known to exist as a family of several subtly different, closely related subspecies—α, βI, βII, γ, δ, ε, ζ and η (Nishizuka 1988). Here, we demonstrate the localization of PKC subspecies and their mRNAs in the rat hippocampal formation by immunocytochemistry and *in situ* hybridization, using light and electron microscopy.

To localize PKC subspecies proteins in the hippocampal formation we have done immunohistochemical studies using four antibodies raised against α, βI, βII and γ-PKC, but not antibodies against other subspecies (Saito et al 1988, 1989, Hosoda et al 1989, A. Ito et al 1990, Kose et al 1990). Figure 2 (see colour plate) shows the distribution of these four PKC subspecies in the frontal sections of rat brain. Immunoreactivity against α, βII and γ-PKC, but not βI-PKC, is more prominent in the hippocampus than in other brain regions, but shows clearly different staining patterns. As summarized in Table 1, α-PKC is located in pyramidal cells of the hippocampus, with its highest density in the CA2 region. α-PKC is also present in interneurons, but not in granule cells of the dentate gyrus. mRNA for βII-PKC is found not only in the CA1 region, but also in the CA3 region; however, intense βII-PKC immunoreactivity is seen only in the CA1 region, and its density is quite low in the CA2–CA3 regions of the dentate gyrus. γ-PKC is localized in pyramidal cells throughout the hippocampus, with the highest concentration in the CA1 region. In the dentate gyrus, granule cells also contain a moderate concentration of γ-PKC.

To determine the distribution of mRNAs encoding PKC subspecies in the rat hippocampal formation we have done *in situ* hybridization histochemistry using oligonucleotide DNA probes. As shown in Table 1, *in situ* hybridization revealed distinct, but partially overlapping, patterns of expression of PKC mRNAs. Signals for the α, βII, γ and ε isoforms are concentrated almost exclusively in the cell layer of the hippocampus and of the dentate gyrus, and are not detected in the strata oriens, radiatum or lacunosum. Signals for δ-PKC and ζ-PKC are weak or absent in the hippocampus and dentate gyrus. The distribution of mRNA shows that α-PKC is expressed at highest concentrations in the CA2 region, and at low concentration in the dentate gyrus. γ-PKC mRNA is more concentrated in hippocampal pyramidal cells than in

**TABLE 1 (*Tanaka*) Location of protein kinase C subspecies in the rat hippocampal formation**

| | *CA1* | *CA2* | *CA3* | *Dentate gyrus* | *Reference* |
|---|---|---|---|---|---|
| messenger RNA | α, βII, γ, ε | α, γ, ε | α, βII, γ, ε | α, γ, ε | unpublished results |
| *Immunoreactivity* | | | | | |
| α-PKC | Pyramidal cells<br>Interneurons | Pyramidal cells | Pyramidal cells<br>Interneurons | Interneurons | Ito et al 1990 |
| βII-PKC | Pyramidal cells | | | | Saito et al 1989<br>Kose et al 1990 |
| γ-PKC | Pyramidal cells | Pyramidal cells | Pyramidal cells | Granule cells | Saito et al 1988<br>Kose et al 1990 |

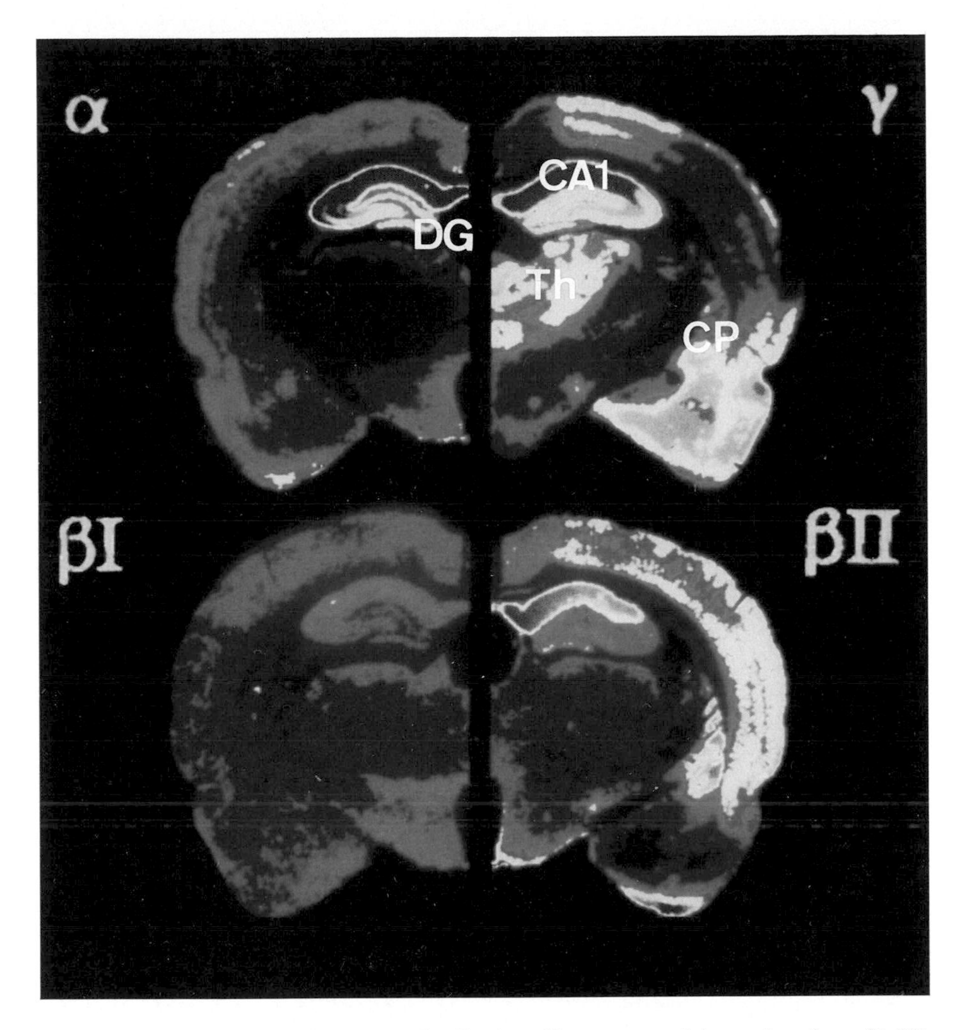


FIG. 2. *(Tanaka)* Colour images of the distribution of immunoreactivity against the α, βI, βII and γ subspecies of protein kinase C (PKC) in frontal sections of rat brain. Immunoreactivity was identified by peroxidase–antiperoxidase immunohistochemistry using polyclonal antisera against synthetic oligopeptides corresponding to the C-terminal sequences of α, βI and βII-PKC, and a monoclonal anti-γ-PKC antibody. In the CA1 and CA3 regions of the hippocampus the highest density *(red)* of α-PKC immunoreactivity was seen, and a moderate *(yellow)* density was observed in the dentate gyrus. In the hippocampus there was no detectable *(blue)* immunoreactivity against βI-PKC; for βII-PKC, the highest level *(red)* of immunoreactivity was observed only in the CA1 region. γ-PKC is found all over the hippocampus, with the highest levels in the CA1 region, and less heavy staining in the dentate gyrus. Th, thalamus; CP, caudoputamen; DG, dentate gyrus.

granule cells of the dentate gyrus. ε-PKC mRNA is quite evenly distributed in the cell layer throughout the hippocampal formation. βII-PKC mRNA is expressed throughout the hippocampus, except in the CA2 region, and a weak signal is also found in the dentate gyrus.

Electron micrographs of CA1 pyramidal cells indicate that immunoreactivity against βII-PKC in pyramidal cell bodies is concentrated in the Golgi region, at the concave face (*trans* region) of the Golgi complex. The cell membrane, nucleus and cytoplasm are immunonegative, except around the Golgi complex. Immunoreactivity against γ-PKC is distributed diffusely in the cytoplasm, being concentrated at the surface of mitochondria, rough endoplasmic reticulum and free ribosomes; the insides of cell organelles are devoid of stain. The nucleus is stained, but much less than the perikaryon, and the nucleolus is unstained.

Higher magnification electron micrographs of CA1 pyramidal cells show that βII-PKC is located around the Golgi complex in proximal apical dendrites with which immunonegative terminals form shaft synapses. Distal dendrites are immunopositive, with strong reactivity along the neurotubules and mitochondria, but most dendritic spines are immunonegative. Interestingly, γ-PKC is found postsynaptically in the cytoplasm of dendrites and dendritic spines, particularly along the neurotubules and the outer membrane of the mitochondria. γ-PKC-positive dendritic trunks and spines were observed to form synapses with immunonegative terminals. γ-PKC appears to be associated with both dendritic shaft synapses and spine synapses. Both these subspecies of PKC may be involved in LTP at the postsynaptic side in the CA1 region. In the CA3 region, γ-PKC is highly associated with the dendritic spines which are surrounded by giant immunonegative mossy fibre terminals. There is a possibility that γ-PKC might be involved in LTP at the postsynaptic side of mossy fibre CA3 synapses.

There are several reports of evidence for a presynaptic contribution of PKC in LTP. In the alveus and the most distal part of the stratum lacunosum-moleculare many myelinated axons are observed to be positive for α, βII and γ-PKC, but no immunopositive terminals have been found in the synapses examined. In the rat hippocampal formation the presynaptic action of PKC may be associated with subspecies of PKC other than α, βII or γ, such as ε-PKC, which is abundant throughout the hippocampal formation.

*Mikoshiba:* Even though the localization of γ-PKC and the $InsP_3$ receptor are very similar in the CA1 region of the hippocampus and in the Purkinje cell, at the level of electron microscopy the distribution differs. The $InsP_3$ receptor is localized exclusively on smooth endoplasmic reticulum, whereas γ-PKC is found in the Golgi apparatus, sometimes in nuclei, or attached to the mitochondrial membrane.

*Tanaka:* It is interesting that γ-PKC is not localized around the calcium stores.

*Mikoshiba:* When we increased the sensitivity of *in situ* hybridization we could detect signals at the dendrites of Purkinje cells. Did you detect signals in the dendrites of Purkinje cells?

*Tanaka:* We cannot detect signals. $Ca^{2+}$/calmodulin-dependent protein kinase II is detectable in the dendrites of hippocampal pyramidal cells but protein kinase C subspecies are not.

*Mikoshiba:* Although $\gamma$-PKC has a signal peptide, the $InsP_3$ receptor does not. The manner of protein synthesis and distribution must be very lifferent.

*Tsien:* Professor Nishizuka, you showed that the $\alpha$ and $\beta$ isoforms of PKC can be activated by arachidonic acid and several other fatty acids, but that diacylglycerol must be present. Can the $\gamma$ subspecies, which Professor Tanaka finds postsynaptically, be activated by arachidonic acid, and to what extent is diacylglycerol required?

*Nishizuka:* It can be activated by arachidonic acid to some extent in the absence of diacylglycerol.

*Tsien:* In that case, the $\gamma$ subspecies, localized in spines, is well positioned to respond to arachidonic acid produced by activation of phospholipase $A_2$ by a calcium influx through NMDA receptors. From everything we know, there is nothing missing for that particular isoform of protein kinase C to be turned on during LTP.

*Nishizuka:* Even the $\gamma$ form can be activated synergistically by arachidonate and diacylglycerol in the absence of added calcium; the $\gamma$ subspecies differs from the others because it can be activated to some extent by arachidonate in the absence of diacylglycerol.

*Tsien:* So even when there is no diacylglycerol production, which may or may not be the case, you would still expect to see some activation of $\gamma$-PKC. Is there much translocation of $\gamma$-PKC?

*Nishizuka:* We don't know.

*Tanaka:* Recently, we found that $\gamma$-PKC is translocated to the cytoskeletal fraction, but not to the cell membrane.

*Nishizuka:* The degree of association with the membrane depends on calcium levels.

## Cross-talk between the phosphatidylinositol and adenylate cyclase signalling systems

*Daly:* We have been faced with a cross-talk puzzle in our lab for two decades. It has proved to be the first example of cross-talk between a phospholipid signalling system and responses of the adenylate cyclase system—at the time we didn't realize it was cross-talk and called it synergism. The original data were reported in 1971 (Huang et al). We were looking at formation of cyclic AMP in brain slices in response to various agents. Control levels of cyclic AMP were very low. Histamine gave a modest response and noradrenaline alone gave no response at all. However, the presence of noradrenaline markedly potentiated the histamine response. Serotonin (5-HT) alone had no effect. We then used a maximal stimulating concentration of adenosine (100 μM) and looked at the effects of combinations of adenosine with histamine, noradrenaline and serotonin. In the case of histamine with adenosine we expected an additive response, indicative of the involvement of separate receptors. Instead, we saw a tremendous potentiation by histamine of the response to adenosine. Noradrenaline, ineffective alone, also markedly potentiated the adenosine response. Similarly, serotonin alone gave no response, but markedly potentiated the adenosine response.

We persevered for 10 years trying to understand these synergisms; we wanted to characterize the receptors, their localization and the mechanism and functional significance. It was relatively easy to characterize the receptors. The adenosine receptor involved is now called an $A_2$ receptor, and it is coupled to adenylate cyclase through a $G_s$ protein in the classical way. The histamine response was found to have two components—one was an $H_2$ histamine receptor, which is known to be coupled to adenylate cyclase through a $G_s$ protein, and the other was an $H_1$ receptor. The augmentation of the response to adenosine by noradrenaline was mediated by an $\alpha_1$ receptor, and serotonin potentiation of the response to adenosine was probably through a 5-$HT_1$ receptor. It was later found that there are other adenylate cyclase responses mediated by receptors that are coupled through $G_s$ to cyclase that can be potentiated by receptors that are not coupled to adenylate cyclase. The responses potentiated were the $A_2$ response, the $H_2$ histamine response, a $\beta_1$ adrenergic response and a vasoactive intestinal peptide (VIP) response. The receptors that potentiated the responses were the $\alpha_1$ adrenergic, the $H_1$ histamine and the 5-$HT_2$ in brain (Daly et al 1980, 1981, Magistretti & Schorderet 1984 and references therein). Potentiation by a $GABA_B$ receptor has also been reported (Karbon & Enna 1985). We also found that depolarization, which triggers influx of sodium, would potentiate the response to adenosine (Shimizu et al 1970, Bruns et al 1980).

What did we find out about mechanisms? We couldn't study mechanisms in membranes—the phenomenon could be demonstrated only in intact cells or in a preparation that we call synaptoneurosomes in which both the presynaptic and postsynaptic entities were re-sealed. We found that the phenomenon was calcium-dependent (Schwabe & Daly 1977).

When we returned to this work in the mid-1980s, others had shown that the receptors that we had found to potentiate the activation of adenylate cyclase were the same receptors that stimulate PtdIns breakdown in brain tissue (Hollingsworth & Daly 1985). It was natural for us to assume that such receptors, linked to PtdIns breakdown, were generating diacylglycerides and thereby activating protein kinase C, and were also generating inositol trisphosphate to release intracellular calcium, and that the protein kinase C activation and/or calcium were interacting with the adenylate cyclase system, augmenting an agonist receptor response to generate a 'synergistic' accumulation of cyclic AMP. In our original brain preparations we couldn't see anything that implicated intracellular calcium in the augmentation of the cyclic AMP responses. However, we found that phorbol esters that activate protein kinase C would enhance the response of adenylate cyclase to adenosine activation in brain preparations (Hollingsworth et al 1985). So, we proposed that there was cross-talk between two systems, where activation of protein kinase C was causing (by phosphorylation) an increased responsiveness of the adenylate cyclase system. The receptors that activate PtdIns breakdown are now known to also couple to phospholipase $A_2$ activation, so others have proposed that arachidonate is perhaps involved (Enna & Karbon 1987). Potentiation of cyclic AMP responses by depolarizing agents appeared to be linked to PtdIns breakdown (Hollingsworth et al 1986a).

We then changed from brain preparations to using cultured cells. Our studies were based primarily on using phorbol esters to directly activate protein kinase C, or using maitotoxin to activate PtdIns breakdown, the latter being used because in many of the cultured cells we couldn't demonstrate a robust receptor-mediated breakdown of PtdIns. In cells that contained only the $\alpha$ isozyme of protein kinase C, either direct activation with phorbol esters or indirect activation through maitotoxin-elicited formation of diacylglycerides resulted in an inhibition of cyclic AMP responses (Hollingsworth & Daly 1987). However, cells, such as pheochromocytoma cells, and brain preparations that contain the $\gamma$ isozyme of protein kinase C showed a potentiation of cyclic AMP responses by phorbol esters, maitotoxin or depolarizing agents (Hollingsworth et al 1986a,b, Gusovsky et al 1989, Karbon et al 1986, Nordstedt & Fredholm 1987). Dr Fabian Gusovsky in my laboratory speculated that the $\gamma$ isozyme was the one that is responsible for the potentiation. He therefore transfected the $\gamma$ isozyme into a fibroblast cell line that contained only $\alpha$-PKC. In the non-transfected cells activation of protein kinase C by phorbol esters resulted in an inhibition of cyclic AMP responses, whereas in the cells transfected with the $\gamma$ isozyme there was a potentiation (Gusovsky & Gutkind 1991). Thus, in this system $\gamma$-PKC does subserve potentiation. In other systems, for example in brain preparations and in other cell lines, a combination of calcium and protein kinase C, or calcium alone, may underlie the potentiation of cyclic AMP responses by agents that activate phospholipid breakdown. This remains a challenge for further research.

*Nishizuka:* Are there any known examples of such 'cross-potentiation' in a single cell?

*Fredholm:* In Jurkat cells, a human T cell leukaemia line, there are two receptors which will stimulate production of cyclic AMP—an adenosine receptor and a prostaglandin receptor. This cell contains $\alpha$-PKC and $\beta$-PKC, which are translocated to the cell membrane by receptor activation or in response to phorbol esters, but not the $\gamma$ form (Kvanta et al 1991). Phorbol ester stimulation of protein kinase C enhances the response to adenosine and decreases the prostaglandin response (Nordstedt et al 1987, 1989). Thus, in one cell line two receptors that stimulate cyclic AMP formation are regulated in diametrically opposite ways by PKC.

*Daly:* I know of a similar example; in one of the fibroblast cell lines phorbol esters enhance the response to prostaglandin, the opposite of the response seen in Jurkat cells, and slightly decrease the response to forskolin (Uzumaki et al 1986). I think the problem with phorbol esters is that you activate all the protein kinase C isozymes and you get a net output signal modulated by different phosphorylation inputs to the adenylate cyclase system. I really feel that calcium is important, despite the fact that we have been unable to demonstrate a role for calcium in brain systems (Hollingsworth et al 1986a, Danoff & Young 1987). However, as with phorbol esters for activation of protein kinase C, I think that using ionophores to change calcium levels is a brute force technique, and that localized rather than global changes in calcium may be important, but at present impossible to investigate experimentally.

*Fredholm:* The response to adenosine can also be enhanced by stimulating T cells with antibodies against CD2 or CD3 (Kvanta et al 1989). This is partly due to activation of PKC, and partly to a mechanism which cannot be blocked by down-regulating PKC or by H-7, but it can be blocked by removing extracellular calcium using BAPTA (Kvanta et al 1990). It would appear that there are two parallel mechanisms activated by the receptor that can enhance the cyclic AMP-mediated signalling. Thus, one way to resolve the apparent conflict of whether PKC is mediating this form of receptor cross-talk is by such a demonstration that PKC is sufficient but not necessary.

*Daly:* In brain slices the most convincing evidence that something in addition to protein kinase C is involved is the fact that if you use phorbol esters to augment the adenylate cyclase response you can block it with staurosporine or H-7, but you can't block the synergism with the receptor agonists (Robinson & Kendall 1990). That's a strong argument that something else is participating in the synergism between receptors that activate phospholipid breakdown and receptors that activate adenylate cyclase via $G_s$ proteins.

## References

Bruns RF, Pons F, Daly JW 1980 Glutamate- and veratridine-elicited accumulations of cyclic AMP in brains slices: a role for factors which potentiate adenosine-responsive systems. Brain Res 189:550–555

Daly JW, McNeal E, Partington C, Neuwirth M, Creveling CR 1980 Accumulations of cyclic AMP in adenine-labelled cell-free preparations from guinea pig cerebral cortex: role of $\alpha$-adrenergic and $H_1$-histaminergic receptors. J Neurochem 35:326–337

Daly JW, Padgett W, Creveling CR, Cantacuzene D, Kirk KL 1981 Cyclic AMP-generating systems: regional differences in activation by adrenergic receptors in rat brain. J Neurosci 1:49–59

Danoff SK, Young JN 1987 Is histamine potentiation of adenosine-stimulated cyclic AMP accumulation in guinea-pig cerebral cortical slices mediated by products of inositol phospholipid breakdown? Biochem Pharmacol 36:1179–1181

Enna SJ, Karbon EW 1987 Receptor-regulation: evidence for a relationship between phospholipid metabolism and neurotransmitter receptor-mediated cAMP formation in brain. Trends Pharmacol Sci 8:21–24

Gusovsky F, Gutkind JS 1991 Selective effects of activation of protein kinase C isozymes on cyclic AMP accumulation. Mol Pharmacol 39:124– 129

Gusovsky F, Yasumoto T, Daly JW 1989 Calcium-dependent effects of maitotoxin on phosphoinositide breakdown and on cyclic AMP accumulation in PC12 and NCB-20 cells. Mol Pharmacol 36:44-53

Hollingsworth EB, Daly JW 1985 Accumulation of inositol phosphates and cyclic AMP in guinea-pig cerebral cortical preparations. Effects of norepinephrine, histamine, carbamylcholine and 2-chloroadenosine. Biochim Biophys Acta 847:207–216

Hollingsworth EB, Daly JW 1987 Inhibition of receptor-mediated stimulation of cyclic AMP accumulation in neuroblastoma-hybrid NCB-20 cells by a phorbol ester. Biochim Biophys Acta 930:272–278

Hollingsworth EB, Sears EB, Daly JW 1985 An activator of protein kinase C (phorbol-12-myristate-13-acetate) augments 2-chloro-adenosine-elicited accumulation of cyclic AMP in guinea-pig cerebral cortical particulate preparations. FEBS (Fed Eur Biochem Soc) Lett 184:339–342

Hollingsworth EB, Sears EB, de la Cruz RA, Gusovsky F, Daly JW 1986a Accumulation of cyclic AMP and inositol phosphates in guinea pig cerebral cortical synaptoneurosomes: enhancement by agents acting at sodium channels. Biochim Biophys Acta 883:15–25

Hollingsworth EB, Ukena D, Daly JW 1986b The protein kinase C activator phorbol-12-myristate-13-acetate enhances cyclic AMP accumulation in pheochromocytoma cells. FEBS (Fed Eur Biochem Soc) Lett 196:131–134

Hosoda K, Saito N, Kose A et al 1989 Immunocytochemical localization of βI-subspecies of protein kinase C in rat brain. Proc Natl Acad Sci USA 86:1393–1397

Huang M, Shimizu H, Daly JW 1971 Regulation of adenosine 3′,5′-phosphate formation in cerebral cortical slices. Interaction among norepinephrine, histamine, serotonin. Mol Pharmacol 7:155–162

Ito A, Saito N, Hirata M et al 1990 Immunocytochemical localization of the $\alpha$ subspecies of protein kinase C in rat brain. Proc Natl Acad Sci USA 87:3195–3199

Ito I, Tanabe S, Kohda A, Sugiyama H 1990 Allosteric potentiation of quisqualate receptors by a nootropic drug aniracetam. J Physiol 424:533–543

Karbon EW, Enna SJ 1985 Characterization of the relationship between $\gamma$-aminobutyric acid B agonists and transmitter-coupled cyclic nucleotide-generating systems in rat brain. Mol Pharmacol 27:53–59

Karbon EW, Shenolikar S, Enna SJ 1986 Phorbol esters enhance neurotransmitter-stimulated cyclic AMP production in rat brain slices. J Neurochem 47:1566–1575

Kose A, Ito A, Saito N, Tanaka C 1990 Electron microscopic localization of $\gamma$ and βII-subspecies of protein kinase C in rat hippocampus. Brain Res 518:209–217

Kvanta A, Nordstedt C, Jondal M, Fredholm BB 1989 Activation of protein kinase C via the T-cell receptor complex potentiates cyclic AMP responses in T-cells. Naunyn-Schmiedeberg's Arch Pharmacol 340:715–717

Kvanta A, Gerwins P, Jondal M, Fredholm BB 1990 Stimulation of T-cells with OKT3 antibodies increases forskolin binding and cyclic AMP accumulation. Cell Signal 2:461–470

Kvanta A, Jondal M, Fredholm BB 1991 Translocation of the $\alpha$-isoforms and β-isoforms of protein kinase C following activation of human T-lymphocytes. FEBS (Fed Eur Biochem Soc) Lett 283:321–325

Magistretti PJ, Schorderet M 1984 VIP and noradrenaline act synergistically to increase cyclic AMP in cerebral cortex. Nature (Lond) 308:280–282

Masu M, Tanabe Y, Tsuchida K, Shigemoto R, Nakanishi S 1991 Sequence and expression of a metabotropic glutamate receptor. Nature (Lond) 349:760–765

Nishizuka Y 1988 The molecular heterogeneity of protein kinase C and its implications for cellular regulation. Nature (Lond) 334:661–665

Nordstedt C, Fredholm BB 1987 Phorbol-12,13-dibutyrate enhances the cyclic AMP accumulation in hippocampal slices induced by adenosine analogues. Naunyn-Schmiedeberg's Arch Pharmacol 335:136–142

Nordstedt C, Jondal M, Fredholm BB 1987 Activation of protein kinase C inhibits prostaglandin- and potentiates adenosine receptor-stimulated accumulation of cyclic AMP in human T-cell leukemia line. FEBS (Fed Eur Biochem Soc) Lett 220:57–60

Nordstedt C, Kvanta A, vanderPloeg I, Fredholm BB 1989 Dual effect of protein kinase-C on receptor stimulated cAMP accumulation in a human T-cell leukemia line. Eur J Pharmacol Mol Pharmacol Sect 172:51–60

Robinson JP, Kendall DA 1990 Prostaglandins and the $\alpha_1$-adrenergic potentiation of neurotransmitter-stimulated cyclic AMP formation in mouse cerebral cortex. Reply from Drs Robinson and Kendall. J Neurochem 54:1083–1084

Saito N, Kikkawa U, Nishizuka Y, Tanaka C 1988 Distribution of protein kinase C-like immunoreactive neurons in rat brain. J Neurosci 8:369– 382

Saito N, Kose A, Ito A et al 1989 Immunocytochemical localization of βII-subspecies of protein kinase C in rat brain. Proc Natl Acad Sci USA 86:3409–3413

Schwabe U, Daly JW 1977 The role of calcium ions in accumulation of cyclic adenosine monophosphate elicited by alpha and beta adrenergic agonists in rat brain slices, J Pharmacol Exp Ther 202:134–143

Shimizu H, Creveling CR, Daly J 1970 Stimulated formation of adenosine 3',5'-cyclic phosphate in cerebral cortex: synergism between electrical activity and biogenic amines. Proc Natl Acad Sci USA 65:1033–1040

Uzumaki H, Yamagoto S, Goto H, Kata R 1986 Potentiation of prostaglandin $E_1$-stimulated cAMP formation by 12-O-tetradecanoyl-phobol-13-acetate in BALB/C mousc 3T3 cells. Biochem Pharmacol 35:835–838

# T lymphocyte activation signals

D. A. Cantrell*, J. D. Graves*, M. Izquierdo*, S. Lucas*, J. Downward†

*Lymphocyte Activation Laboratory and †Signal Transduction Laboratory, Imperial Cancer Research Fund, PO Box 123, 44 Lincoln's Inn Fields, London WC2A 3PX, UK

*Abstract.* Activation of T lymphocytes results in immediate biochemical changes including increases in intracellular calcium levels, activation of protein kinase C (PKC) and changes in tyrosine phosphorylation. In T cells recent studies have indicated that activation of the guanine nucleotide-binding proteins $p21^{ras}$ is mediated by PKC, which suggests that the $p21^{ras}$ proteins may regulate intracellular signalling events downstream of PKC. The $p21^{ras}$ proteins can be activated in T cells by signals generated by triggering of the T cell antigen receptor (TCR), the CD2 antigen and the interleukin 2 receptor. Experiments using a PKC pseudosubstrate inhibitor indicate that PKC does not mediate TCR-induced activation of $p21^{ras}$. These results imply that an alternative signal transduction pathway not involving PKC can regulate the activity of $p21^{ras}$ proteins in T cells.

*1992 Interactions among cell signalling systems. Wiley, Chichester (Ciba Foundation Symposium 164) p 208–222*

T lymphocyte activation results in the induction of the expression of new genes, and ultimately in the induction of T cell effector functions and T cell proliferation (reviewed in Crabtree 1989). Proliferation is mediated via interactions between the T cell growth factor interleukin 2 (IL-2) and its cellular receptor (IL-2R) (Cantrell & Smith 1984). Accordingly, a primary focus in T cell activation studies has been the elucidation of the receptors and intracellular signalling mechanisms that control the expression of the IL-2 gene and the genes encoding the IL-2 receptor subunits.

The physiological ligand for T cell activation is antigen presented to the T lymphocyte in association with major histocompatibility molecules by another cell. Multiple receptors on the T cell surface mediate the interaction between the T cell and the antigen-presenting cell. The antigen specificity of T cell activation indicates that the critical receptor in the activation process is the T cell antigen receptor (TCR). However, other receptors, non-specific for that antigen, such as the adhesion molecules LFA-1 and CD2, and the CD4 and CD8 antigens, also play important roles in T cell activation (Bierer et al 1989).

## Biochemical changes initiated during T cell activation

In studies examining the intracellular signalling requirements for regulation of the IL-2 and IL-2 receptor genes attention has been focused both at the cell membrane and in the nucleus. At the T cell membrane it is known that triggering of receptors such as the TCR and CD2 antigens results in the activation of at least two signal transduction pathways. The activation of an inositol lipid-specific phospholipase C (PLC) by the TCR and CD2 is well documented (Imboden & Stobo 1985, Pantaleo et al 1987). This event leads to the hydrolysis of inositol phospholipids, liberating two second messengers—inositol polyphosphates, which regulate intracellular calcium, and diacylglycerols, the regulators of protein kinase C (PKC). Also, the TCR and CD2 antigens activate a tyrosine kinase, such that within seconds of T cell activation phosphorylation of proteins on tyrosine residues is induced (Klausner & Samelson 1991). Candidates for the TCR- and CD2-regulated tyrosine kinases include the c-*fyn* gene product, which can be co-immunoprecipitated with the TCR complex, and Lck, which physically associates with CD4 and CD8 molecules. It is also recognized that phosphatases may have a critical role in the regulation of tyrosine phosphorylation in T cells. T cells express the CD45 molecule, the cytoplasmic domain of which is a phosphotyrosine phosphatase. Interactions that have been described between the TCR or CD2 antigens and CD45 may explain certain of the regulatory effects of the TCR and CD2 molecules on the tyrosine phosphorylation events that occur after these receptors have been triggered (Alexander & Cantrell 1989).

## Coupling of the T cell receptor to phospholipase C

Originally it was hypothesized that in T cells the signalling pathways involved in the regulation of PLC and tyrosine kinases by the TCR were independent and parallel. However, it is now recognized that there is cross-talk between these two signalling systems. It has been demonstrated that T cells express the phospholipase C$\gamma$ isotype that has been shown to be regulated by phosphorylation on tyrosine residues in other cell types, such as fibroblasts. (Klausner & Samelson 1991, Park et al 1991). Furthermore, experiments with two inhibitors of tyrosine kinases have indicated that tyrosine phosphorylation is required for TCR and CD2 regulation of inositol phospholipid metabolism (Klausner & Samelson 1991, Mustelin et al 1990). Finally, TCR-mediated phospholipase C activation requires the surface expression of the phosphotyrosine phosphatase CD45 (Koretzky et al 1990). Although these findings clearly implicate tyrosine phosphorylation as an important regulator of TCR coupling to PLC, they do not establish whether the TCR regulates PLC via a direct mechanism involving solely PLC tyrosine phosphorylation.

It is clear that T cells express a PLC that can be activated via guanine necleotide-biding proteins (G proteins) (Graves et al 1990). This conclusion is based on experiments in permeabilized T cells which show that GTP[S], a non-hydrolysable analogue of GTP, is a potent activator of inositol lipid hydrolysis. It appears, however, that the coupling between the TCR and this G protein is not analogous to that between the receptors and G proteins involved in the hormonal control of adenylate cyclase. Most notably, there is no synergy between TCR agonists and GTP[S] with regard to PLC activation, which indicates that the TCR does not regulate guanine nucleotide exchange on the PLC-linked G proteins (Graves & Cantrell 1991). Nevertheless, there is evidence that GDP, which antagonizes G protein functions, can inhibit TCR-induced inositol lipid metabolism (Graves & Cantrell 1991), so a G protein may be involved in coupling between PLC and the TCR. It is possible of course that the inhibitory effect of GDP on TCR functions does not result from GDP blocking G protein activity. The $\zeta$ chain of the TCR complex has a putative nucleotide-binding site (Weissman et al 1988). It remains to be established whether this binding site interacts with guanine nucleotides, and whether such an interaction has any effect on the functions of the TCR.

## The role of $[Ca^{2+}]_i$, PKC and protein tyrosine phosphorylation in T cell activation

The observation that activation of T cells resulted in increases in intracellular calcium levels, stimulation of PKC and stimulation of tyrosine kinases has inevitably generated the hypothesis that these signalling pathways are responsible for the subsequent changes in gene expression. The use of pharmacological agents such as calcium ionophores and phorbol esters has shown that increases in $[Ca^{2+}]_i$, combined with PKC stimulation, can regulate IL-2 and IL-2R gene expression in T cells (Weiss et al 1984, Cantrell et al 1988, Desai et al 1990). Similarly, T cells transfected with v-*src*, a constitutionally activated tyrosine kinase, express the IL-2 gene (O'Shea et al 1991). However, although it has been demonstrated that these different signalling pathways are involved in regulation of lymphokine genes, it is not known how the signals that are induced on the cytoplasmic face of the T cell membrane are transduced to the cell nucleus where they control gene expression. One strategy to explore this issue has been to study the nuclear events controlling gene expression and attempt to identify the *trans*-activating factors involved (Crabtree 1989). For example, a tissue-specific and activation-dependent enhancer −326 to −52 base pairs 5′ of the transcription initiation site of the IL-2 gene has been identified (Durand et al 1988). In the IL-2 gene promoter–enhancer region there are sequences typical of those with which the transcription factors AP-1, NF-$\kappa$B and Oct-1 interact. There are also DNA sequences that interact with T cell-specific transcription factors; most notably, there is a sequence which interacts with a protein complex, present only in the nuclei of activated T cells, termed NFAT-1 (nuclear factor of activated T cells). The emphasis of T cell activation studies is shifting accordingly, with attention now being focused on the role of T cell receptors and their signal transduction pathways in the regulation and function of these distinct *trans*-activating factors (Durand et al 1988).

## Signalling mechanisms downstream of PKC

As discussed, there have been considerable advances in our understanding of the events at and near the T cell that accompany activation. However, more distal biochemical signals, particularly those downstream of kinases such as PKC, are poorly understood. Recent studies have shown that PKC can induce

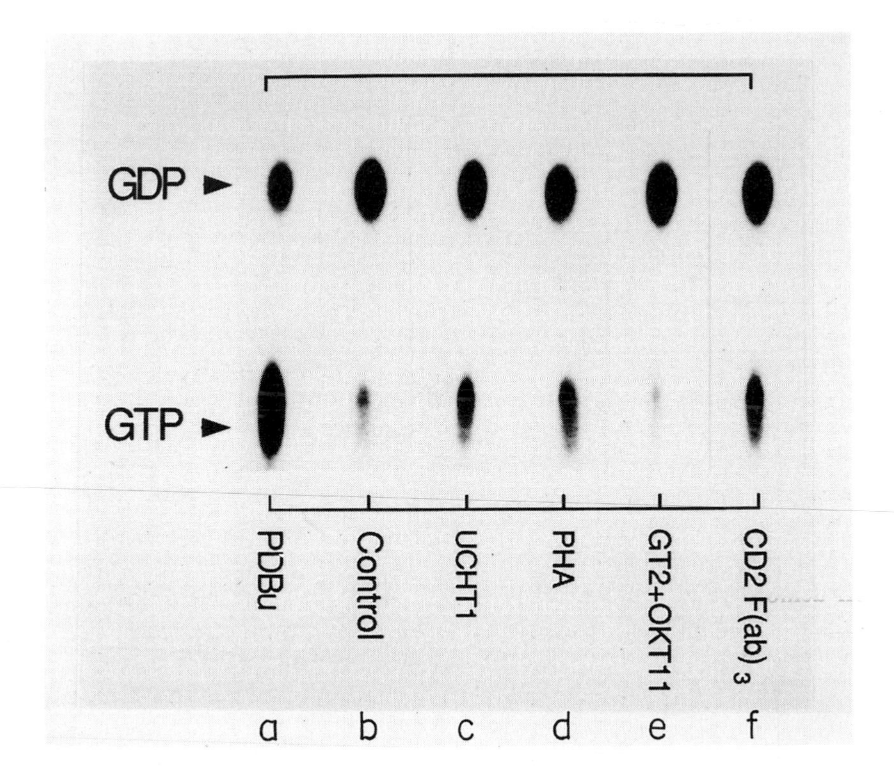


FIG. 1. Effect of T cell activation on guanine nucleotide bound to p21$^{ras}$. Thin-layer chromatogram of the guanine nucleotides eluted from immunoprecipitates of p21$^{ras}$ from $^{32}$P-orthophosphate-labelled T cells from human peripheral blood. Immunoprecipitation was with the anti-p21$^{ras}$ monoclonal antibody Y13-259. Cells were (b) unstimulated or stimulated for 10 minutes with (a) the PKC activator phorbol dibutyrate (PDBu, 50 ng/ml), (c) a TCR-agonistic anti-CD3 antibody, UCHTI, (10 μg/ml), (d) phytohaemagglutinin (PHA; 10 μg/ml), (e) non-cross-linked intact agonistic anti-CD2 antibodies GT2 and OKT11 (10 μg/ml of each antibody) and (f) a CD2 agonist, an anti-CD2 bispecific antibody (1 μg/ml) The position at which GTP and GDP standards ran is indicated.

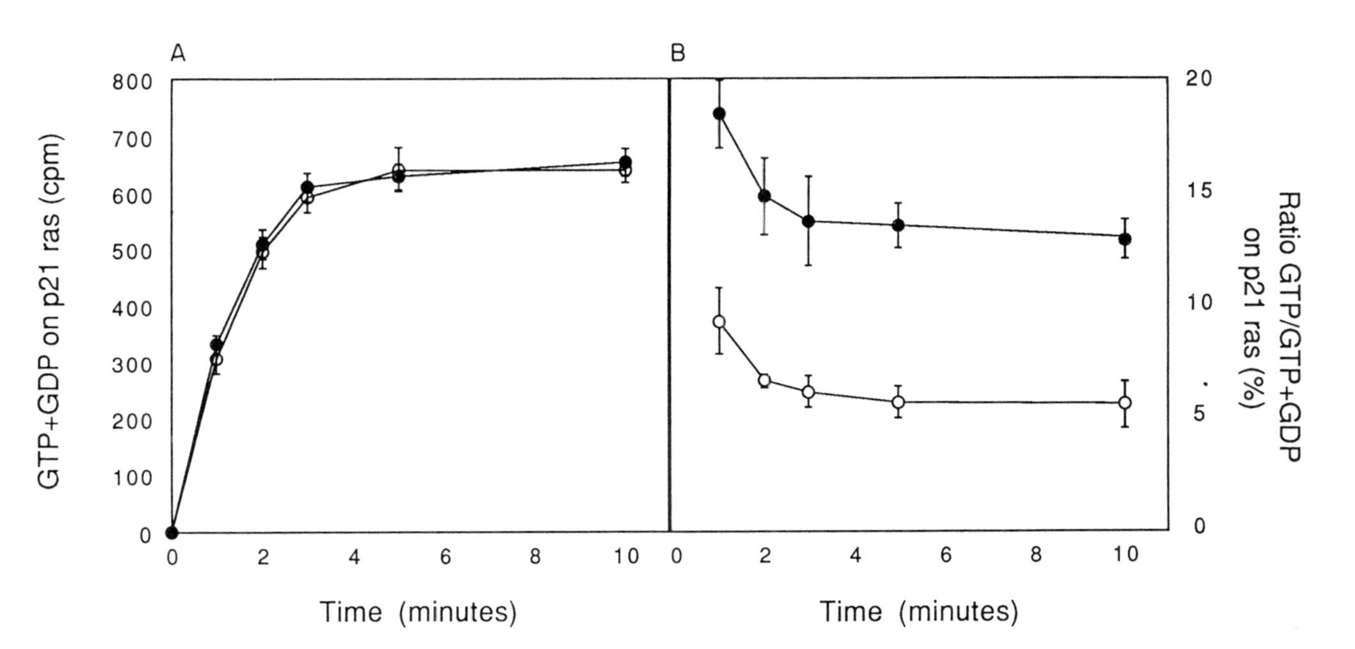


FIG. 2. Effect of PKC activation on p21$^{ras}$ activation in permeabilized cells. (A) Time course showing the amount of [$\alpha$-$^{32}$P]-labelled GTP plus GDP bound to p21$^{ras}$ in permeabilized T cells treated at 37 °C with 50 ng/ml phorbol dibutyrate (●) or unstimulated (○). (B) The same data as in (A), but with [$\alpha$-$^{32}$P]GTP shown as % of total labelled nucleotide bound to p21$^{ras}$. Human T lymphocytes were permeabilized as described by Downward et al (1990) by addition of 0.4 IU/ml of streptolysin-0 in 10mM Pipes buffer pH 7.2 containing 120 mM KCl, 30mM NaCl, 1 mM ATP, 5 mM free $Mg^{2+}$, 100 nm free $Ca^{2+}$, 5 μCi of [$\alpha$-$^{32}$P]GTP (3000 Ci/mmol); for each assay point $2\times10^7$ cells in 0.5 ml, were used. At the end of the incubation p21$^{ras}$ was recovered by immunoprecipitation with antibody Y13-259. Error bars indicate SD.

serine phosphorylation of the c-Raf kinase and phosphorylation of MAP (microtubule-associated protein) kinase on tyrosine and threonine residues (reviewed in Klausner & Samelson 1991). The interpretation of these observations is that PKC's effects on gene expression may be mediated by signals generated by a cascade of kinases. In T cells there is also evidence that the proteins encoded by the *ras* proto-oncogenes may be mediators of the action of PKC (Downward et al 1990). The *ras* proto-oncogenes encode proteins ($p21^{ras}$) that bind GTP and catalyse its hydrolysis to GDP (Barbacid 1987, Bos 1989). These proteins are able to stimulate cell growth and transformation when GTP is bound ('active') but not when GDP is bound ('inactive'). In T cells phorbol dibutyrate, which activates PKC, can induce the accumulation of $p21^{ras}$–GTP complexes (Fig. 1, Control and PDBu tracks). Hence, in unstimulated T cells less than 5% of $p21^{ras}$ is GTP-bound, whereas in phorbol dibutyrate-stimulated cells about 80% of $p21^{ras}$ is GTP-bound (Fig. 1). The mechanism for control of $p21^{ras}$ by PKC seems to involve a decrease in the intrinsic GTPase activity of $p21^{ras}$, which allows active $p21^{ras}$–GTP complexes to accumulate. Thus, activation of PKC does not influence the rate of guanine nucleotide exchange onto $p21^{ras}$ (Fig. 2A), but it does lower the rate of GTP hydrolysis by $p21^{ras}$ (Fig. 2B). The way in which activated PKC brings about the reduction of $p21^{ras}$ GTPase activity appears to involve it inhibiting a $p21^{ras}$ GTPase-activating protein, presumably GAP (Downward et al 1990). The molecular basis for regulation of GAP by PKC remains to be characterized. In T Lymphocytes PKC has multiple roles in the activation response (Alexander & Cantrell 1989). Firstly, it acts as a positive signal in the regulation of the expression and functioning of the *trans*-activating factors that regulate gene expression. Secondly, it is involved in controlling the expression and functioning at the cell surface of receptors such as the TCR and CD2 antigens that control T cell activation. Future studies will establish whether these functions of PKC are mediated via the $p21^{ras}$ proteins.

## The role of PKC in TCR and IL-2R regulation of $p21^{ras}$

Triggering of the TCR and CD2 antigens can also activate the $p21^{ras}$ proteins (i.e., lead to the accumulation of $p21^{ras}$–GTP complexes) (Fig.1, lanes c–f) and, as with PKC, TCR and CD2 agonists appear to stimulate $p21^{ras}$ by inhibiting $p21^{ras}$ GTPase activity (Graves et al 1991). Because triggering of the TCR and CD2 molecules also activates PKC, it is possible that the regulatory effect of these receptors is PKC-mediated. To examine this possibility we looked at the effect of a PKC pseudosubstrate inhibitor on $p21^{ras}$ activation via the TCR. The ability of this inhibitor to regulate TCR-mediated activation of PKC is shown in Fig. 3A, and has been described previously (Graves et al 1990). As predicted, the PKC pseudosubstrate inhibitor could prevent phorbol ester-induced $p21^{ras}$ activation but was unable to prevent TCR-mediated stimulation of $p21^{ras}$ (Fig. 3B). These results suggest that the TCR may use a signalling

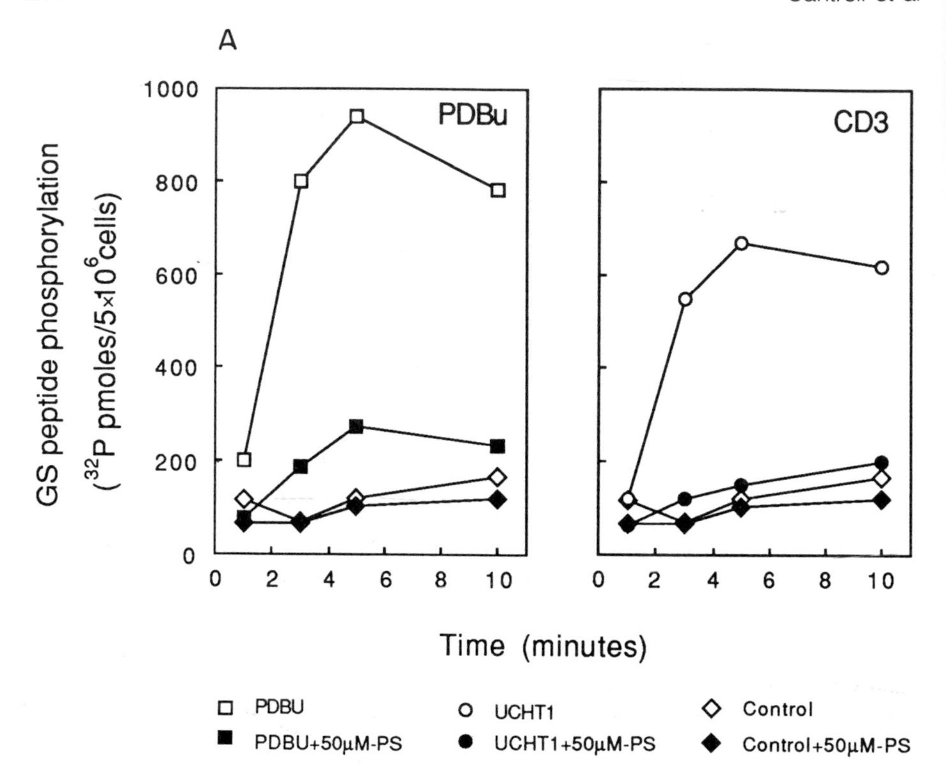


FIG. 3(A). Effect of a PKC pseudosubstrate peptide inhibitor on phorbol dibutyrate (PDBu)- and CD3-induced stimulation of PKC. The incorporation of $^{32}$P from [$\gamma$-$^{32}$P]ATP into the PKC substrate peptide GS was measured in T cells permeabilized with streptolysin-0 as described in Fig. 2. Assays were carried out for the indicated time in the absence of any stimuli or in the presence of 50 ng/ml PDBu (*left*) or 10 μg/ml of the CD3 agonist UCHTI (*right*), either with (closed symbols) or without (open symbols) the PKC pseudosubstrate inhibitor, PS, at 50 μM. A 250 μM concentration of peptide GS was used for each assay. Peptide GS has the sequence Pro-Leu-Ser-Arg-Thr-Leu-Ser-Val-Ala-Ala-Lys-Lys and its properties as a selective PKC substrate in permeabilized T cells have been described (Alexander et al 1990). The PKC pseudosubstrate inhibitor has the sequence Arg-Phe-Ala-Arg-Lys-Gly-Ala-Leu-Arg-Gln-Lys-Asn-Val.

pathway distinct from that used by PKC to control p21$^{ras}$ activity in T cells. There is further evidence to suggest that PKC is not the only intracellular regulator of p21$^{ras}$ in T cells. For example, interactions between IL-2 and the IL-2 receptor induce T cell proliferation, but IL-2 does not activate PKC (reviewed in Farrar et al 1990. However, the results in Fig. 4 show that IL-2 can activate p21$^{ras}$. At present there is no information on the mechanism of non-PKC-mediated regulation of p21$^{ras}$. In particular, we have yet to determine whether IL-2 regulates p21$^{ras}$ through an effect on p21$^{ras}$ GTPase activity, or

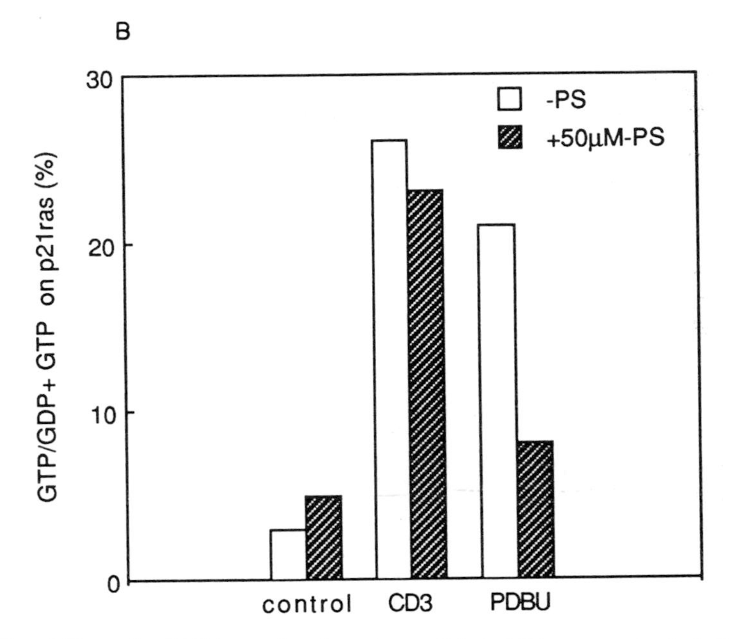


FIG. 3(B) Effect of a PKC pseudosubstrate peptide inhibitor on phorbol dibutyrate and CD3-induced p21$^{ras}$ stimulation. Date show GTP as %(GTP + GDP) from p21$^{ras}$ proteins immunoprecipitated from T cells (permeabilized as described in Fig. 2) incubated in the absence of any stimuli or in the presence of 50 ng/ml PDBu or 10 μg/ml UCHTI, either with (hatched bars) or without (open bars) 50 μM of the PKC pseudosubstrate inhibitor (PS). The results show that the PKC pseudosubstrate inhibited phorbol dibutyrate-induced p21$^{ras}$ stimulation, but had no effect on CD3-induced p21$^{ras}$ activation.

whether it controls guanine nucleotide exchange on the p21$^{ras}$ proteins. The salient point to emerge from our studies is that the p21$^{ras}$ proteins are regulatory targets for receptors that control the $G_0/G_1$ stage of the T cell cycle (i.e., the TCR and CD2 antigens) and also for the IL-2R, which controls T cell progression through $G_1$ into the S phase of the cell cycle, ultimately controlling T cell mitosis. This could mean that a prolonged stimulation of p21$^{ras}$ activity throughout $G_0/G_1$ is necessary for T cell growth.

## Conclusions

Knowledge about the immediate membrane proximal signals generated during activation of T lymphocytes has increased over the last few years. Despite this, many of the molecular details regarding the regulation of T cell activation have yet to be defined. It is recognized that changes in $[Ca^{2+}]_i$, PKC activity and tyrosine kinase activity are involved in T cell activation, but an understanding

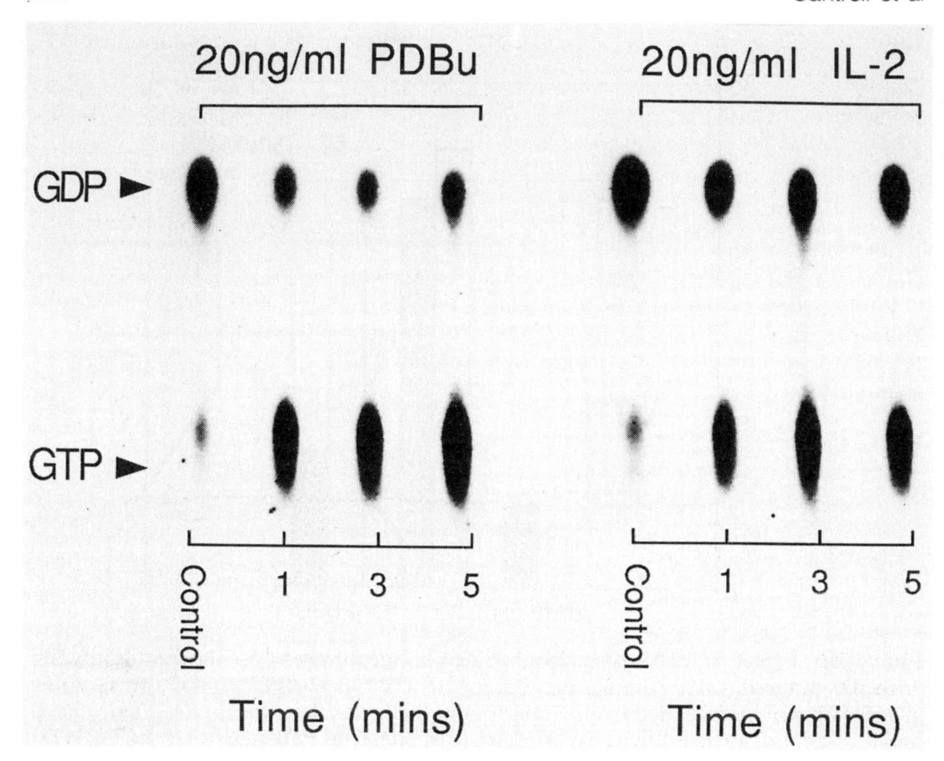


FIG. 4. IL-2 activates $p21^{ras}$ in T lymphocytes. Thin-layer chromatogram of the guanine nucleotides eluted from $p21^{ras}$ immunoprecipitates from $^{32}P$-labelled IL-2R-positive T cells prepared as described by Cantrell & Smith (1984). Immunoprecipitation was with the anti-$p21^{ras}$ monoclonal antibody Y13-259. Cells were unstimulated or stimulated for the indicated time with recombinant IL-2 (20 ng/ml). The position at which GTP and GDP standards ran is indicated. The effect of phorbol dibutyrate is shown for comparison.

of the distal events regulated by these different signalling pathways and their role in gene regulation is only beginning to emerge. What is evident, however, is that activation of T cells is a complex process which involves the integration of multiple receptors and their respective signal transduction pathways.

## References

Alexander D, Cantrell DA 1989 Kinases and phosphatases in T cell activation. Immunol Today 10:200–205

Alexander D, Graves J, Lucas S, Cantrell D, Crumpton M 1990 A method for measuring protein kinase C activity in permeabilised T cells by using peptide substrates. Biochem J 268:303–308

Barbacid M 1987 ras genes. Annu Rev Biochem 56:779–827

Bierer BE, Sleckman BP, Ratnofsky SE, Burakoft SJ 1989 The role of the CD2, CD4 and CD8 molecules in T cell activation. Annu Rev Immunol 7:579–599

Bos JL 1989 Ras oncogenes in human cancer: a review. Cancer Res 49:4662–4869

Cantrell DA, Smith KA 1984 The interleukin-2 T-cell system: a new cell growth model. Science (Wash DC) 224:1312–1316

Cantrell DA, Collins MKL, Crumpton MJ 1988 Autocrine regulation of lymphocyte proliferation: differential induction of IL-2 and IL-2 receptor. Immunology 65:343–349

Crabtree GR 1989 Contingent regulatory events in T cell activation. Science (Wash DC) 243:355–361

Desai DM, Newton ME, Kadlecek T, Weiss A 1990 Stimulation of the phosphatidyl inositol pathway can induce T cell activation. Nature (Lond) 348:66–69

Downward J, Graves JD, Warne PH, Rayter S, Cantrell DA 1990 Stimulation of p21$^{ras}$ upon T cell activation. Nature (Lond) 346:719–722

Durand DB, Shaw JP, Bush MR, Replogle RE, Belagajue R, Crabtree GR 1988 Characterisation of antigen receptor response elements within the IL-2 enhancer. Mol Cell Biol 8:1715–1724

Farrar WL, Linnekin D, Brini AT, Kelvin DJ, Michiel DF 1990 The interleukin-2 receptor complex: structure, gene regulation and signal transduction. In: Cambier J (ed) Ligands, receptors and signal transduction in regulation of lymphocyte function. American Society for Microbiology, Washington DC

Graves J, Cantrell D 1991 An analysis of the role of guanine nucleotide binding proteins in antigen receptor/CD3 antigen coupling to phospholipase C. J Immunol 146:2102–2107

Graves JD, Lucas SC, Alexander DR, Cantrell DA 1990 Guanine nucleotide regulation of inositol lipid hydrolysis and CD3 antigen phosphorylation in T cells. Biochem J 265:407–410

Graves J, Downward J, Rayter S, Warne P, Tutt AL, Glennie M, Cantrell DA 1991 CD2 antigen mediated activation of p21$^{ras}$ in human T lymphocytes. J Immunol 146:3709–3712

Imboden JB, Stobo JD 1985 Transmembrane signalling by the T cell antigen receptor J Exp Med 161:446–456

Klausner RD, Samelson LE 1991 T cell antigen receptor activation pathways: the tyrosine kinase connection. Cell 64:875–878

Koretzky GA, Picus J, Thomas ML, Weiss A 1990 Tyrosine phosphatase CD45 is essential for coupling the T cell antigen receptor to the phosphatidylinositol pathway. Nature (Lond) 346:66–69

Mustelin T, Coggeshall KM, Isakov N, Altman A 1990 T cell antigen receptor regulation of phospholipase C requires tyrosine phosphorylation. Science (Wash DC) 247: 1584–1587

O'Shea JJ, Ashwell JD, Bailey TL, Cross SL, Samelson LE, Klausner RD 1991 Expression of v-src in a murine T-cell hybridoma results in constitutive T-cell receptor phosphorylation and interleukin-2 production. Proc Natl Acad Sci USA 88:1741–1746

Pantaleo G, Olive D, Poggi A, Kozumbo WJ, Moretta L, Moretta A 1987 Transmembrane signalling via the T11-dependent pathway of human T cell activation. Eur J Immunol 17:55–60

Park DJ, Rho HW, Rhee SG 1991 CD3 stimulation causes phosphorylation of phospholipase Cγ1 on serine and tyrosine residues in a human T cell line. Proc Natl Acad Sci USA 88:5453–5456

Weiss A, Wiskocil RL, Stobo JD 1984 The role of T3 surface molecules in the activations of human T cells: a two stimulus requirement for IL-2 production reflects events happening at a pre-translational level. J Immunol 133:123–128

Weissman AM, Hou P, Orloff DG et al 1988 Molecular cloning of the T cell receptor zeta chain: distinction from the molecular CD2 complex. Proc Natl Acad Sci USA 85:9709–9713

## DISCUSSION

*Krebs:* Is there direct evidence that $p21^{ras}$ mediates the various processes that you have mentioned?

*Cantrell:* Many experiments in which mutated *ras* genes have been transfected into cells have suggested that ras controls growth and transformation and changes in the membrane and the cytoskeleton. It's very difficult to extrapolate that data from fibroblast-type systems to the T cell. In the T cell there is no evidence that *ras* proteins control membrane trafficking, but the experiments haven't been done. Interestingly, the protein kinase C regulation of $p21^{ras}$ that we see in the T lymphocyte definitely does not happen in fibroblast cells such as Swiss 3T3 cells. Many of the *ras* transfection experiments have been done in fibroblast-type cells, where the regulatory pathways that I have described don't seem to operate. We have even looked in other haemopoietic cells, such as myeloid cells into which IL-2 receptors had been transfected, and those cells also seem to completely lack this PKC *ras* regulation response. It happens in T cells and B cells, and in fairly immature thymocytes, but has not so far been found in other cell lineages.

*Ui:* Why is the active site of $p21^{ras}$ so stable in comparison with other heterotrimeric G proteins? This seems to be a unique property of p21.

*Cantrell:* We have no idea. We think the regulation of the GAP-like activity in the T cell is a stable regulation; we think nucleotide is continually exchanging onto $p21^{ras}$, but because the GAP-like protein is continually inactivated by PKC stimulation it is able to continue inhibiting activation of $p21^{ras}$ GTPase activity. We have maintained the stimuli in all our experiments and we haven't actually looked to see what happens when the stimuli are removed; I imagine that the $p21^{ras}$–GTP complexes would decay quickly, but I could be wrong.

*Fredholm:* Agonistic anti-CD2 and anti-CD3 antibodies increase GTPase activity in membranes from leukaemic T cells (Kvanta et al 1990); could this be related to the GTPase activity is that expressed by $p21^{ras}$?

*Cantrell:* That is not possible in T cells because we see *inhibition* of $p21^{ras}$ GTPase activity by anti-CD2 and CD3 antibodies.

*Fredholm:* Do you think that our findings could mean that the T cell receptor can influence GTPases other than $p21^{ras}$?

*Cantrell:* We never have been able to see an overall effect on GTPase activity in T cell membranes in response to activation via CD2 and CD3. We now know that there are at least three G proteins in the T cell membrane (Harvey, Kirsten and N-Ras) whose GTPase activity is inhibited by these stimuli, so to see a stimulatory effect the GTPase activity of many more molecules of GTP-binding proteins would need to be stimulated. To see changes in p21*ras* GTPase activity we have to use immunoprecipitation techniques and focus on the *ras* proteins.

*Michell:* Are you looking at a pool of all the *ras* types? Do you have any information about which forms are involved?

*Cantrell:* Our antibody (Y13-259) recognizes the Harvey, Kirsten and N-Ras forms in T cells. It appears that we are studying an effect on all three proteins.

*Michell:* Have you any idea whether there is any cell-selective activity?

*Cantrell:* We have done an experiment using an antibody (Y13-238) that recognizes the Kirsten Ras and the Harvey Ras but not N-Ras. This antibody had the same effect as Y13-259, so we have no evidence about which of those three types is or are involved. We were interested in whether the Kirsten Ras was the main target because it is a protein kinase C substrate, but that doesn't seem to be the case.

*Michell:* What are the relative abundances of the three *ras* proteins in T cells?

*Cantrell:* We haven't really looked at that in great detail.

*Carpenter:* I gather that your results with ionomycin suggest that calcium does not have a role. Have you tried ionomycin in combination with low TPA, to see if there is cooperativity?

*Cantrell:* No, we haven't. I imagine there would be, because ionomycin acts in synergy with low doses of phorbol dibutyrate to activate protein kinase C. However, it would be difficult to say whether the effect was due simply to the synergy in activating protein kinase C or to an independent pathway. We certainly don't think that calcium *alone* can regulate p21*ras*, but there must be some synergy.

*Kato-Homma:* Do you think that the changes in GAP activity are more important than changes in other factors, such as guanine nucleotide exchange factor, GEF, or guanine nucleotide dissociation factor, GDI?

*Cantrell:* There are no differences in nucleotide exchange under any conditions we have looked at in permeabilized cells. I must stress though that we are working with permeabilized cells and we may have lost some control mechanism. The exchanges of GTP on Ras that we measure in permeabilized cells were very, very fast, whereas exchange is very slow with isolated Ras, taking about 40 min to exchange GTP. Unfortunately, exchange is too fast in permeabilized cells for us to do good kinetic measurements, so there might be changes occurring that we can't observe.

*Hunter:* Have you looked at the state of tyrosine phosphorylation of GAP, or its associated p190 or p62 in these cells?

*Cantrell:* We don't have any idea about p190 or p62. In the Jurkat cell line CD3 activation induces a very low level, 1–2%, tyrosine phosphorylation of GAP. Phorbol esters apparently do not induce tyrosine phosphorylation of GAP. The effects of IL-2 on tyrosine phosphorylation of GAP are not known.

So, there is some evidence for tyrosine phosphorylation of GAP but it doesn't correlate with the activation of Ras that we have seen.

*Daly:* Do you know how reversible the effect on nucleotide exchange is? It might be worthwhile looking at that because if the effect on nucleotide exchange were not readily reversible, you could stimulate the cells with IL-2 and then permeabilize them and look at exchange.

*Cantrell:* Actually, we have tried pre-stimulating the cells with IL-2 before permeabilization; IL-2 still fails to work (unpublished results).

*Tsien:* You permeabilized with streptolysin-0. The holes are large so I take it that you use some sort of calcium-buffering system.

*Cantrell:* The permeabilized cells are buffered to 100 nM calcium, which is roughly the cytosolic level in our cells. Protein kinase C regulation seems to be maintained perfectly well in our system and we can get good receptor stimulation of phospholipase C, inositol phosphate production and subsequent protein kinase C activation.

*Tsien:* Given that PKC inhibitors don't have an effect, would it make sense to try inhibitors of calmodulin-dependent kinases, as lymphocytes contain these enzymes?

*Cantrell:* We have tried these experiments in permeabilized cells at many different calcium concentrations, and have seen no regulatory effect of calcium other than that we could attribute to an effect of protein kinase C regulation.

*Tsien:* Could a sceptic invoke localized changes in calcium pools and say you hadn't buffered the calcium rigorously enough?

*Cantrell:* The calcium buffering is vigorous. The problem with a calmodulin kinase inhibitor is that we wouldn't know how to assay that it was working, because we don't have a substrate for it.

*Tsien:* You would certainly need that control if there were no kinase inhibitor effect.

*Cantrell:* I suppose I am predicting that we would see a negative effect.

*Tsien:* If it failed to act you wouldn't know anything, but a positive effect of the inhibitor might be informative.

*Daly:* Dr Cantrell, do you know how much of the different kinases are retained in the permeabilized cells? Could you have lost a kinase that IL-2 activates?

Cantrell: That's certainly possible; we don't even know whether we have destroyed the affinity of the IL-2 receptor for its ligand. We have clearly lost something from the IL-2 mechanism, but we don't know whether the receptor has become uncoupled. You can make no predictions about what will be lost and what will be retained in these permeabilized cells.

*Hunter:* In fibroblasts after microinjection of monoclonal antibody 259 one sees inhibition of serum-induced mitogenesis (Mulcahy et al 1985), indicating that p21$^{ras}$ is apparently essential for all signalling pathways! Can you inject that antibody into your permeabilized cells?

*Cantrell:* No we can't—the antibodies will not go in.

*Hunter:* Have you tried more vigorous permeabilization methods?

*Cantrell:* We have spent a lot of time using different detergents, looking for methods to maintain signal transduction pathways and get antibodies into the cells. When we managed to get the antibody in, signal transduction capacity was low, whereas if we maintained signal transduction capacity the antibodies wouldn't go in. It's not because the holes are too small; with streptolysin-0 permeabilization, the antibody should theoretically be able to enter the cell, but practically, it doesn't.

*Hunter:* Has anyone microinjected T cells?

*Cantrell:* Not that I know of.

*Hunter:* Is it possible?

*Cantrell:* I would imagine it's possible. One problem would be that most of the assays for what ensues would require many cells.

*Hunter:* Induction of c-*fos* can be seen using fluorescence of individual microinjected cells.

*Cantrell:* That might work.

*Tsien:* Richard Lewis uses whole-cell voltage clamping of T cells to study calcium oscillations in individual cells in real time (Lewis & Cahalan 1989). Also, Gerald Crabtree is developing single-cell assays for gene expression.

*Cantrell:* Obviously, at the moment we are trying to develop as many single-cell assays as possible because ultimately you need those to be able to answer many of the outstanding questions.

*Michell:* If you could microinject a reasonable number of cells in a microscope field you could simply watch them and see if they turned into blasts.

*Nishizuka:* I am unclear whether CD2 causes hydrolysis of phosphatidylinositol.

*Cantrell:* It definitely does, in a non-G protein-linked fashion, probably through tyrosine phosphorylation (unpublished results).

*Fredholm:* Does that require CD3?

*Cantrell:* We don't know.

*Fredholm:* By using mutant Jurkat cells it can be shown that the effect of agonistic antibodies to CD2 on $InsP_3$ formation is lost if there is no CD3 present (Kvanta et al 1990).

*Cantrell:* We have done a lot of experiments but in all the $CD3^-$ Jurkat cells we have, CD2 still activates protein kinase C. We have not been able to find cells with levels of CD3 sufficiently low that CD2 will not induce the first signal transduction response. You don't need many CD3 molecules on the surface for CD2 to still function.

*Tsien:* Do immunologists have any idea why lymphocytes have evolved such a complicated system involving the production of an autocrine factor and expression of receptors for the autocrine factor to switch the cells on? There would seem to be other, much simpler ways for them to be switched on.

*Cantrell:* Immunologists use the term 'autocrine' very loosely. We really mean that one population of cells makes the growth factor; we don't necessarily think that one cell produces the growth factor and then responds to it. The kinetics of IL-2 receptor production in the primary cell population show that the cells make IL-2 first, with production being complete within 12 h of activation, but receptors are not seen on the surface until about 24–48 h. That doesn't seem a logical way for a truly autocrine system to operate and the system is better geared up for cell interactions, probably in a very local environment, for the T cell to make a growth factor to which another T cell responds.

*Tsien:* I see—if you think in terms of coordinated actions of different colonies the functional logic of the system becomes more evident.

*Cantrell:* If one cell makes a growth factor for another cell, this provides a way for one T cell to control the proliferation of other cells. It's interesting that a much lower threshold of antigenic stimulation is needed to induce T cells to express IL-2 receptors than is needed to initiate IL-2 production. A low affinity antigen could probably induce a lot of T cells to express receptors at some stage, but a much stronger signal is required to make the cells produce IL-2.

T cell proliferation is very tightly controlled. There are very few examples of adult T cell leukaemias—the only case I can think of is the HTLV tumours. Apart from those, the system seems almost perfectly controlled; there can be immune deficiencies but not lymphoproliferative adult T cell tumours.

## References

Kvanta A, Gerwins P, vanderPloeg I, Nordstedt C, Jondal M, Fredholm BB 1990 Stimulation of the T-cell receptors CD2 and CD3 with OKT3 and OKT11 antibodies activates a common pertussis toxin-insensitive G protein. Eur J Pharmacol Mol Pharmacol Sect 189:363–372

Lewis RS, Cahalan MD 1989 Mitogen-induced oscillations of cytosolic $Ca^{2+}$ and transmembrane $Ca^{2+}$ current in human leukemic cells. Cell Regul 1:99–112

Mulcahy LS, Smith MR, Stacey DW 1985 Requirement for *ras* proto-oncogene function during serum-stimulated growth of NIH 3T3 cells. Nature (Lond) 313:241–243

# Growth factor phosphorylation of PLC-$\gamma_1$

Graham Carpenter*†, S. M. Teresa Hernández-Sotomayor*, Shunzo Nishibe*, Gordon Todderud*†, Marc Mumby‡ and Matthew Wahl*

*Departments of Biochemistry* and Medicine (Dermatology)†, Vanderbilt University School of Medicine, Nashville, Tennessee 37232-0146 and Department of Pharmacology‡, University of Texas Southwestern Medical Center, Dallas, Texas 75235-0941, USA*

*Abstract.* The hydrolysis of phosphatidylinositol 4,5-bisphosphate has a central role in many signalling pathways. One of the phospholipase C (PLC) isozymes that mediates this reaction is a direct substrate for the tyrosine kinase activity of several growth factor receptors. Growth factors elicit increases in both the phosphoserine and the phosphotyrosine content of the PLC-$\gamma_1$ isozyme. PLC-$\gamma_1$ contains three tyrosine phosphorylation sites, which have been identified as residues 771, 783 and 1254. Phosphorylation of tyrosine residues is sufficient to increase the catalytic activity of PLC-$\gamma_1$, though other proteins may modulate this activation. However, the role of growth factor-enhanced phosphorylation of serine residues on PLC-$\gamma_1$ remains obscure. *In vitro* studies of PLC-$\gamma_1$, recovered from growth factor-treated cells, indicate that activation by tyrosine phosphorylation is not due to increased sensitivity to $Ca^{2+}$, a required co-factor, but is reflected in altered kinetic constants, i.e. $V_{max}$ and, to a lesser extent, $K_m$.

*1992 Interactions among cell signalling systems. Wiley, Chichester (Ciba Foundation Symposium 164) p 223–240.*

Epidermal growth factor (EGF), the first mitogenic growth factor isolated (Cohen 1962), has served as a model for growth factor action, and the information gained has been generally applicable to a variety of mitogenic molecules. Although the exact physiological roles of endogenous EGF in the intact animal remain elusive, substantial progress has been made in understanding the mechanism of action of EGF at the cellular level.

Results that define the signal transduction scheme for EGF are clearly rudimentary at this point, but elements of the pathway are being identified and understood as biochemical entities. The receptor for EGF has been isolated (Cohen et al 1980), cDNA cloning has revealed its complete sequence (Ullrich et al 1984) and mutational studies have demonstrated that the tyrosine kinase cytoplasmic domain of the receptor must be active for ligand binding to the extracellular domain to produce biological responses (Chen et al 1987, Honegger et al 1987). Therefore, tyrosine phosphorylation, both of the receptor molecule itself (autophosphorylation) and of other cellular proteins, constitutes the initial biochemical step in the signalling pathway. Several proteins have been identified as tyrosine phosphorylation substrates and have functions that are plausible in a signal transduction scheme. These include PLC-$\gamma_1$ (Wahl et al 1989a,b, Margolis et al 1989, Meisenhelder et al 1989), the GTPase-activating protein (GAP) for the *ras* gene product (Molloy et al 1989, Ellis et al 1990) and the 85 kDa subunit of phosphatidylinositol 3-kinase (Kaplan et al 1987). Interestingly, these three proteins share a common structural motif—the presence of two SH2 (*src* homology) domains and one SH3 domain (Koch et al 1991). Although two serine/threonine kinases, MAP (microtubule-associated protein) kinase and the *raf* gene product, are reported to show growth factor-increased levels of tyrosine phosphate, it is not clear that these are direct substrates of growth factor receptors (Carpenter & Cohen 1990).

Phospholipase C activation is a common feature in the mechanism of action of a wide variety of hormones. Hydrolysis of its substrate, a minor membrane phospholipid, phosphatidylinositol 4,5-bisphosphate ($PtdInsP_2$), produces two second messenger molecules—diacylglycerol, an activator of protein kinase C, and inositol 1,4,5-trisphosphate, which mobilizes intracellular stores of $Ca^{2+}$. There are several PLC isozymes which mediate this step, and their molecular properties have been reviewed (Rhee et al 1989). Only the PLC-$\gamma_1$ isozyme, however, mediates the action of growth factor tyrosine kinase receptors.

## Phosphorylation of PLC-$\gamma_1$ in cells

Studies in intact cells have shown that EGF (Wahl et al 1989a, Margolis et al 1989, Meisenhelder et al 1989), platelet-derived growth factor (PDGF) (Meisenhelder et al 1989, Wahl et al 1989b), fibroblast growth factor (Burgess et al 1990) and nerve growth factor (Kim et al 1991) elicit increased tyrosine phosphorylation of PLC-$\gamma_1$. In all cases, both phosphoserine and phosphotyrosine levels on PLC-$\gamma_1$ are increased by mitogen treatment. Significantly, neither insulin (Meinsenhelder et al 1989, Nishibe et al 1990a) nor colony stimulating factor 1 (Downing et al 1989) provoke phosphorylation of PLC-$\gamma_1$, but they do activate tyrosine kinase receptors. This suggests that determinants exist within the tyrosine kinase molecules that allow selection of appropriate molecules as substrates.

A second significant feature of PLC-$\gamma_1$ tyrosine phosphorylation is that a very large fraction, over 50%, of the total molecules are phosphorylated rapidly after mitogen treatment of cells (Margolis et al 1989, Meisenhelder et al 1989, Wahl et al 1990). By comparison, the tyrosine phosphorylation of GAP may involve only 5% of the total GAP molecules (Moran et al 1991). The tyrosine phosphate on PLC-$\gamma_1$ is rapidly dephosphorylated when the mitogenic agent is removed, suggesting that tyrosine phosphatases are very active towards this substrate (Wahl et al 1989a, Meisenhelder et al 1989). Rapid reversal upon removal of the activating agent is a characteristic expected of signal transduction elements.

Several groups (Meisenhelder et al 1989, Margolis et al 1989, Kumjian et al 1989) have detected formation of complexes between PLC-$\gamma_1$ and activated EGF or PDGF receptors, and one report (Kumjian et al 1991) has presented experimental evidence to suggest that formation of a receptor–PLC-$\gamma_1$ complex is part of the mechanism for increasing activity of this phospholipase.

## Analysis of PLC-$\gamma_1$ phosphorylation *in vitro*

That the EGF receptor directly phosphorylates PLC-$\gamma_1$ could be determined only by experiments *in vitro* with purified components. In the absence of this demonstration it may be argued that another tyrosine kinase, activated by the EGF receptor kinase, might be the enzyme which actually phosphorylates the phospholipase.

It has been shown that the purified EGF receptor will phosphorylate purified PLC-$\gamma_1$ (Nishibe et al 1989, Meisenhelder et al 1989), but not the related PLC-β or PLC-δ isozymes (Nishibe et al 1989). In contrast to experiments in intact cells, the *in vitro* system produces only tyrosine phosphorylation of PLC-$\gamma_1$; no phosphoserine is formed, because there are no serine kinases for PLC-$\gamma_1$ present. The identity of the growth factor-activated serine kinase that phosphorylates PLC-$\gamma_1$ in intact cells is unknown, as are the sites of serine phosphorylation on PLC-$\gamma_1$. *In vitro* phosphorylation experiments (Kim et al 1990) indicate that three tyrosine residues (771, 783 and 1254) on bovine PLC-$\gamma_1$ are highly phosphorylated by the purified (human) EGF receptor, while a fourth, tyrosine 472, is phosphorylated to a lesser extent. Attempts to sequence PLC-$\gamma_1$ isolated from human cells treated with EGF produced three tryptic phosphopeptides (Wahl et al 1990). However, only two of these could be sequenced; these contained phosphotyrosine corresponding to tyrosines 771 and 1254 in the bovine PLC-$\gamma_1$ molecule. The chromatographic behaviour of the third phosphopeptide, which did contain phosphotyrosine, led us to believe that it contained the tyrosine 783 phosphorylation site.

Tyrosines 771 and 783 lie in a region of 40 residues that acts as a spacer between the second SH2 domain and the single SH3 domain in PLC-$\gamma_1$, while tyrosine 1254 is located near the carboxy terminus of PLC-$\gamma_1$. The mechanism by which these individual tyrosine phosphorylation sites influence the catalytic activity of PLC-$\gamma_1$ is not yet known. SH2 motifs, however, have been shown to interact with certain phosphotyrosine-containing proteins (Matsuda et al 1990) and the isolated SH2 motifs from PLC-$\gamma_1$ are capable of forming a complex with the activated EGF receptor (D. Anderson et al 1990). This interaction apparently occurs at the carboxy terminus of the EGF receptor, which contains the receptor's autophosphorylation sites (Margolis et al 1990).

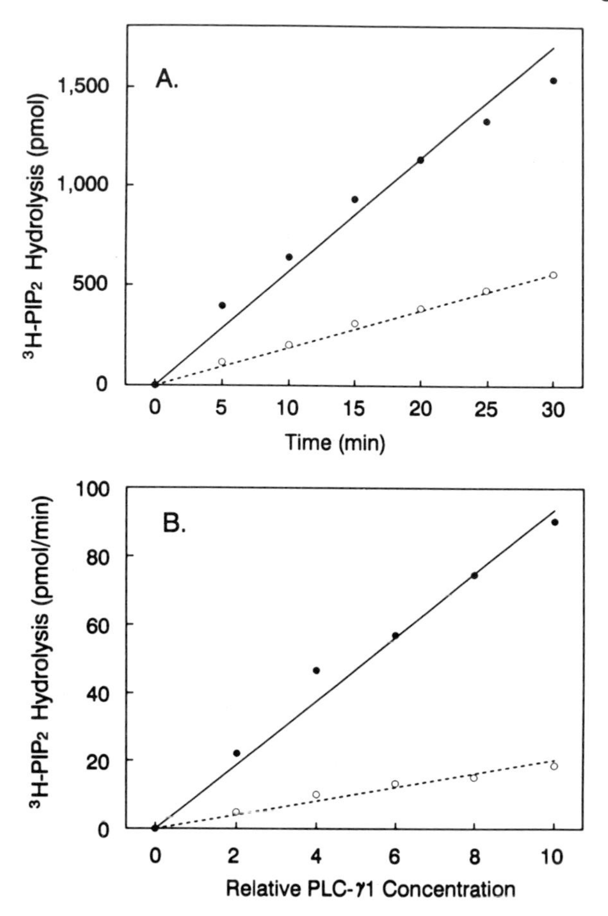


FIG. 1. Hydrolysis of phosphatidyl[$^3$H]inositol 4,5-bisphosphate ($^3$H-PIP$_2$) by PLC-$\gamma_1$ immunoprecipitates recovered from EGF-treated or untreated A-431 cells. A-431 cells were treated with or without EGF for 5 min at 37 °C. Cytosolic extracts were prepared and PLC-$\gamma_1$ was immunoprecipitated. PLC activity in immunoprecipitates from control (○) or EGF-treated (●) cells was assayed with respect to time (panel A) and the amount of immunoprecipitate (panel B). The data in A represent the activity of PLC-$\gamma_1$ immunoprecipitated from 100 μg cytosol measured after incubation of reaction mixtures for the indicated periods of time. In panel B, the amount of Ptd[$^3$H]InsP$_2$ hydrolysed by PLC-$\gamma_1$ immunoprecipitated from 25, 50, 75, 100 or 125 μg of cytosol was measured after incubation of reaction mixtures for 15 min. Each reaction mixture contained 100 μM Ptd[$^3$H]Ins(4,5)P$_2$, 0.14% octylglucoside and 0.05% Triton X-100. Data are from one representative experiment.

## Activation of PLC-$\gamma_1$ by phosphorylation

As with any phosphoprotein, a major challenge is to demonstrate a functional consequence due to the phosphorylation of amino acid residues. We (Nishibe

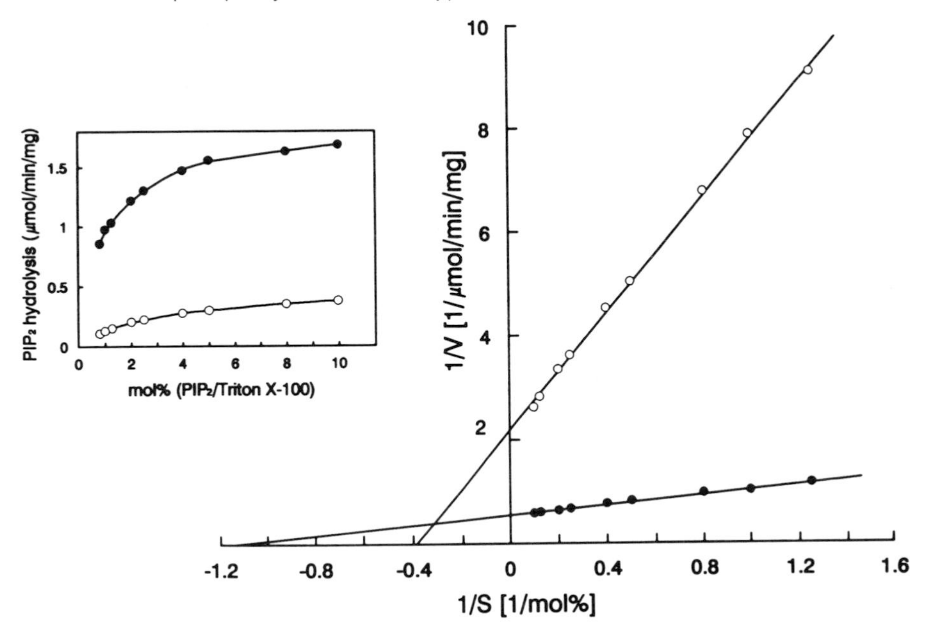


FIG. 2. Kinetic analysis of PLC-$\gamma_1$ activity immunoprecipitated from A-431 cells treated with (●) or without (○) EGF for 5 min at 37 °C. PLC-$\gamma_1$ activity was measured with a constant amount of Ptd[$^3$H]Ins(4,5)$P_2$ ($PIP_2$) (100 µM), resuspended initially in 1.4% octylglucoside (final concentration in the reaction mixtures was 0.14%). Increasing concentrations of Triton X-100 (0.063–0.78%) were added to the reaction mixture to alter the surface concentration of Ptd[$^3$H]Ins(4,5)$P_2$. The ratio of Ptd[$^3$H]Ins(4,5)$P_2$ to Triton X-100 is indicated as mol%. The amount of PLC-$\gamma_1$ protein in the immunoprecipitates was measured by immunoblot analysis with anti-PLC-$\gamma_1$ monoclonal antibodies, with purified bovine brain PLC-$\gamma_1$ being used as the standard. Approximately 0.06% of total cytosolic protein was recovered as PLC-$\gamma_1$ protein from both EGF-treated (59 ng/100 µg) and untreated 63 ng/100 µg) cells. The direct substrate concentration dependence of PLC-$\gamma_1$ activity recovered from control (○) or EGF-treated (●) cells is shown in the inset.

et al 1990b) have modified the *in vitro* assay for PLC activity by adding a modest amount of Triton X-100 to create mixed micelles of detergent and PtdIns$P_2$ using immunoprecipitated PLC-$\gamma_1$ from control or EGF-treated cells as the source of enzyme. The data in Fig. 1 show that under these assay conditions PLC-$\gamma_1$ from growth factor-treated cells is several-fold more active than the enzyme obtained from control cells. These data also show that in the *in vitro* assay enzyme activity is linear with respect to time or enzyme concentration. With this assay we have begun to compare and analyse the kinetic parameters of PLC-$\gamma_1$ obtained from EGF-treated or control A-431 cells (a human epidermoid carcinoma cell line). The results, as shown in Fig. 2, demonstrate that the growth factor-activated enzyme displays a higher $V_{max}$ and $K_m$ toward

**TABLE 1 Dephosphorylation of the autophosphorylated EGF receptor by purified tyrosine phosphatases**

| | *% Dephosphorylation*[a] | |
|---|---|---|
| *Phosphatase* | *30 min* | *60 min* |
| None | 5 | 5 |
| CD45 | 52 | 81 |
| TCPTPase | 25 | 39 |

[a]Purified human EGF receptor was autophosphorylated at room temperature and a molar excess of unlabelled ATP was added to prevent further incorporation of radioactivity from [$^{32}$P]ATP. Aliquots of buffer or one of the purified tyrosine phosphatases were then added and the incubation was continued for an additional 30 min or 60 min. The amount of $^{32}$P-labelled receptor remaining was determined by SDS-PAGE, autoradiography, excision of the receptor bands from the gel and Cherenkov counting. Percentage dephosphorylation was calculated relative to the amount of radioactivity present in the EGF receptor before the addition of unlabelled ATP.

$PtdInsP_2$ than the enzyme from control cells. The exact interpretation of these kinetic parameters is, however, complicated by the fact that catalysis occurs at a lipid-water interface (Hendrickson & Dennis 1984).

In subsequent experiments we have used purified serine and tyrosine phosphatases to dephosphorylate the growth factor-activated species of PLC-$\gamma_1$ and have determined how these treatments effect the activity of the activated enzyme. Two tyrosine phosphatases, CD45 and T cell phosphatase (TCPTPase), and one serine phosphatase (PP2A) were used. As reported elsewhere (Nishibe et al 1990b), TCPTPase removed phosphate from tyrosine residues on PLC-$\gamma_1$ and produced a 70% decrease in enzyme activity. TCPTPase did not decrease the level of phosphoserine on PLC-$\gamma_1$. This indicates that phosphotyrosine is necessary for maintenance of the activated form of the enzyme. Interestingly, CD45 did not dephosphorylate PLC-$\gamma_1$, nor did it reduce enzyme activity. Under nearly identical conditions almost the opposite results were obtained with the autophosphorylated EGF receptor (Table 1). With the EGF receptor as the substrate, CD45 was much more active than the TCPTPase. These results, although limited in scope, do support the notion that tyrosine phosphatases are selective towards substrates.

To address the question of whether tyrosine phosphorylation actually activates PLC-$\gamma_1$, we have shown that addition of purified EGF receptor and ATP to PLC-$\gamma_1$ immunoprecipitated from untreated cells results in tyrosine phosphorylation of PLC-$\gamma_1$ and a 2.5-fold increase in enzyme activity (Nishibe et al 1990b). This result argues for a direct role of tyrosine phosphate in the creation of the activated state of PLC-$\gamma_1$. Another group (Goldschmidt-Clermont et al 1991) has reported that in the presence of profilin, a $PtdInsP_2$-binding protein, the tyrosine phosphorylated form of PLC-$\gamma_1$ is more active than the non-phosphorylated enzyme.

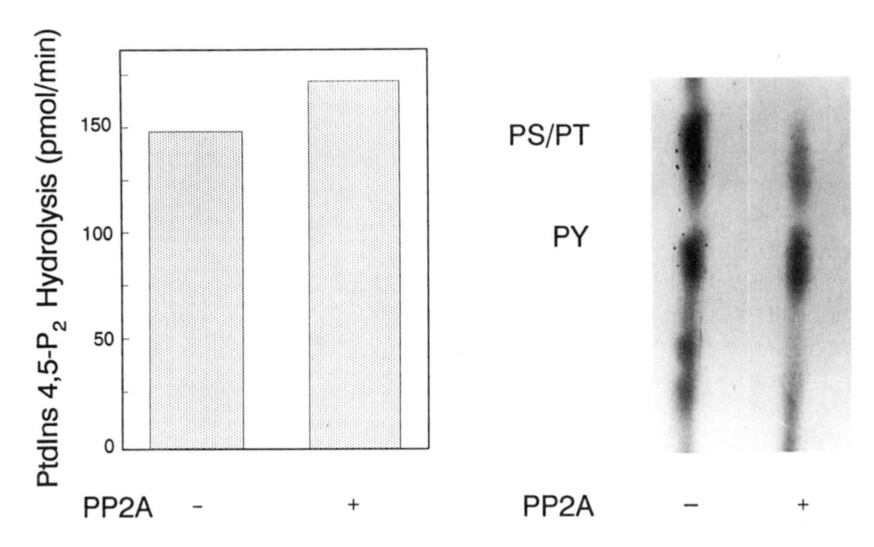


FIG. 3. Effect of serine dephosphorylation of activated PLC-$\gamma_1$ recovered from EGF-treated A-431 cells. (*Left*) *In vitro* enzyme activity. Immunoprecipitates of PLC-$\gamma_1$ were prepared from EGF-treated A-431 cells as previously described (Nishibe et al 1990b). The immunoprecipitates were then incubated at 30 °C for 30 min with or without (buffer only) 0.2 µg purified catalytic subunit of the serine phosphatase PP2A. The PLC-$\gamma_1$ precipitates were then washed and phospholipase C activity was measured. (*Right*) Phosphoamino acid analysis. A similar protocol was used to measure the phosphoamino acid content of PLC-$\gamma_1$ from EGF-treated cells before (−) and after (+) PP2A treatment. In this instance, however, the cells were labelled with $^{32}PO_4^{3-}$ for 4 h and exposed to EGF for 5 min. Phosphoamino acid analysis was performed after SDS–PAGE, autoradiography and elution of the radiolabelled PLC-$\gamma_1$ band (Wahl et al 1989a). PS/PT, phosphoserine/phosphotheronine; PY, phosphotyrosine.

Considerably less is understood about the role of growth factor-enhanced phosphorylation of PLC-$\gamma_1$. As shown in Fig. 3, treatment of the activated form of PLC-$\gamma_1$ with the serine phosphatase PP2A removes a substantial amount of the phosphoserine from PLC-$\gamma_1$, but does not significantly alter enzyme activity. This result is interesting with regard to the data obtained with MAP kinase. Activated MAP kinase from EGF-treated cells was inactivated by treatment with either a tyrosine phosphatase or a threonine phosphatase (N.G. Anderson et al 1990), suggesting that growth factor-stimulated levels of both phosphotyrosine and phosphothreonine were necessary to activate that enzyme.

We have also determined whether the formation of phosphoserine on PLC-$\gamma_1$ could be a permissive but prerequisite event for tyrosine phosphorylation and activation of PLC-$\gamma_1$. As shown in Fig. 4, pretreatment of PLC-$\gamma_1$ with PP2A did not alter subsequent activation of PLC-$\gamma_1$ by EGF receptor-catalysed tyrosine phosphorylation. In this experiment, PLC-$\gamma_1$ was obtained from

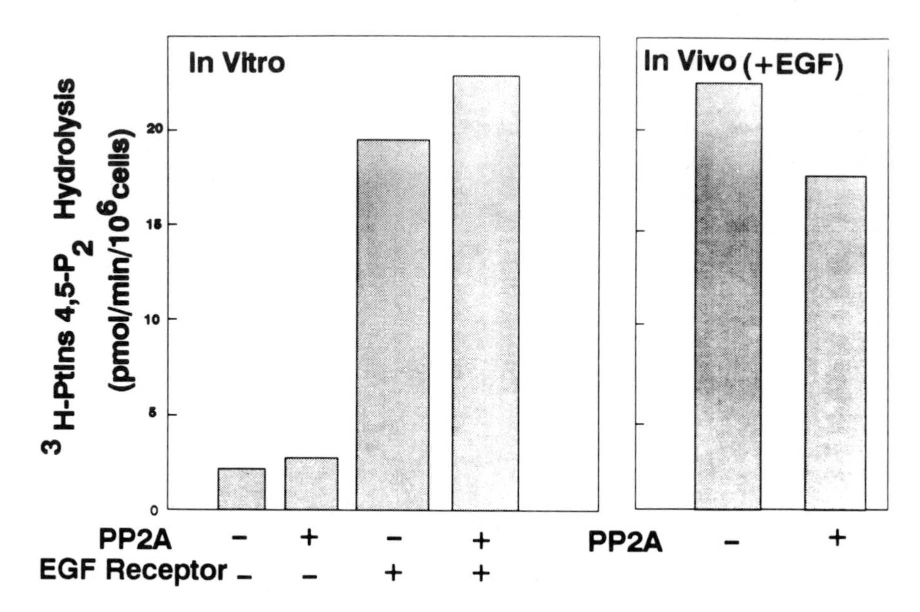


FIG. 4. Capacity of tyrosine phosphorylation to increase the activity of PLC-$\gamma_1$ after dephosphorylation of phosphoserine residues. (*Left*) PLC-$\gamma_1$ was immunoprecipitated from untreated A-431 cells and incubated with or without 0.2 μg PP2A at 30 °C for 30 min. The immunoprecipitates were then washed and incubated with purified EGF receptor and ATP for 60 min at 0 °C. After a second wash, PLC activity in the immunoprecipitates was measured (Nishibe et al 1990b). (*Right*) For comparison of the PLC-$\gamma_1$ activity levels produced by EGF receptor activation *in vitro* with those resulting from EGF treatment of cells, an equivalent number of EGF-treated A-431 cells were lysed and PLC-$\gamma_1$ was immunoprecipitated. The precipitates were then treated with or without PP2A, as above, and PLC activity was measured. The +PP2A value is not significantly different from the −PP2A value.

untreated cells and did contain a reasonable level of phosphoserine (Wahl et al 1989a,b, Meisenhelder et al 1989). At the moment, we are unable to define the functional significance of increased phosphoserine on PLC-$\gamma_1$ in growth factor-treated cells. However, in intact cells exposed to growth factors all tyrosine kinase substrates described to date have increased levels of phosphoserine/threonine as well as phosphotyrosine.

## Conclusions

PLC-$\gamma_1$ has been shown by both experiments with intact cells and reconstitution studies *in vitro* to be a direct tyrosine kinase substrate. Data have been presented to show that phosphotyrosine formation leads to an increase in the specific activity of this enzyme. The results do not, however, resolve two additional questions. First, it is very possible that within the intracellular environment other cellular components have a significant role. Likely candidates in this regard include activated receptors, $PtdInsP_2$-binding proteins, phosphoserine and phosphotyrosine phosphatases, and other molecules whose functions and identities are not yet known. The second issue concerns the role of PLC activity in the mitogenic response. Growth factors, like all hormones, produce a spectrum of distinct responses depending on the physiological nature of the target cell. At this point, there is a lack of experimental evidence demonstrating how PLC activation actually participates in a response to EGF, mitogenic or non-mitogenic.

## *Acknowledgements*

The authors thank Sue Carpenter for preparation of the manuscript and collaborators (Sue Goo Rhee, Nicholas Tonks) who have shared critical reagents. Also, we acknowledge support for these studies from the NIH (CA43720, HL31107, GM07374) and the American Cancer Society (RD304).

## References

Anderson D, Koch CA, Grey L, Ellis C, Moran MF, Pawson T 1990 Binding of SH2 domains of phospholipase Cγ1, GAP, and *src* to activated growth factor receptors. Science (Wash DC) 250:979–982

Anderson NG, Maller JL, Tonks NK, Sturgill TW 1990 Requirement for integration of signals from two distinct phophorylation pathways for activation of MAP kinase. Nature (Lond) 343:651–653

Burgess WH, Dionne CA, Kaplow J et al 1990 Characterization and cDNA cloning of phospholipase C-γ a major substrate for heparin-binding growth factor 1 (acidic fibroblast growth factor)-activated tyrosine kinase. Mol Cell Biol 10:4770–4777

Carpenter G, Cohen S 1990 Epidermal growth factor. J Biol Chem 265:7709–7712

Chen WS, Lazar CS, Poenie M, Tsien RY, Gill GN, Rosenfeld MG 1987 Requirement for intrinsic protein tyrosine kinase in the immediate and late actions of the EGF receptor. Nature (Lond) 328:820–823

Cohen S 1962 Isolation of a mouse submaxillary gland protein accelerating incisor eruption and eyelid opening in the new-born animal. J Biol Chem 237:1555–1562

Cohen S, Carpenter G, King L 1980 Epidermal growth factor-receptor-kinase interactions. Co-purification of receptor and epidermal growth factor-enhanced phosphorylation activity. J Biol Chem 255:4834–4842

Downing JR, Margolis BL, Zilberstein A et al 1989 Phospholipase Cγ, a substrate for PDGF receptor kinase, is not phosphorylated on tyrosine during the mitogenic response to CSF-1. EMBO (Eur Mol Biol Organ) J 8:3345–3350

Ellis C, Moran M, McCormick F, Pawson T 1990 Phosphorylation of GAP and GAP-associated proteins by transforming and mitogenic tyrosine kinases. Nature (Lond) 343:377–381

Goldschmidt-Clermont PJ, Kim JW, Machesky LM, Rhee SG, Pollard TD 1991 Regulation of phospholipase C-γ1 by profilin and tyrosine phosphorylation. Science (Wash DC) 251:1231–1233

Hendrickson HS, Dennis EA 1984 Kinetic analysis of the dual phospholipid model for phospholipase $A_2$ action. J Biol Chem 259:5734–5739
Honegger AM, Szapary D, Schmidt A et al 1987 A mutant epidermal growth factor receptor with defective protein tyrosine kinase is unable to stimulate proto-oncogene expression and DNA synthesis. Mol Cell Biol 7:4568–4571
Kaplan DR, Whitman M, Schaffhausen B et al 1987 Common elements in growth factor stimulation and oncogenic transformation: 85kd phosphoprotein and phosphatidylinositol kinase activity. Cell 50:1021–1029
Kim JW, Sim SS, Kim U-H et al 1990 Tyrosine residues in bovine phospholipase C-$\gamma$ phosphorylated by the epidermal growth factor receptor *in vitro*. J Biol Chem 265:3940–3943
Kim U-H, Fink D, Kim HS et al 1991 Nerve growth factor stimulates phosphorylation of phospholipase C-$\gamma$ in PC12 cells. J Biol Chem 266:1359–1362
Koch CA, Anderson D, Moran MF, Ellis C, Pawson T 1991 SH2 and SH3 domains: regulatory elements that control the interactions of signalling proteins. Science (Wash DC) 252:668–675
Kumjian DA, Wahl MI, Rhee SG, Daniel TO 1989 Platelet-derived growth factor (PDGF) binding promotes physical association of PDGF receptor with phospholipase C. Proc Natl Acad Sci USA 86:8232–8236
Kumjian DA, Barnstein A, Rhee SG, Daniel TO 1991 Phospholipase C$\gamma$ complexes with ligand-activated platelet-derived growth factor receptors. An intermediate implicated in phospholipase activation. J Biol Chem 266:3973–3980
Margolis B, Rhee SG, Felder S et al 1989 EGF induces tyrosine phosphorylation of phospholipase C-II: a potential mechanism for EGF receptor signalling. Cell 57:1101–1107
Margolis B, Li N, Koch A et al 1990 The tyrosine phosphorylated carboxyterminus of the EGF receptor is a binding site for GAP and PLC-$\gamma_1$. EMBO (Eur Mol Biol Organ) J 9:4375–4380
Matsuda M, Mayer GJ, Fukui Y, Hanafusa H 1990 Binding of transforming protein, $P47^{gag\text{-}crk}$, to a broad range of phosphotyrosine-containing proteins. Science (Wash DC) 248:1537–1539
Meisenhelder J, Suh P-G, Rhee SG, Hunter T 1989 Phospholipase C$\gamma$ is a substrate for the PDGF and EGF receptor protein-tyrosine kinases *in vivo* and *in vitro*. Cell 57:1109–1122
Molloy CJ, Bottaro DP, Fleming TP, Marshal MS, Gibbs JB, Aaronson SA 1989 PDGF induction of tyrosine phosphorylation of GTPase activating protein. Nature (Lond) 342:711–714
Moran MF, Polakis P, McCormick F, Pawson T, Ellis C 1991 Protein-tyrosine kinases regulate the phosphorylation, protein interactions, subcellular distribution, and activity of $p21^{ras}$ GTPase-activating protein. Mol Cell Biol 11:1804–1812
Nishibe S, Wahl MI, Rhee SG, Carpenter G 1989 Tyrosine phosphorylation of phospholipase C-II *in vitro* by the epidermal growth factor receptor. J Biol Chem 264:10335–10338
Nishibe S, Wahl MI, Wedegaertner PB, Kim JJ, Rhee SG, Carpenter G 1990a Selectivity of phospholipase C phosphorylation by the epidermal growth factor receptor, the insulin receptor, and their cytoplasmic domains. Proc Natl Acad Sci USA 87:424–428
Nishibe S, Wahl MI, Hernández-Sotomayor SMT, Tonks NK, Rhee SG, Carpenter G 1900b Increase of the catalytic activity of phospholipase C-$\gamma$1 by tyrosine phosphorylation. Science (Wash DC) 250:1253–1256
Rhee SG, Suh P-G, Ryu S-H, Lee SY 1989 Studies of inositol phospholipid-specific phospholipase C. Science (Wash DC) 244:546–550

Ullrich A, Coussens L, Hayflick JS et al 1984 Human epidermal growth factor receptor cDNA sequence and aberrant expression of the amplified gene in A-431 epidermoid carcinoma cells. Nature (Lond) 309:418–424
Wahl MI, Nishibe S, Suh P-G, Rhee GS, Carpenter G 1989a Epidermal growth factor stimulates tyrosine phosphorylation independently of receptor internalization and extracellular calcium. Proc Natl Acad Sci USA 86:1568–1572
Wahl MI, Olashaw NE, Nishibe S, Rhee SG, Pledger WJ, Carpenter G 1989b Platelet-derived growth factor induces rapid and sustained tyrosine phosphorylation of phospholipase C-$\gamma$ in quiescent BALB/c 3T3 cells. Mol Cell Biol 9:2934–2943
Wahl MI, Nishibe S, Kim JW, Kim H, Rhee SG, Carpenter G 1990 Identification of two epidermal growth factor-sensitive tyrosine phosphorylation sites of phospholipase C-$\gamma$ in intact HSC-1 cells. J Biol Chem 265:3944–3948

## DISCUSSION

*Parker:* Can I play devil's advocate and ask about the activation of PLC-$\gamma_1$ on tyrosine phosphorylation? The effect you see is really a suppression of the inhibition of the enzyme's activity by Triton X-100. You have shown that there is both an increase in $V_{max}$ and a shift in the apparent $K_m$, although in your explanation of the effect of Triton you suggested that there's a surface dilution, which would affect the $K_m$ value but not the increase in $V_{max}$. To what extent is this a real activation rather than a protection from inhibition?

*Carpenter:* We are trying to work through all the conditions involved in this assay and I don't think that we are yet at a point at which I can respond accurately to your question. The kinetic analysis seems to be more complicated than we had envisioned and there is still a long way to go. Trying to decide which parameter, association with vesicles or catalysis on the micelle surface, we are really looking at in the $K_m$ and $V_{max}$ values that I showed is difficult. The kinetic constants also depend on the ratio of Triton to $PtdInsP_2$. Goldschmidt-Clermont et al (1991) recently showed essentially the same phenomenon; profilin decreased the activity of the basal PLC-$\gamma_1$ enzyme but did not decrease the activity of the tyrosine phosphorylated form. What both profilin and Triton seem to be doing is sequestering the substrate, making it less accessible to the enzyme. Clearly, there is no Triton in the cell, but there may be something else that sequesters the enzyme in the membrane.

*Kasuga:* Do you think tyrosine phosphorylation of PLC-$\gamma$ has an effect other than activation, such as the detachment of the enzyme from the receptor?

*Carpenter:* There is some evidence suggesting that the presence of the receptor may further augment the activation of the phosphorylated enzyme. Tom Daniel argues this on the basis of kinetic analysis in intact cells (Kumjian et al 1991). Gordon Gill has a mutant EGF receptor, the behaviour of which suggests that PLC-$\gamma_1$ activity is augmented by formation of a complex between the receptor and the phosphorylated enzyme (Vega et al 1990). We have some evidence *in vitro* that is consistent with that idea. I don't know how long the phosphorylated enzyme exists as a complex in the cell with the receptor, but it is possible that such a complex may be physiologically significant.

*Krebs:* In the experiments with fibroblasts EGF, PDGF and FGF elicited tyrosine phosphorylation of PLC-$\gamma$, but insulin and colony-stimulating factor did not. Are comparable numbers of receptors for these various factors present in the fibroblasts?

*Carpenter:* I don't know. We did a similar experiment using a cell line, provided by Jonathan Whittaker, that overexpresses the insulin receptor, and could not detect PLC-$\gamma_1$ phosphorylation in those cells either (Nishibe et al 1990). Tony Hunter did the same experiment in Swiss 3T3 cells and got a negative result (Meisenhelder et al 1989).

*Hunter:* In our experiments there were relatively low numbers of insulin receptors.

*Carpenter:* We also failed to see significant phosphorylation *in vitro* using the purified insulin receptor.

*Fischer:* Do you know whether the IGF-1 receptor behaves differently from the insulin receptor?

*Carpenter:* We recently looked at that in collaboration with Renato Baserga (unpublished work). The results were negative.

*Fischer:* Were the experiments on dephosphorylation by CD45 and the T cell phosphatase done with isolated, phosphorylated PLC-$\gamma_1$, or in the presence of the EGF receptor?

*Carpenter:* The enzyme was separated from the receptor before the dephosphorylation assays.

*Fischer:* The presence of the EGF receptor might make a difference; the two molecules might interact through the SH2 domains.

*Carpenter:* There are technical complications in measuring kinase activity at the same time as lipase activity. For example, the $Mn^{2+}$ or $Mg^{2+}$ required for kinase activity are inhibitory towards phospholipase activity.

*Fischer:* You could allow phosphorylation to proceed, and then add EDTA, for example, to block the kinase activity, before adding the phosphatase. I think your results might be quite different.

*Carpenter:* In what way?

*Fischer:* You might find that the truncated form of the T cell enzyme is more active than CD45 or the full length enzyme.

*Tsien:* The interpretation of Lineweaver-Burk analysis is complicated in a system such as yours where you are not dealing with freely soluble enzymes. Could you tell us a little more about the $V_{max}$ effect?

*Carpenter:* I think the $V_{max}$ effect may be a result of the phosphorylated enzyme binding better to the vesicles, such that there are more activated enzymes bound to vesicles. I showed the results to illustrate that the kinetic parameters seem to differ between phosphorylated and non-phosphorylated PLC-$\gamma_1$ species. The interpretation of the kinetic data is, however, very complex.

*Michell:* You have a working assay system whereby you can see a difference that is dependent on phosphorylation. Have there been any careful studies done that take into account the fact that $PtdInsP_2$ in its natural environment in the plasma membrane is a rare lipid, looking at the interaction of the enzyme with a lipid mixture containing the proportion of $PtdInsP_2$ found in a normal plasma membrane? Does the enzyme see that situation differently, according to its phosphorylation state?

*Carpenter:* The closest to that is probably the work by Tom Pollard, which involved large unilamellar vesicles containing $PtdInsP_2$ and other phospholipids (Goldschmidt-Clermont et al 1991). The specific enzyme activity measured was about 1% of the activity we measure. I don't know whether that difference is due to the form of the substrate or to some other factor.

*Michell:* What happened to the phosphorylated form in that case?

*Carpenter:* It is more active than the non-phosphorylated enzyme in the presence of profilin.

*Tsien:* PLC-$\gamma$ can be phosphorylated on both serine and tyrosine, but when you were working *in vitro* with the purified EGF receptor the enzyme was phosphorylated only on tyrosine. Could this mean that in cells there is a sequence of kinase actions, the EGF receptor first acting to produce phosphotyrosine, which stimulates the PLC-$\gamma$, followed by another enzyme phosphorylating serine residues? Do you think this would have functional significance, given that serine phosphorylation isn't absolutely necessary for phospholipase activity?

*Carpenter:* I think the serine phosphorylation of PLC-$\gamma_1$ has a function, but we haven't been smart enough to detect it yet.

*Krebs:* Serine/threonine phosphorylation seems to occur with all of the receptor tyrosine kinases; they often become heavily phosphorylated on serine residues. With the EGF receptor, phosphorylation by protein kinase C is definitely inhibitory. Have the serine/threonine phosphorylations involving other receptor-type tyrosine kinases been clearly implicated in functional changes?

*Carpenter:* I think all of the tyrosine kinase substrates are also phosphorylated in an agonist-dependent way on either Ser or Thr. I believe that only for MAP kinase has a function been ascribed for these non-tyrosine phosphorylations. I believe that PKC will phosphorylate the insulin receptor, but I don't think that it phosphorylates the PDGF receptor at all. The EGF receptor is the only case where the function of PKC-dependent serine phosphorylation of a receptor tyrosine kinase is known.

*Hunter:* There has been a report that PKC phosphorylation of the insulin receptor decreases its activity (Bollag et al 1986), but I am not sure that is generally believed.

*Tsien:* In other signalling systems, such as ion channels, there is often an orderly sequence of activation followed by inactivation. With the the enzyme systems that we are discussing, is it known whether the tyrosine phosphorylation has to happen first in order for the target to then become vulnerable to serine phosphorylation?

*Carpenter:* Sue Goo Rhee has mutated each of the three tyrosine phosphorylation sites (Kim et al 1991). Mutation of Tyr-771 did not alter responsiveness to PDGF, but mutation of Tyr-783 completely abolished it. After mutation of Tyr-1254, which is located near the C terminus, the response to PDGF was reduced by 50%. It seems that two of the tyrosine phosphorylation sites are involved in the activation of the enzyme.

*Tsien:* What effect did the mutations have on Ser/Thr phosphorylation?

*Hunter:* The combined mutation of tyrosines 1254 and 783 led to a loss of serine phosphorylation as well, and also to a loss of receptor association; you could argue that the whole protein is inactive as a result.

*Parker:* I believe there was also a decrease in the basal PtdIns turnover with that mutant.

*Carpenter:* We have never been able to separate the tyrosine phosphorylation and the serine phosphorylation events. If you incubate cells at 0 °C and add the ligand both serine and tyrosine phosphorylation are observed. We haven't found any condition that produces only one type of phosphorylation.

*Tsien:* One might try to mutate the enzyme, to see whether ablation of the tyrosine sites prevents the serine phosphorylation.

*Hunter:* There was some hint that the double mutant PLC-$\gamma$ might be a dominant negative, but the appropriate experiment of co-expressing the mutant with the wild-type enzyme was not done. These results show that tyrosine phosphorylation is important for activation *in vivo* regardless of the difficulties with the *in vitro* assays.

*Nishizuka:* What percentage of the total PLC activity in your cells is due to PLC-$\gamma$?

*Carpenter:* Immunoprecipitation of all the PLC-$\gamma_1$ from A-431 extracts reduces the total PLC activity by about 50%. The remaining activity cannot be ascribed to PLC-$\gamma_2$, so I think there must be PLC-$\beta$ isozymes.

*Kasuga:* Is there any data to indicate which phosphorylated tyrosine on the EGF receptor interacts with the SH2 domain of PLC-$\gamma$?

*Carpenter:* I don't think that's known. A C-terminal fragment of the receptor containing three or four autophosphorylation sites seems to interact with PLC-$\gamma_1$.

*Hunter:* Rusty Williams has synthesized phosphotyrosine-containing peptides from the PDGF receptor and has found differential binding of PLC-$\gamma$ to those phosphopeptides (unpublished). I can't tell you which one has the highest affinity for PLC-$\gamma$ but there's clearly some specificity in the binding of SH2 domains.

*Fischer:* Joseph Schlessinger has mutated the tyrosyl residues at the C-terminus of the EGF receptor and is looking at the interaction, not with PLC-$\gamma$ but with constructs containing the SH2 and SH3 domains of PLC-$\gamma$ (unpublished work). I believe he has found that mutation of some of the tyrosyl residues has no effect whereas mutation of others blocks the interaction.

*Parker:* One has to be careful about interpretation of the interactions of those SH2 domain constructs with phosphorylated receptors. p85 $\alpha$ and $\beta$ proteins in monomeric form will interact stoichiometrically with both the PDGF receptor and the EGF receptor. The heterodimeric, active kinase complex of p85 and p110 does not bind stoichiometrically to the EGF receptor, but it binds nevertheless to the PDGF receptor (Otsu et al 1991). There must be steric considerations or some other restriction in this interaction.

*Carpenter:* We have done experiments with mutant receptors in which one, two or three of the autophosphorylation sites are altered. When one or two sites are mutated there is no decrease in PLC-$\gamma$ phosphorylation; only when all three are changed is there a drastic decrease in PLC-$\gamma$ phosphorylation.

*Michell:* Have you compared tyrosine phosphorylation and activation of purified PLC-$\gamma_1$ with that obtained in the immunoprecipitates from A-431 cells?

*Carpenter:* The immunoprecipitated PLC-$\gamma_1$ gave a phosphorylation stoichiometry of two, whereas purified PLC-$\gamma_1$ gave a phosphorylation stoichiometry of one. The purified enzyme (obtained from Sue Ghoo Rhee) also yielded a much smaller extent of apparent activation.

*Michell:* Did you look for differences in the pattern of phosphorylation of the various sites between immunoprecipitated and purified enzymes, to see whether you could correlate this with the difference in stoichiometry?

*Carpenter:* We didn't have enough purified protein to do that, and are now purifying some PLC-$\gamma_1$ ourselves to do that experiment. I am worried that the brain enzyme we had might have been regionally denatured in the SH region, leading to the low phosphorylation activity.

*Cantrell:* There have been a number of reports that protein kinase C can regulate phospholipase C activity in cells through feedback mechanisms. Do you have any ideas about a mechanism for that?

*Carpenter:* PKC will phosphorylate the EGF receptor in our system and attenuate its kinase activity. We have not seen direct phosphorylation of PLC-$\gamma$ by PKC.

*Michell:* Many of the situations where PKC attenuates responses involve phospholipase activation via G protein-coupled pathways. These involve different species of phospholipase.

*Carpenter:* That's correct. There is evidence that PKC may phosphorylate PLC-$\beta$, but there's no evidence that it does so for PLC-$\gamma$.

*Rasmussen:* I believe that cyclic AMP-dependent protein kinase will phosphorylate PLC-$\gamma$.

*Carpenter:* There is some evidence to that effect from Rhee's laboratory (Kim et al 1989).

*Cantrell:* Do you think there is any cross-talk between the $\gamma$ and $\beta$ families of phospholipase C isozymes?

*Carpenter:* I do wonder what the cell does when you activate these two different enzymes. Does one enzyme hydrolyse the substrate more effectively than the other? Do they communicate in some way? Would they have an additive effect on $PtdInsP_2$ hydrolysis, or a synergistic effect? These might be interesting experiments to do.

*Tsien:* There might be a competitive effect.

*Michell:* One of the problems we have is to explain how different phospholipases select $PtdInsP_2$ out of multiple pools. Another interesting dimension is about to be added to that problem. Len Stephens has some data on f-Met-Leu-Phe-driven and G protein-mediated inositol lipid 3-kinase activation (Stephens et al 1991). He concludes, in contrast to the groups of Cantley and of Majerus, that $PtdInsP_2$ is probably the sole substrate of the 3-kinase and that in the f-Met-Leu-Phe-stimulated neutrophil the fluxes of $PtdInsP_2$ usage through phosphoinositidase C and through inositol lipid 3-kinase are probably in a ratio of 4:1.

*Carpenter:* I think the pool is a very small fraction of the total.

*Michell:* I don't think we really know that. Experiments on $PtdInsP_2$ depletion in intensely stimulated cells suggest that there is turnover of a large fraction of the cell's $PtdInsP_2$ pool per minute. In some cells the pool is reduced in size by 60 or 70%, so whatever is present in the cell is used quickly and must therefore be replenished very quickly.

*Carpenter:* Do you think that the 3-kinase activity is sufficient to represent a competitive pathway?

*Michell:* Because we don't know whether the f-Met-Leu-Phe receptor is ever able to selectively regulate these two pathways, this is very difficult to judge.

*Nishizuka:* Professor Carpenter, I have no experience with the measurement of PLC-$\gamma$. In your crude membrane system, how accurately and specifically can it be measured?

*Carpenter:* The antibody seems to be the crucial reagent in this.

*Tsien:* As a control you could take a known amount of another PLC, such as PLC-$\beta$, and subject it to the same immunoprecipitation procedure to see how specific the assay is.

*Carpenter:* Rhee's results suggest that the antibody that we use does not cross react with PLC-$\beta$.

*Exton:* Our experience with the same antibody is that there is no cross-reactivity.

*Hunter:* We have mapped the epitopes for some of the anti-PLC-$\gamma$ monoclonal antibodies and they are clearly against regions that are unique to PLC-$\gamma_1$. Unless there is a complex between different PLCs in the cell the assay is probably specific.

Professor Carpenter, is the PLC-$\gamma$ activity that you measure *in vitro* sufficient to account for that detected in the cell?

*Carpenter:* Yes.

*Michell:* Is it actually higher? In most circumstances if you simply extrapolate data from highly active enzyme preparations of isolated phospholipases back to the situation in the intact cell you discover if the activity *in situ* matched that in broken cell preparations, then everything would disappear in seconds!

*Carpenter:* Our calculations indicate that activity *in vitro* is a little lower than in the cell.

*Hunter:* The basal level is much higher than it ought to be.

*Carpenter:* That's probably true, and there is some fluctuation in the extent of activation that we obtain *in vitro*. We sometimes observe as much as a 5–10-fold activation. It isn't clear to me whether that is due to differences in the proportion of co-precipitating proteins, or to fluctuations in the basal level of enzyme activity.

*Michell:* I am not sure we are talking about the same thing. I am comparing the measurement of lipid hydrolysis in a stimulated cell (for example, by measuring inositol phosphate signals) with assay of an isolated enzyme or broken cell preparation *in vitro*.

*Carpenter:* We are not talking about the same thing.

## References

Bollag GE, Roth RA, Beaudoin J, Mochly-Rosen D, Koshland DE Jr 1986 Protein kinase C directly phosphorylates the insulin receptor in vitro and reduces its protein-tyrosine kinase activity. Proc Natl Acad Sci USA 83:5822–5824

Goldschmidt-Clermont PJ, Kim JW, Machesky LM, Rhee SG, Pollard TD 1991 Regulation of phospholipase C-$\gamma$1 by profilin and tyrosine phosphorylation. Science (Wash DC) 251:1231–1233

Kim U-H, Kim JW, Rhee SG 1989 Phosphorylation of phospholipase C-$\gamma$1 by cAMP-dependent protein kinase. J Biol Chem 264:20167–20170

Kim HK, Kim JW, Zilberstein A et al 1991 PDGF stimulation of inositol phospholipid hydrolysis requires PLC-$\gamma$1 phosphorylation on tyrosine residues 783 and 1254. Cell 65:435–441

Kumjian DA, Barnstein A, Rhee SG, Daniel TO 1991 Phospholipase C$\gamma$ complexes with ligand-activated platelet-derived growth factor receptors. An intermediate implicated in phospholipase activation. J Biol Chem 266:3973–3980

Meisenhelder J, Suh P-G, Rhee SG, Hunter T 1989 Phospholipase C$\gamma$ is a substrate for the PDGF and EGF receptor protein-tyrosine kinases *in vitro* and *in vitro*. Cell 57:1109–1122

Nishibe S, Wahl MI, Wedegaertner PB, Kim JJ, Rhee SG, Carpenter G 1990 Selectivity of phospholipase C phosphorylation by the epidermal growth factor receptor, the insulin receptor, and their cytoplasmic domains. Proc Natl Acad Sci USA 87:424–428

Otsu M, Hiles I, Goot I et al 1991 Characterization of two related 85 kd proteins containing SH2 and SH3 domains that bind to receptor tyrosine kinases and polyoma virus middle T antigen/pp60c-src complexes, and associate with phosphatidylinositol 3-kinase. Cell 65:91–104

Stephens L, Hughes KT, Irvine RF 1991 Pathway of phosphatidylinositol 3,4,5-trisphosphate synthesis in activated neutrophils. Nature (Lond) 337:181–184

Vega QC, Cochet C, Rhee SG, Kao J, Gill GN 1990 Structural features of the epidermal growth factor receptor are required for $Ca^{2+}$ and inositol phosphate regulation. J Cell Biol 111:95a

# Protein tyrosine kinases belonging to the *src* family

Kumao Toyoshima*, Yuji Yamanashi†, Kazushi Inoue*, Kentaro Semba†, Tadashi Yamamoto† and Tetsu Akiyama*

**Research Institute for Microbial Diseases, Osaka University, Suita, Osaka 565 and †Institute of Medical Science, University of Tokyo, Minato-ku, Tokyo 108, Japan*

*Abstract.* There are nine non-receptor-type protein tyrosine kinases that show a high level of similarity in their primary structures and in the structures of their functional domains. Together, they are called the *src* family. They seem to have common sites specific for oncogenic activation. Recent findings suggest that the kinases are closely associated with cell surface molecules and that they mediate extracellular signals through the activation of their tyrosine kinase activity. They appear to act more on the differentiated phenotype than in haemopoietic cell proliferation. Possible functions of the products of the *lck*, *fyn*, *lyn* and *fgr* genes in lymphocytes and monocytes are discussed.

*1992 Interactions among cell signalling systems. Wiley, Chichester (Ciba Foundation Symposium 164) p 240–253*

Protein tyrosine kinases (PTKs) are divided into two major categories—the receptor type and the non-receptor type. Kinases in the former category, represented by the product of the *erbB*/EGF receptor gene or the *fms*/CSF (colony-stimulating factor)-1 receptor gene, are encoded by several gene families. Similarly, the non-receptor tyrosine kinases are also encoded by at least four families; namely *src*, *abl*, *fps/fes* and *src*-kinase. The last enzyme is a novel PTK responsible for phosphorylation of Tyr-527 of $p60^{c\text{-}src}$, the product of the c-*src* gene (Okada & Nakagawa 1989). PTKs in each family show structural similarity, as well as sharing genomic structure (such as that of the splicing junctions of the coding region; Toyoshima et al 1987). However, the structure of the regions that regulate gene expression are specific to each gene. These similarities and specificities suggest that the members have common mechanisms for regulation of their enzymic activities, but have expression patterns that differ, possibly according to their particular function (Toyoshima et al 1990).

The biological significance of receptor-type PTKs has been widely discussed by many research groups, but the functions of the non-receptor type PTKs in normal states have been clarified only recently. The *src* gene, the representative of the *src* family, was known to be expressed in neuronal cells (Sudoh 1988), possibly playing a role in differentiation (Bjelfman et al 1990), though the pathway of the signal is still unknown. The first evidence for a link between an external signal and a non-receptor type PTK was the association of $p56^{lck}$ with CD4/CD8 antigen in T cells (Rudd et al 1988, Veillete et al 1988).

Here, we describe significant aspects of the structure of the *src* family and discuss possible functions of some members of this family in haemopoietic systems.

## Structural similarity in the *src* family kinases

Nine members of the non-receptor type PTK family are reported to have highly conserved structure, typified by that of $p60^{c\text{-}src}$ (Fig. 1). These kinases have molecular masses around 55 to 60 kDa and have a glycine residue at position 2, which is myristoylated after removal of the N-terminal methionine. This myristic acid and several amino acids next to Gly-2 are believed to be involved in the localization of the enzyme in the plasma membrane (Shulz et al 1985). In fact, PTKs of this family are known to associate with membrane fractions; this is in contrast to the cytosolic or nuclear localization of other non-receptor type PTKs, such as Fps and Abl.

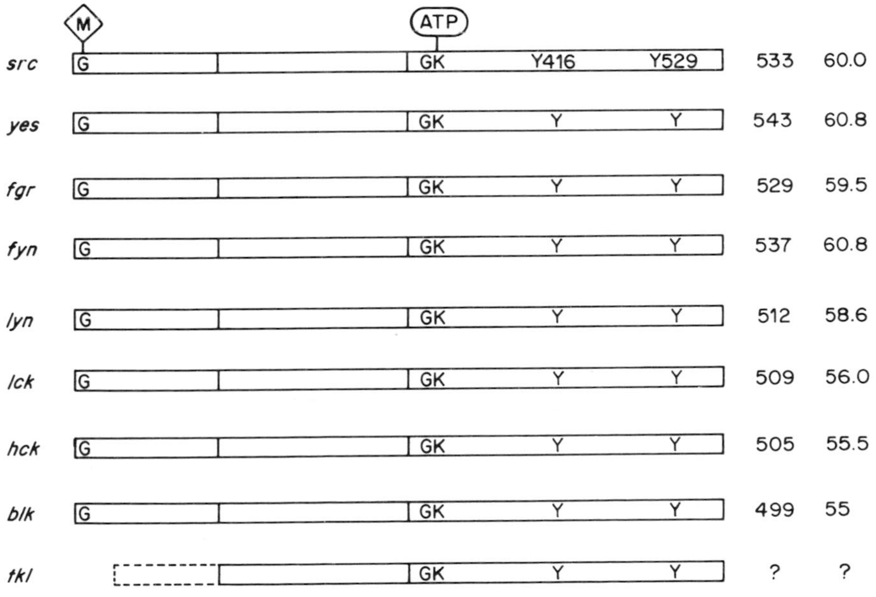

FIG. 1. Schematic structures of six non-receptor-type protein tyrosine kinases. M, myristic acid; P, phosphate; G, glycine; K, lysine; Y, tyrosine. Numbers (*right*) indicate the number of amino acids and molecular mass in kDa (*far right*) of each kinase.

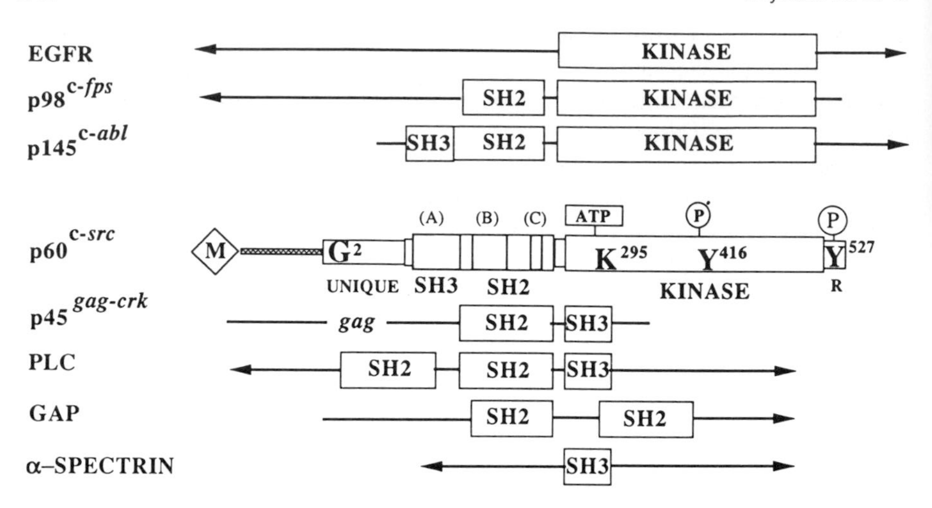


FIG. 2. A model of the functional domains of the *src* family tyrosine kinases. Additional proteins that have regions homologous to the modulatory domain are illustrated. EGFR, epidermal growth factor receptor; PLC, phospholipase C-$\gamma_1$; GAP, GTPase-activating protein.

Following the N-terminal sequence is a region of about 50 to 80 amino acids that is unique to each member—almost no homology is evident in this region. The difference in molecular size of the kinases is mostly due to the variation in the number of amino acids in this region. The region's function is still not clear, but it is postulated to play a role in association of the kinase with a surface molecule which is able to receive extracellular signals. The next region of the sequence, about 150 amino acids, has moderate similarity among members of the *src* family, and includes *src*-homologous (SH) regions. These SH regions were originally found as conserved regions in other non-receptor-type PTKs such as the *abl* and *fps* gene products. This region is also conserved in a variety of cytoplasmic molecules that are thought to interact with PTKs (Fig. 2), and is considered to be the site that interacts with cytoplasmic proteins to modulate signals. The C-terminal half of the molecule is a highly conserved region; sequence identity is about 80 to 90%. This region includes the catalytic domain, which phosphorylates tyrosine residues of target proteins, and the C-terminal regulatory domain, which down-regulates the catalytic activity when the tyrosine at the most C-terminal position is phosphorylated.

## Similarity and dissimilarity of genomic structures

The genomic organization of the coding regions of c-*src* is identical in the chicken and human (Takeya & Hanafusa 1983, Anderson et al 1985). When the human c-*fgr* clone was analysed, the splicing junctions of exons 4 to 12 were found to coincide with those of the c-*src* gene. Similar coincidences of splicing junctions were also found in the c-*yes* gene (Nishizawa et al 1985, 1986). Possible splicing junctions found in other members of the *src* family were also in the same locations as those of the c-*src* gene, except in the unique region. In contrast, splicing junctions in the c-*erbB*2 gene were completely different from those of the *src* family, even within the catalytic domain, suggesting that the nine members of the *src* family arose by duplication of a preformed ancestral gene.

In contrast, regulatory elements located upstream of the *src* family genes are specific to each member of the family. This difference of regulatory elements may explain the unique pattern of each expression of *src* family genes in different tissues and organs.

## Oncogenic activation of the *src* family genes

The c-*src* gene, integrated into a retroviral genome, acquires cell-transforming capacity only after mutation. v-*src*, v-*yes* and v-*fgr* were found as oncogenes of acutely oncogenic retroviruses. All of these, including S1 and S2, new isolates containing activated c-*src* (Ikawa 1986), have lost a sequence encoding the C-terminal portion of their regulatory domains through recombination with viral or other cellular sequences. Thus, they have invariably lost a tyrosine residue at the most C-terminal position (corresponding to Tyr-527 of the c-*src* gene product). Substitution of phenylalanine for Tyr-527 also activated tyrosine kinase activity and added transforming capacity to $p60^{c\text{-}src}$. In addition, substitution of an amino acid residue in the modulatory domain (Tyr-90 to Phe; Tyr-92 to Phe; Arg-95 to Glu, Lys or Trp) or in the kinase domain (Thr-338 to Ile; Glu-378 to Gly; Ile-441 to Phe), or deletions including part or almost all of the modulatory domain also confer transforming activity on $p60^{c\text{-}src}$. These mutations are also known to activate the tyrosine kinase activity of the enzyme. Activation of other members of the family has not been tested extensively; deletion of C-terminal sequences including the tyrosine residue corresponding to Tyr-527 of $p60^{c\text{-}src}$, or substitution of the same tyrosine by phenylalanine, activates both tyrosine kinase and transforming activities in c-Yes, c-Fgr, Fyn, Lck and Hck. Deletions in the modulatory domain and substitution of an amino acid in the kinase domain in Fyn are also reported to activate transforming capacity (Semba et al 1990). These mutations and substitutions are summarized by Cooper (1990) and Semba & Toyoshima (1990). A precise analysis of mutations within the modulatory domain of $p60^{src}$ was published recently by Hirai & Varmus (1990).

## *lck* and *fyn* expression in T cells

*lck* was identified as a gene encoding a novel tyrosine kinase, $p56^{lck}$, that is abundantly expressed in mouse LSTRA (T cell lymphoma) cells (Marth et al 1985, Voronova & Sefton 1986). Its human counterpart was soon found to be expressed in T lymphocytes. The dramatic decrease of p56$^{lck}$ in T cells during their activation suggests that it plays an important role in this process (Marth et al 1987). The protein appears to be functionally and physically associated with CD4 or CD8 in normal T cells, and cross-linking of CD4 activates the protein tyrosine kinase activity of p56$^{lck}$ (Veillette et al 1988). Activation of T cells occurs when the T cell receptor is stimulated by a specific antigen and the CD4 antigen interacts with a class II major histocompatibility complex (MHC) molecule. The association of CD4 with p56$^{lck}$ is critical in this process (Glaichenhaus et al 1991).

High expression of *fyn* was observed in T cells of mice with the autosomal recessive mutation *lpr* (lymphoproliferation) or *gld* (generalized lymphoproliferative disease). These homozygous mice with autoimmune disease have polyclonal proliferation of CD4$^-$/CD8$^-$ T cells and lymph node enlargement (Katagiri et al 1989). Expression of *fyn* is low in T cells of normal mice that have the same genetic background except for the *lpr* or *gld* gene. p56$^{lck}$ is expressed at similar levels in diseased and normal mice. The expression of *fyn* in negatively selected CD4$^-$/CD8$^-$ T cells was increased 10-fold about two hours after stimulation by treatment with anti-CD3ε, concanavalin A, 12-*O*-tetradecanoyl phorbol-13-acetate (TPA) or A23187 (a calcium ionophore), while *lck* expression remained constant. In addition, p59$^{fyn}$ and CD3 were co-precipitated by anti-CD3. Thus, p56$^{lck}$ associated with CD4/CD8 aids in the recognition of MHC molecules and p59$^{fyn}$ associated with CD3 is involved in the response to growth-stimulating treatment. CD45, a common lymphocyte antigen that has phosphotryosine phosphatase activity, is thought to activate p56$^{lck}$ tyrosine at position 505 (Ostergaard et al 1989). Thus, at least three components may act in the regulation of T cell growth in response to a stimulus from outside the cell.

### Association of p56$^{lyn}$ with the B cell antigen receptor complex

*lyn* is preferentially expressed in B cells in the lymphoid system (Yamanashi et al 1989). p56$^{lyn}$ is not readily precipitated as an immune complex by known antibodies to lymphocyte surface molecules. However, mild treatment with digitonin of WEHI-231 cells, a mouse B cell line, allowed us to co-immunoprecipitate p56$^{lyn}$ with surface membrane-bound IgM. This loose association between p56$^{lyn}$ and the IgM can be expected because B cell antigen receptors have only a few amino acids in their cytoplasmic regions (Rogers et al 1980). WEHI-231 cells produce two types of μ chain—a membrane form (μM) and a soluble form (μS). μM was co-immunoprecipitated with p56$^{lyn}$ by anti-Lyn, and IgM-associated molecules of 30 to 40 kDa were co-immunoprecipitated with p56$^{lyn}$ by anti-IgM (Yamanashi et al 1991). Cross-linkage of membrane IgM up-regulates the tyrosine kinase activity of p56$^{lyn}$, and phosphorylation at tyrosine residues of at least 10 molecules increases within

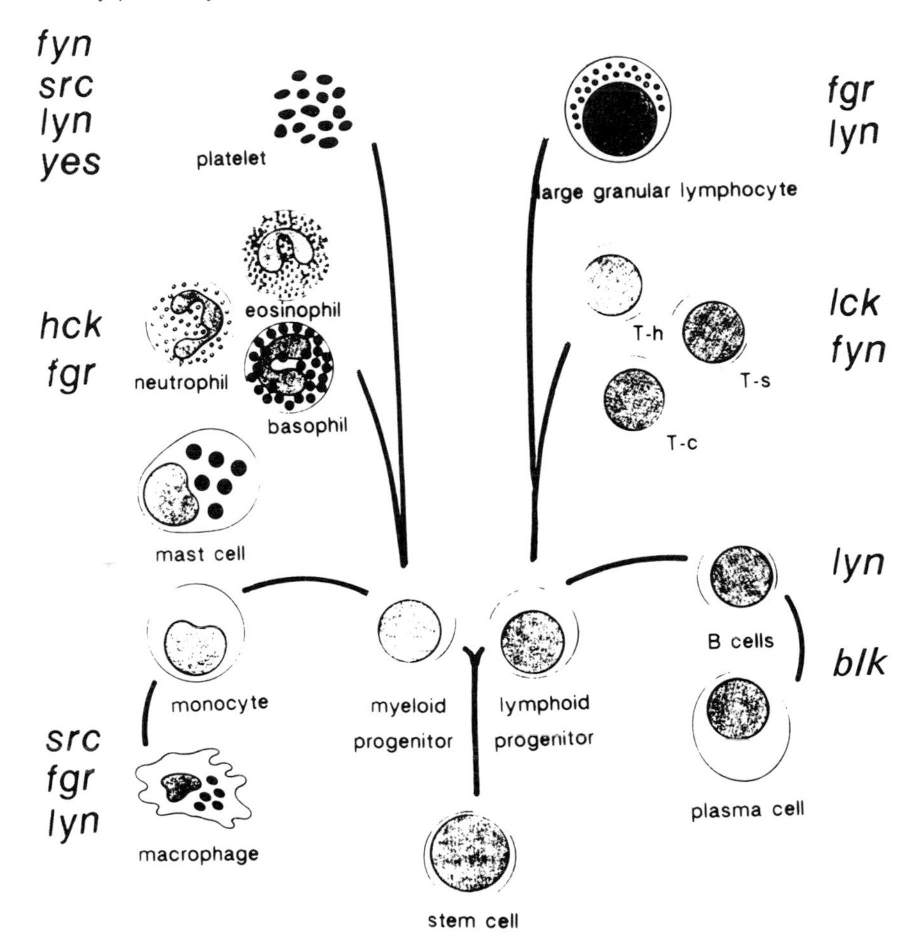


FIG. 3. Expression of cytoplasmic tyrosine kinases of the *src* family in haemopoietic cells.

2–10 min after cross-linkage. As a result of this treatment $p56^{lyn}$ was down-regulated instantly, reaching a minimum level about 30 min after stimulation, whereas $p53^{lyn}$, which was also associated with IgM, was not down-regulated. These results, together with the recent finding that $p56^{lyn}$ and membrane IgD can be co-immunoprecipitated (Y. Yamanashi, T. Yamamoto & K. Toyoshima, unpublished work), suggest that $p56^{lyn}$ plays an important role in antigen-dependent B cell differentiation. Although there is no evidence, CD45 and $p55^{blk}$ (Dymecki et al 1990) may also play an important role in B cell proliferation and differentiation (Fig. 3). The connection between an 80 kDa tyrosine kinase described by Campbell & Sefton (1990) and $p56^{lyn}$ is unclear at present.

**Possible functions of p58$^{fgr}$ in monocytes**

p58$^{c\text{-}fgr}$ is known to be expressed in B cells infected with Epstein–Barr virus. Recently, we reported that the enzyme is specifically expressed in peripheral monocytes, granulocytes and natural killer cells. When HL-60, a promyelocytic cell line, was treated with retinoic acid or TPA to induce myelomonocytic differentiation, p58$^{c\text{-}fgr}$ accumulated in differentiated cells (Inoue et al 1990, Kawakami et al 1988), suggesting that p58$^{c\text{-}fgr}$ is more closely associated with differentiation than with proliferation. The c-*fgr* gene and a mutant encoding phenylalanine instead of Tyr-523, in expression vector pCO, were transfected into NIH-3T3 (contact-inhibited NIH Swiss mouse embryo) cells. The mutant induced cell transformation, recognizable by morphological changes as well as the capacity of the cells to form colonies on soft agar. The cells transfected with non-mutated c-*fgr*, which were selected by resistance to G418 (a neomycin derivative), were indistinguishable from untransfected or vector-transfected NIH-3T3 clones. Markers for differentiated myelomonocytic cells were tested for in transfected NIH-3T3 cells. Peroxidase, a representative enzyme of the mature granulocyte, was negative in all transfectants, but tests for 2-naphthyl butyrate esterase, a maker of mature monocytes, were strongly positive in clones transfected with normal c-*fgr*. The esterase reaction was completely inhibited by sodium fluoride, a specific inhibitor of the monocyte enzyme. The level of enzyme is dependent on the level of c-*fgr* expression. In contrast, cell clones transformed by mutant *fgr* showed only a minor population of esterase-positive cells. c-*src*, c-*yes*, *fyn* and *lyn* expression in NIH-3T3 cells was not correlated with esterase expression. The difference between normal and activated c-*fgr* genes with regard to esterase induction may be related to differences in phosphorylation of target proteins. (K. Inoue, W. Bussabah, T. Akiyama & K. Toyoshima, unpublished work).

*Acknowledgements*

Most of the work in this study was supported by a Grant in Aid for Special Project Research on Cancer Bioscience from the Ministry of Education, Science and Culture, Japan. We also thank Ms Yoko Sugiyama for preparation of the manuscript.

**References**

Anderson, SK, Griffs CP, Tanaka A, Kung H-J, Fujita DJ 1985 Human cellular *src* gene: nucleotide sequence and derived amino acid sequence of the region coding for the carboxyl-terminal two-thirds of pp 60$^{c\text{-}src}$. Mol Cell Biol 5:1122–1129

Bjelfman C, Meyerson G, Cartwright CA, Mellstrom K, Hammerling U, Parlman S 1990 Early activation of endogenous pp60$^{src}$ kinase activity during neuronal differentiation of cultured human neuroblastoma cells. Mol Cell Biol 10:361–370

Campbell M-A, Sefton BM 1990 Protein tyrosine phosphorylation is induced in murine B lymphocytes in response to stimulation with anti-immunoglobulin. EMBO (Eur Mol Biol Organ) J 9:2125–2131

Cooper JA 1990 The *src*-family of protein-tyrosine kinases. In: Kemp B, Alewood PF (eds) Peptides and protein phosphorylation. CRC Press, Boca Raton, FL, p 85–113

Dymecki SM, Niederhuber JE, Desiderio SV 1990 Specific expression of a tyrosine kinase gene, *blk*, in B lymphoid cells. Science (Wash DC) 247:332–336

Glaichenhaus N, Shastri N, Littman DR, Turner JM 1991 Requirement for association of $p56^{lck}$ with CD4 in antigen-specific signal transduction in T cells. Cell 64:511–520

Hirai H, Varmus HE 1990 Site-directed mutagenesis of the SH2- and SH3-coding domains of c-*src* produces varied phenotypes, including oncogenic activation of $p60^{c\text{-}src}$. Mol Cell Biol 10:1307–1318

Ikawa S, Hagino-Yamagishi K, Kawai S, Yamamoto T, Toyoshima K 1986 Activation of the cellular *src* gene by transducing retrovirus. Mol Cell Biol 6:2420–2428

Inoue K, Yamamoto T, Toyoshima K 1990 Specific expression of human c-*fgr* in natural immunity effector cells. Mol Cell Biol 10:1789–1792

Katagiri T, Urakawa K, Yamanashi Y et al 1989 Overexpression of *src* family tyrosine-kinase $p59^{fyn}$ in $CD4^-CD8^-$T cells of mice with lympho-proliferative disorder. Proc Natl Acad Sci USA 86:10064–10068

Kawakami T, Kawakami Y, Aaronson SA, Robbins KC 1988 Acquisition of transforming properties by *FYN*, a normal *SRC*-related human gene. Proc Natl Acad Sci USA 85:3870–3874

Marth JD, Peet R, Krebs EG, Perlmutter RM 1985 A lymphocyte-specific protein-tyrosine kinase gene is rearranged and overexpressed in the murine T cell lymphoma LSTRA. Cell 43:393–404

Marth JD, Lewis DB, Wilson CB, Gearen ME, Krebs EG, Perlmutter RM 1987 Regulation of $pp56^{lck}$ during T cell activation: functional implications for the *src*-like protein tyrosine kinases. EMBO (Eur Mol Biol Organ) J 6:2727–2734

Nishizawa M, Semba K, Yamamoto T, Toyoshima K 1985 Human c-*fgr* gene does not contain coding sequence for actin-like protein. Jpn J Cancer Res 76:155–159

Nishizawa M, Semba K, Yoshida MC, Yamamoto T, Sasaki M, Toyoshima K 1986 Structure, expression and chromosomal location of the human c-*fgr* gene. Mol Cell Biol 6:511–517

Okada M, Nakagawa H 1989 A protein tyrosine kinase involved in regulation of $pp60^{c\text{-}src}$ function. J Biol Chem 264:20886–20893

Ostergaard HL, Shackelford DA, Hurley TR et al 1989 Expression of CD45 alters phosphorylation of the *lck*-encoded tyrosine protein kinase in murine lymphoma T-cell lines. Proc Natl Acad Sci USA 86:8959–8963

Rogers J, Early P, Carter C et al 1980 Two mRNAs with different 3′ ends encode membrane-bound and secreted forms of immunoglobulin μ chain. Cell 20:303–312

Rudd CE, Trevillyan JM, Dasgupta JD, Wong LL, Schlossman SF 1988 The CD4 receptor is complexed in detergent lysates to a protein-tyrosine kinase (pp58) from human T lymphocytes. Proc Natl Acad Sci USA 85:5190–5194

Semba K, Toyoshima K 1990 Structure and function of *src* family kinases. In: Carney D, Sikora K (eds) Genes and cancer. Wiley, Chichester p 73–83

Semba K, Kawai S, Matsuzawa Y, Yamanashi Y, Nishizawa M, Toyoshima K 1990 Transformation of chicken embryo fibroblast cells by avian retroviruses containing the human *fyn* gene and its mutated genes. Mol Cell Biol 10:3095–3104

Shulz AM, Henderson LE, Oroszlan S, Carber EA, Hanafusa H 1985 Amino terminal myristylation of the protein kinase $p60^{src}$, a retroviral transforming protein. Science (Wash DC) 227:427–429

Sudoh M 1988 Expression of proto-oncogenes in neural tissues. Brain Res Rev 13:391–403
Takeya T, Hanafusa H 1983 Structure and sequence of the cellular gene homologous to the RSV *src* gene and the mechanism for generating the transforming virus. Cell 32:881–890
Toyoshima K, Semba K, Nishizawa M et al 1987 Non-receptor type protein-tyrosine kinases closely related to *src* and *yes* compose a multigene family. In: Aaronson SA, Bishop JM, Sugimura T, Terada M, Toyoshima K, Vogt PK (eds) Oncogenes and cancer. Japan Scientific Societies Press, Tokyo, p 203–210
Toyoshima K, Yamanashi T, Katagiri T, Inoue K, Semba K, Yamamoto T 1990 Characterization and functional allotment of proto-oncogenes belonging to the *src* family. In: Knudson AG, Stanbridge EJ Jr, Sugimura T, Terada M, Watanabe S (eds) Genetic basis for carcinogenesis. Japan Scientific Societies Press, Tokyo, p 111–117
Veilette A, Bookman MA, Horak EM, Bolen JB 1988 The CD4 and CD8 T cell surface antigens are associated with the internal membrane tyrosine-protein kinase $p56^{lck}$. Cell 55:301–308
Voronova AF, Sefton BM 1986 Expression of a new tyrosine protein kinase is stimulated by retrovirus promoter insertion. Nature (Lond) 319:682–685
Yamanashi Y, Mori S, Yoshida M et al 1989 Selective expression of a protein-tyrosine kinase $p56^{lyn}$, in hematopoietic cells and association with production of human T-cell lymphotropic virus type 1. Proc Natl Acad Sci USA 86:6538–6542
Yamanashi Y, Kakiuchi T, Mizuguchi J, Yamamoto T, Toyoshima K 1991 Association of B cell antigen receptor with protein tyrosine kinase *lyn*. Science (Wash DC) 251:192–194

## DISCUSSION

*Cantrell:* Do B lymphocytes express *fyn*?

*Toyoshima:* I do not think so. If they do, the expression is very low.

*Fischer:* Do you know what proportion of Fyn is linked to the T cell receptor in T lymphocytes?

*Toyoshima:* I don't know exactly, although we could co-immunoprecipitate CD3 and Fyn with anti-CD3. Expression of *fyn* in unstimulated T cells is very low compared with that of *lck*.

*Hunter:* Larry Samelson, under the best conditions, finds that the fraction of Fyn that is associated with the T cell receptor is about 30% (Samelson et al 1990). The fraction of the T cell receptor associated with Fyn is also about 30%.

*Fischer:* That's not very high.

*Hunter:* The association is not tight either. Digitonin, a mild detergent, has to be used to detect these complexes.

*Cantrell:* I was interested in your results on the regulation of *fyn* expression by CD3. CD3 can regulate tyrosine phosphorylation. Do you see any changes in the effects of CD3 on tyrosine phosphorylation that correlate with its effects on *fyn* expression?

*Toyoshima:* We have not examined an immediate tyrosine phosphorylation in response to our anti-CD3 antibody (which is an anti-ε monoclonal antibody).

*Cantrell:* CD3 can clearly regulate tyrosine phosphorylation before an increase in *fyn* expression is seen, because most results show that it activates phosphorylation within five seconds. The phospholipase C activation would certainly be happening very rapidly. I wondered what the function of the increased *fyn* expression is. Does Fyn bind to another receptor? Has it been found associated with anything else in the T cell?

*Toyoshima:* We are currently looking at that. Our work with the anti-CD3 antibody is preliminary.

*Hunter:* By how much do the levels of the protein change in response to stimulation of CD3?

*Toyoshima:* The level of protein increases up to ten-fold.

*Hunter:* There are also changes in the level of *lck* mRNA when you activate T cells; it first decreases and then it increases above the original level (Marth et al 1987).

*Toyoshima:* With anti-CD3 stimulation the increase in *lck* mRNA is two-fold, at most, whereas *fyn* mRNA is increased ten-fold.

*Hunter:* There is already expression of *fyn*, so the levels do not start at zero.

*Parker:* In your experiments on the association of the p56 and p53 forms of Lyn with IgM, when you treated cells you saw disappearance of p56 only. How do you explain the fact that both seem to be in the complex, but only one disappears?

*Toyoshima:* We do not have an answer to that question at this time. The type of association probably differs.

*Cantrell:* Did those experiments involve a Western blot analysis of the level of Lyn?

*Toyoshima:* Yes.

*Hunter:* Was the antibody directed against the internal region?

*Toyoshima:* The major part of the unique domain is produced as a fused protein with bacterial protein, and this is used for immunization. The antibody is therefore polyclonal, directed against the majority of the unique domain.

*Hunter:* Roger Perlmutter's group saw an apparent loss of Lck after activation (Marth et al 1987), but it turned out that was because it became highly phosphorylated and could no longer be detected by the antibody. Is Lyn really disappearing, or might it be becoming undetectable?

*Toyoshima:* The disappearance of Lyn was checked by examining a total lysate.

*Hunter:* But if phosphorylation affected antibody binding you might fail to detect the protein, even if it were still present.

*Krebs:* But Professor Toyoshima is using a polyclonal anti-protein antiserum.

*Hunter:* It's directed against the unique region, which might be phosphorylated, as it is in other members of the family.

*Cantrell:* I have a general question about tyrosine phosphorylation of phospholipase C. Is there any specificity of the different lymphocyte tyrosine kinases for phospholipase C? Will Lck phosphorylate phospholipase C as well as Fyn?

*Carpenter:* Activation of the T cell receptor leads to phosphorylation of PLC-$\gamma_1$, presumably by Fyn (Park et al 1991, Secrist et al 1991, Weiss et al 1991).

*Cantrell:* CD4 does regulate phospholipase C, so perhaps Lck and Fyn regulate tyrosine phosphorylation of phospholipase C.

*Carpenter:* In B cells activation of IgM, presumably through Lyn, leads to phosphorylation of PLC-$\gamma_1$ also (Carter et al 1991).

*Toyoshima:* When we stimulate WEHI-231 cells with anti-IgM a band closely resembling PLC is phosphorylated, but we haven't checked what protein it is.

*Hunter:* In fibroblasts transformed by v-*src* you do not see constitutive phosphorylation of PLC-$\gamma$ (unpublished work), which might imply that it's not a good substrate for *src* family members, which generally have similar specificities.

*Parker:* Isn't there a constitutive turnover of PtdIns? The original observations with the v-*src* temperature-sensitive mutant imply that there is some sort of stimulated turnover (Diringer & Friiss 1977).

*Hunter:* There is an increased level of diacylglycerol but not of $InsP_3$. David Foster repeated those experiments very carefully and saw release of diacylglycerol from phosphatidylcholine but no increased hydrolysis of PtdIns. That is actually consistent with PLC-$\gamma$ not being tyrosine phosphorylated.

*Cantrell:* I understand that in T cells transfected with v-*src* there are no changes in inositol lipid metabolism.

*Carpenter:* Analogously, the CSF-1 receptor, which is similar to the PDGF receptor, does not phosphorylate PLC-$\gamma_1$, whereas the PDGF receptor does.

*Hunter:* I think one has to be careful. It may well be that Lyn and Fyn really can phosphorylate PLC-$\gamma$, but someone ought to try the experiment *in vitro*, at least, to see if it's possible.

*Cantrell:* We have heard some suggestions that in the B cell IgM might be regulating phospholipase C via tyrosine phosphorylation, yet the published results from permeabilized B cells have clearly established that a G protein-linked system connects IgM to phospholipase C. Does anyone have any thoughts about how these two ideas can be reconciled?

*Hunter:* Perhaps IgM uses both systems.

*Cantrell:* I was hoping someone would say that!

*Carpenter:* In hepatocytes there is evidence that EGF-induced PtdIns turnover is sensitive to pertussis toxin (Johnson & Garrison 1987). That's difficult to explain on the basis of a simple model of tyrosine phosphorylation only.

*Exton:* This is a single unconfirmed report, but it's from a very reliable laboratory.

*Parker:* Can I ask a general question about the tyrosine-to-phenylalanine substitutions in receptor molecules? For *neu*, the idea was that mutation might create an inhibitor that is no longer phosphorylated, and therefore does not dissociate from the active site. Has it been shown for any peptide substrates or competitive inhibitors containing tyrosine or phenylalanine that the phenylalanine analogues are as effective at binding at the active site as the tyrosine equivalents?

*Hunter:* Ron Kahn's group did a series of experiments with an insulin receptor autophosphorylation site peptide (Shoelson et al 1989). They had a phenylalanine-substituted peptide and a 4-methoxyphenylalanine-substituted peptide; these both acted as competitive inhibitors of receptor autophosphorylation with about equal potency.

*Carpenter:* I think they also saw an inhibition of the EGF receptor with those peptides.

*Hunter:* That's correct; they weren't specific for the insulin receptor but he did test the sort of thing that Peter Parker was asking about. I can't remember whether the $K_i$ values were as good as the $K_m$ values.

*Cantrell:* Does anyone know if there are ways in which G proteins could regulate tyrosine phosphorylation, by inducing the activation of tyrosine kinases or changing the activity of tyrosine phosphatases?

*Parker:* There are examples of that. Angiotensin II action in WB cells is an example where a seven-transmembrane-domain receptor-driven event leads to a tyrosine phosphorylation (Huckle et al 1990). Whether that's the type of tyrosine phosphorylation that you are talking about, a MAP kinase tyrosine phosphorylation rather than a Src- or Lck-driven phosphorylation, is not clear.

*Exton:* Vasopressin can also induce tyrosine phosphorylation.

*Carpenter:* Dr Fischer, is there any evidence that G proteins might be involved in the control of tyrosine phosphatase activity? I wonder whether enhanced tyrosine phosphorylation by vasopressin might be mediated by a G protein effect on a tyrosine phosphatase rather than a tyrosine kinase.

*Fischer:* That's a good question. We have no evidence either way.

*Exton:* Might phosphorylation of a protein by protein kinase C make it a better substrate for tyrosine kinases? This would explain the tyrosine phosphorylation induced by calcium-mobilizing agonists.

*Fischer:* You can increase the level of tyrosine phosphorylation with okadaic acid or PKC. There are many ways in which this could be achieved. Serine/threonine phosphorylation could activate a tyrosine kinase or inactivate a tyrosine phosphatase; or, it could make a substrate more or less susceptible to a tyrosine kinase or phosphatase.

*Exton:* I think it is questionable that there is direct communication between tyrosine kinases and G protein; the data do not support that idea unequivocally and can be interpreted in other ways.

*Hunter:* In principle, there is no reason why G proteins shouldn't be able to regulate tyrosine kinases; they regulate other types of enzymes.

*Nishizuka:* Is there any evidence of tyrosine kinases in nuclei?

*Hunter:* There has been a report that the type IV c-*abl* gene product is partially localized in the nucleus (Van Etten et al 1989). Several years ago it was shown that in v-*abl*-transformed 3T3 cells there is a subset of DNA-binding proteins that are tyrosine phosphorylated (Bell et al 1987). That was found by isolating DNA-binding proteins from $^{32}P$-labelled cells and using anti-phosphotyrosine antibodies.

*Fischer:* David Baltimore's data suggests that the c-*abl* gene product is translocated from the nucleus to the cytoplasm or the membrane when it becomes oncogenic (Van Etten et al 1989).

*Hunter:* The mutation that makes c-*abl* transforming prevents its translocation into the nucleus. There's also been a recent report that Fer, which is a member of the *fps/fes* family of tyrosine kinases, has a nuclear localization (Hao et al 1991).

## References

Bell JC, Mahadevan LC, Colledge WH, Frackelton AR Jr, Sargent MG, Foulkes JG 1987 Abelson-transformed fibroblasts contain nuclear phosphotyrosyl-proteins which preferentially bind to murine DNA. Nature (Lond) 325:552–554

Carter RH, Park DJ, Rhee SG, Fearon DT 1991 Tyrosine phosphorylation of phospholipase C induced by membrane immunoglobulin in B lymphocytes. Proc Natl Acad Sci USA 88:2745–2749

Diringer H, Friis RR 1977 Changes in phosphatidylinositol metabolism correlate to growth state of normal and Rous sarcoma virus-transformed Japanese quail cells. Cancer Res 37:2979–2984

Hao QL, Ferris DK, White G, Heisterkamp N, Groffen J 1991 Nuclear and cytoplasmic location of the FER tyrosine kinase. Mol Cell Biol 11:1180–1183

Huckle WR, Prokop CA, Dy RC, Herman B, Earp S 1990 Angiotensin II stimulates protein-tyrosine phosphorylation in a calcium-dependent manner. Mol Cell Biol 10:6290–6298

Johnson RM, Garrison JC 1987 Epidermal growth factor and angiotensin II stimulate formation of inositol 1,4,5- and inositol 1,3,4-trisphosphate in hepatocytes. Differential inhibition by pertussis toxin and phorbol 12-myristate 13-acetate. J Biol Chem 262:17285–17293

Marth JD, Lewis DB, Wilson CB, Gearn ME, Krebs EG, Perlmutter RM 1987 Regulation of $pp56^{lck}$ during T-cell activation: functional implications for the src-like protein tyrosine kinases. EMBO (Eur Mol Biol Organ) J 6:2727–2734

Park DJ, Rho HW, Rhee SG 1991 CD3 stimulation causes phosphorylation of phospholipase Cγ1 on serine and tyrosine residues in a human T-cell line. Proc Natl Acad Sci USA 88:5453–5456

Samelson LE, Philips AF, Luong ET, Klausner RD 1990 Association of the fyn protein-tyrosine kinase with the T-cell antigen receptor. Proc Natl Acad Sci USA 87: 4358–4362

Secrist JP, Karnitz L, Abraham RT 1991 T-cell antigen receptor ligation induces tyrosine phosphorylation of phospholipase Cγ1. J Biol Chem 266:12135–12139

Shoelson SE, White MF, Kahn CR 1989 Nonphosphorylatable substrate analogs selectively block autophosphorylation and activation of the insulin receptor, epidermal growth factor receptor, and pp60v-src kinases. J Biol Chem 264:7831–7836

Van Etten RA, Jackson P, Baltimore D 1989 The mouse type IV c-*abl* gene product is a nuclear protein, and activation of transforming ability is associated with cytoplasmic localization. Cell 58:669–678

Weiss A, Koretzky G, Schatzman RC, Kadlecek T 1991 Functional activation of the T-cell antigen receptor induces tyrosine phosphorylation of phospholipase Cγ1. Proc Natl Acad Sci USA 88:5484–5488

# General discussion II

*Parker:* I would like to ask a general question about cross-talk at the receptor level. There have been reports about heterodimeric complexes formed between the EGF receptor and Neu, for example (Wada et al 1990). Have mix-and-match experiments been done, expressing the lysine mutant of Neu in cells containing the EGF (epidermal growth factor) receptor, to see if there are changes in the ability of the cells to grow in soft agar in response to EGF, for example?

*Hunter:* I don't know if that has been done—it should be.

Has anyone got any good ideas about the function of phosphorylation of phosphoinositides at the 3 position by PtdIns 3-kinase? PtdIns 3-phosphates are not subject to hydrolysis by the usual PLCs.

*Michell:* That's correct—no one has yet managed to find phospholipases that will cleave these lipids.

*Hunter:* But they are degraded by specific phosphatases.

*Michell:* It has been shown that in f-Met-Leu-Phe-stimulated neutrophils there is rapid accumulation of PtdIns(3,4,5)$P_3$. The labelling patterns and kinetics are consistent with the direct formation of PtdIns(3,4,5)$P_3$, which is then stepwise dephosphorylated, via PtdIns(3,4)$P_2$ to PtdIns3P (Stephens et al 1991). I don't know how the enzymes that remove the phosphates at positions 4 and 5 on that pathway compare with those that dephosphorylate those positions on PtdIns4P and PtdIns(4,5)$P_2$.

*Parker:* The kinase that can work on the 3-phosphate may be different from the one that works on the 4-phosphate (Yamamoto et al 1990). I don't think the phosphatase experiments have been reported.

*Michell:* The work of Stephens et al (1991) is only one example of G protein-driven phosphorylation of PtdIns(4,5)$P_2$ to PtdIns(3,4,5)$P_3$. If this is a general phenomenon, with regulation being at that particular step, rather than at the further phosphorylation of PtdIns3P and/or PtdIns(3,4)$P_2$, interesting questions would arise as to the significance of the reports of kinases that drive these reactions.

*Cantrell:* Are there different kinases that can work on the 4 position of PtdIns?

*Hunter:* There are at least two; if there are two there could easily be many.

*Tsien:* It appears that there are a lot of different types of calcium oscillations, with different mechanisms. $InsP_3$ receptors can participate in oscillations, but there are cells, such as sympathetic neurons, which largely lack $InsP_3$ responses but can display oscillations dominated by calcium-induced calcium release. There are diverse mechanisms that could mesh together in different ways. We are still at the point of establishing which mechanism is active in any particular cell. More work needs to be done at the physiological level. There are no direct measurements of the changes in calcium levels in the store in response to messengers such as $InsP_3$.

*Michell:* In any cell in which there can, in principle, be calcium-induced calcium release, anything that generates a focal production of calcium at one point in the cell is liable to initiate waves.

*Tsien:* The Ref-52 fibroblasts that my brother Roger Tsien works on lack the calcium-induced calcium release mechanism and yet display $Ca^{2+}$ oscillations. This defines one extreme case of a cell with $InsP_3$-sensitive stores but no calcium-induced calcium release. Most cells probably lie in between this and the other extreme. How the two stores interact hasn't really been worked out to my satisfaction. Micheal Berridge has an appealing story about how the $InsP_3$-sensitive calcium stores can trigger the calcium-induced calcium release stores in cells where the two coexist, but this is probably not a universal explanation.

*Thomas:* The data Dr Carpenter and Dr Fischer presented suggested that tyrosine phosphatases will turn out to be much more specific than serine/threonine phosphatases.

*Fischer:* The fact that there is a real plethora of tyrosine phosphatase isoforms suggests that these will be localized within specific cellular compartments or that they will have distinct substrate specificities; both may be the case. This is being looked at. Peptides with variations around the tyrosyl residues are being used to examine differences in substrate specificity. I am sure there is some kind of trafficking going on within the cell but we don't know very much about that as yet.

*Carpenter:* Dr Thomas, what led you to suggest that the serine phosphatases are less specific?

*Thomas:* Your findings *in vitro*, which showed that CD45 and the placental tyrosine phosphatase differentially recognize the EGF receptor and PLC-$\gamma$.

*Carpenter:* All our *in vitro* experiments are done with only the catalytic subunit of phosphatase 2A. The regulatory subunit, which probably affects specificity *in vivo*, was not present.

*Fischer:* Such *in vitro* conditions are very artificial. There are phosphatase-binding proteins, such as the Type I glycogen-binding protein, that direct enzymes towards one substrate or another and participate in catalysis and control (Cohen & Cohen 1989).

*Thomas:* Would you not agree that there are potentially more tyrosine phosphatases than serine/threonine phosphatases?

*Fischer:* That's probably true. Using the polymerase chain reaction Matt Thomas has found nearly 30 protein tyrosine phosphatase isoforms in the sea squirt (Matthews et al 1991).

*Hunter:* I am not sure that such experiments have been done for the Ser/Thr phosphatases.

*Fischer:* You are correct. David Brantigan and Ned Lamb have shown that the type 1 phosphatase can be translocated during the cell cycle towards various compartments.

*Tsien:* Given that not enough work has been done on the Ser/Thr phosphatases let's start from the opposite point of view. You know what phosphatase 2A does, because you use it routinely in biochemical experiments to verify the prior formation of a serine/threonine phosphate group. If you measured the amount of phosphatase 2A in a particular cell type and then extrapolated from its *in vitro* properties, you could estimate a lower limit for the rapidity of reversal of any particular phosphorylation reaction. Presumably, extra phosphatases of unknown character could only add to the rate of dephosphorylation; so, calculations based on estimated levels and activities of the pertinent kinases would set an upper limit on the steady-state level of phosphorylation. These are simple-minded calculations, certainly, but they would provide a starting point.

In the glycogen breakdown pathway, which everyone thinks of as a paradigm, are the calculated levels of phosphorylation of by phosphorylase kinase roughly comparable to those actually measured?

*Fischer:* For the phosphatases, in particular, there is control by binding of effectors to the substrate molecule. For example, you can calculate the activity of phosphorylase kinase and phosphorylase phosphatase, but the phosphatase reaction is totally blocked by AMP, so, within the cell, conversion of phosphorylase *a* to phosphorylase *b* can be blocked totally by the nucleotide.

*Tsien:* You seem to be saying that the calculation is more complicated than it first appears; but, is it so complicated that it can't be done? Ultimately, you have to return to the cell to see if all the known factors are adequate to account for measured phosphorylation levels.

*Krebs:* The phosphatase inhibitor okadaic acid gives one the ability to manipulate phosphatase action within intact cells. It seems to me that a lot of the results that have been obtained with okadaic acid are reassuring, in that a number of the predictions that had been made by Phil Cohen about phosphatase 2A and phosphatase 1 have been validated.

*Hunter:* That is basically because okadaic acid works on two of the major classes of phosphatase that he had defined. However, there could be many more members of each of those classes than are currently known.

*Krebs:* Enzymologists have always been obsessed with the measurement of $V_{max}$ and $K_m$ values *in vitro* and using them to predict what would happen in the cell. I am sure much of such work was a great over-simplification, because it didn't take into account the localization of enzymes in the cell and a great variety of other factors. However, this was a starting point and did prove to be useful in studies involving serine/threonine phosphorylation. With tyrosine phosphorylation and dephosphorylation, however, there's a startling lack of that type of information for the various kinases and phosphatases. One difficulty is that tyrosine kinases are generally membrane-bound rather than existing as readily accessible soluble enzymes. Also, there is relatively little information concerning the natural substrates of tyrosine kinases and phosphatases. A third problem is the frequent use of immunoprecipitates as a source of enzyme. Quantitative data obtained using immunoprecipitates is obviously suspect.

## References

Cohen P, Cohen PTW 1989 Protein phosphatases come of age. J Biol Chem 264:21435–21438

Matthews RJ, Flores E, Thomas ML 1991 Protein tyrosine phosphatase domains from the Protochordate *Styela plicata*. Immunogenetics 33:33– 42

Stephens L, Hughes KT, Irvine RF 1991 Pathway of phosphatidylinositol 3,4,5-trisphosphate synthesis in activated neutrophils. Nature (Lond) 337:181–184

Wada T, Qian X, Greene ML 1990 Intermolecular association of the p185neu protein and EGF receptor modulates EGF receptor function. Cell 61:1339–1347

Yamamoto K, Graziani A, Carpenter G, Cantley L, Lapetina EG 1990 A novel pathway for the formation of phosphatidylinositol 3,4-bisphosphate. J Biol Chem 265:22086–22089

# Index of contributors

*Non-participating co-authors are indicated by asterisks. Entries in bold type indicate papers; other entries refer to discussion contributions.*

*Indexes compiled by John Rivers*

# Subject index